Forest Landscape Restoration

WORLD FORESTS

Series Editors

World Forests Description

As forests stay high on the global political agenda, and forest-related industries diversify, cutting edge research into the issues facing forests has become more and more transdisciplinary. With this is mind, Springer's World Forests series has been established to provide a key forum for research-based syntheses of globally relevant issues on the interrelations between forests, society and the environment.

The series is intended for a wide range of readers including national and international entities concerned with forest, environmental and related policy issues; advanced students and researchers; business professionals, non-governmental organizations and the environmental and economic media.

Volumes published in the series will include both multidisciplinary studies with a broad range of coverage, as well as more focused in-depth analyses of a particular issue in the forest and related sectors. Themes range from globalization processes and international policies to comparative analyses of regions and countries.

For further volumes:
http://www.springer.com/series/6679

John Stanturf • David Lamb • Palle Madsen
Editors

Forest Landscape Restoration

Integrating Natural and Social Sciences

Springer

Editors
John Stanturf
Center for Forest Disturbance Science
US Forest Service
Athens, GA, USA

David Lamb
Center for Mined Land Rehabilitation
Queensland University
Brisbane, QLD, Australia

Palle Madsen
Forest & Landscape
University of Copenhagen
Frederiksberg, Denmark

ISSN 1566-0427 ISSN 1566-0427 (electronic)
ISBN 978-94-007-5325-9 ISBN 978-94-007-5326-6 (eBook)
DOI 10.1007/978-94-007-5326-6
Springer Dordrecht Heidelberg New York London

Library of Congress Control Number: 2012953276

Printed on acid-free paper

Springer is part of Springer Science+Business Media (www.springer.com)

Contents

Part I
Introduction

Chapter 1
What Is Forest Landscape Restoration?

David Lamb, John Stanturf, and Palle Madsen

1.1 Introduction

The extent and distribution of global forests is a matter of considerable concern. The overall rate of deforestation remains high although recent reports suggest it is finally beginning to decline (FAO 2011). But this hides regional differences. In temperate regions net forest cover is increasing because of afforestation and natural expansion of forests. By contrast, net forest cover in most tropical regions continues to decline and few of the remaining forests are being managed on a sustainable basis (Asner et al. 2010; Foley et al. 2005). This means that more and more tropical countries are changing from being exporters of forest products to being importers. Across the globe most deforestation has been carried out to create agricultural lands but a large proportion of these lands have subsequently been abandoned (Ramankuty and Foley 1999).

In upland mainland Southeast Asia, for example, Fox and Vogler (2005) note that as much as 49% of these new agricultural lands are reported to have been subsequently abandoned and become shrub, brush or other forms of secondary forest. At the same time the Millennium Ecosystem Assessment found 60% of the world's

D. Lamb (✉)
Center for Mined Land Rehabilitation, University of Queensland,
Brisbane, Australia
e-mail: d.lamb@uq.edu.au

J. Stanturf
Center for Forest Disturbance Science, US Forest Service, Athens, GA, USA
e-mail: drdirt48@gmail.com

P. Madsen
Danish Center for Forest, Landscape and Planning, University of Copenhagen,
Copenhagen, Denmark
e-mail: pam@life.ku.dk

J. Stanturf et al. (eds.), *Forest Landscape Restoration: Integrating Natural and Social Sciences*, World Forests 15, DOI 10.1007/978-94-007-5326-6_1,

ecosystem services to have been degraded (MEA 2005). Much deforestation has been unplanned and has generated a series of socio-economic and ecological problems (Chomitz 2007). This is especially true within the tropics. Despite the wealth generated by logging tropical forests many poor people remain living in and around these landscapes and there have been widespread losses of biodiversity and substantial losses of soil.

But a number of changes are now underway that will also affect the extent to which existing forests are protected and the likelihood that any deforested land will be restored. These include a rising demand for food as populations continue to increase and, especially in the humid tropics, development of plantation crops such as palm oil (Gerber 2011). Both will place pressure in existing forests and on the availability of land for reforestation and afforestation. Countervailing forces are apparent in the increasing level of concern in many communities about environmental issues which makes forest restoration more attractive and a general drift of rural people to urban areas which, in some areas at least is allowing forests to regrow.

But perhaps the greatest unfolding change likely to affect the world's forests in future are the changes that will occur as a result of global warming (Bolte et al. 2009; Lindner et al. 2002; Liu et al. 2010). Although it is currently difficult to specify local impacts with any great confidence the broad global trends are reasonably clear (IPPC 2007). These mean there will be changes in the temperature, rainfall patterns and water resources and thus changes in the distribution of many plant and wildlife species (Beaumont et al. 2007; Iverson et al. 2004; Saxon et al. 2005). These will lead to changes in the distribution of agricultural crops and changes in the distribution of pests and diseases that affect them (Berry et al. 2006). Likewise, many of the existing protected areas will become unable to conserve the species for which they were originally established (Milad et al. 2011). These events will probably encourage further deforestation in some places but may act as stimuli for some form of restoration in other places including former farmlands and shrublands.

For those concerned with finding ways to restore degraded or under-used lands there are several challenges. While there are areas of extensive forest cover, especially in the boreal zone, much of the world's remaining forests exist in a landscape mosaic together with other land uses, particularly agriculture. Where property rights are well established and the rule of law prevails, restoration at a landscape scale will be constrained by the diversity of landowners. Additional limitations apply in the developing world where tenurial rights are sometimes ambiguous, governance issues abound, and corruption is often present (Kolstad and Søreide 2009).

One challenging task will be finding ways of restoring forest cover that suit the ecological constraints of particular sites as well as the socio-economic circumstances of the landowner or land user. These forms of forest restoration will have to be resilient enough to cope with the range of future uncertainties and also sufficiently economically attractive to persuade landowners to embrace them. The second task will be to find ways of implementing this restoration at an appropriately large or landscape scale. Both tasks need some explanation and this chapter firstly reviews how to carry out forest 'restoration' at a site level and then considers the most effective ways of undertaking forest restoration at a landscape scale. By doing this the chapter seeks to provide a background and a context for the more detailed and location-specific studies that follow in subsequent chapters.

1.2 What Is Forest Restoration?

Many people see the task of overcoming degradation as one of forest restoration yet consensus is lacking on terminology (Stanturf 2005). Perceptions of degradation and naturalness are social constructs (Emborg et al., Chap. 7, this volume) without universally accepted meaning. In the present context we shall regard a degraded forest as one with a reduced capacity to supply specified goods and services. This may be because of changes to the composition, structure or productivity of the forest caused by previous usage or by a catastrophic natural event such as a storm, landslip or a tsunami. Degraded lands are those whose ecosystems have suffered a persistent loss in their productivity caused by losses in soil fertility, changes in fire regimes, modifications to microclimate or because of invasive species. Over-abundant populations of herbivores such as deer can also cause degradation.

Even here we have to recognize that this definition must be somewhat flexible to account for circumstances where fertilizers have been added, introduced species have become naturalized, or fire in adapted communities has become a problem because suppression has altered fuel loads to dangerous levels. All of these events can limit the extent to which the supply of certain goods or services can be re-established at a site. There are, of course, often degrees of degradation and some have drawn a distinction between marginal, fragile and degraded with degraded lands being most severely affected while marginal or fragile lands might have lost some of their productivity but still be useable for agricultural purposes (Hudson and Alcántara-Ayala 2006; Biot et al. 1995).

Degradation can be overcome by restoration but definitions of restoration are also contentious (Hobbs et al. 2011). The Society for Ecological Restoration (SER 2004) defines 'restoration' as the process of assisting the recovery of an ecosystem that has been degraded, damaged or destroyed. The implication is that it is an intentional process which aims to accelerate the recovery of an ecosystem with respect to its structure (i.e., species composition, cover, physiognomy) functional properties (e.g., productivity, energy flow, nutrient cycling), and exchanges with surrounding landscapes and ecosystems. While not explicit in the SER definition of restoration, others have argued for the need to use reference sites to define restoration goals (Clewell and Rieger 1997), with the implication that only restoration to some historic condition is really ecological restoration.

This has proved controversial since it can present difficulties for those working in areas that have evolved after a long period of human occupancy and management and that may have involved certain grazing, burning or harvesting regimes. What should be done once these traditional management systems are abandoned? Should one try to re-establish the cultural landscape and its forests by re-establishing the former management regimes or should one seek to re-establish the supposed 'original' ecosystem assuming, of course, that condition can even be known (Aronson et al. 1995)? This means that historical settings will continue to be useful in some situations but not in others. But there are also other difficulties as well including:

- *environmental conditions at the site have altered*: the physical attributes of the site may have changed (e.g. soil fertility has been reduced by erosion or increased by agricultural fertilizers) and the sites are no longer suitable habitats for the original tree species or other biota.
- *the landscape has changed*: deforestation or farmland drainage may have altered hydrological regimes or local microclimates. Likewise, fire regimes may have been altered by fire suppression or grazing.
- *the target is unknown*: where deforestation is complete, or where humans have occupied the landscape for long periods, there may be no record, let alone examples, of the original forests or of the wildlife that occupied them. Even early records can be misleading since the site may have subsequently changed over time.
- *some species changes may be hard to reverse*: some of the original plants and animals are now extinct while the populations of others may have grown and become over-abundant; exotics may have colonised the area and become naturalized and impossible to eradicate.
- *the cost*: all forms of restoration can be expensive and many landowners may be only interested in simple forms of reforestation or afforestation involving commercially attractive species that can be harvested for a financial gain. Others may be reluctant to invest in what, to them, is a new land use about which they have only limited knowledge without substantial compensation. Still others may be concerned they may lose rights to subsidies for the present agricultural land-use and enter an unsubsidized land-use like forestry. Such a loss could affect the overall capital value of the property.

Add to these the complications arising out of the impacts of climate changes and it is clear that it may not always be possible to restore the original forest ecosystems even if one wished to.

In practice, the type of restoration adopted at a particular site is likely to depend not only on the degree of degradation that has occurred but also on the objectives of the land manager and on the resources available to them. Broadly speaking there are three possible alternatives for those interested in some form of forest restoration. The first might appeal to landowners primarily concerned with planting trees in order to generate a financial return from harvesting forest products or perhaps, where there is a carbon trading scheme by simply maximising biomass accumulation to sequester carbon. In such cases their main objective will be to increase productivity and they may be quite prepared to use a single commercially attractive native or exotic species to achieve this purpose (Fig. 1.1). This approach returns forest cover and regains some forest functions but does not strictly qualify as restoration according to the SER definition. This might be seen as the traditional approach to afforestation and is the one most favoured by industrial forestry companies primarily concerned with timber production. It might also be attractive to those wishing to accelerate the sequestration of atmospheric carbon.

Fig. 1.1 A young monoculture of *Anthocephalus chinensis* in Sabah, Malaysia

A second group of managers could be more interested in improving conservation or functional outcomes rather than maximising financial returns. Some may even wish to restore, in so far as they understand what was there before, the pre-disturbance forests. This group may attempt this through plantings at deforested sites as well as by protecting and managing natural forest regrowth (Elliott et al. 2006; Parrotta and Knowles 2001). This is illustrated in Fig. 1.2. Their emphasis will be on restoring as many as possible of the native tree species and restoring the habitats of wildlife species. In some temperate forests in particular the task may involve manipulating the density of trees in existing planted forests to allow additional species to become established (Hahn et al. 2005). This is illustrated in Fig. 1.3. Such undertakings can occur on both privately and publicly owned land. This obviously does qualify as restoration according to the SER definition although it may be many years before anything approaching pre-disturbance conditions are achieved even after an appropriate successional trajectory has been established.

A third group will be those wishing to achieve some elements of both of these objectives. This may be because they wish to generate an income as well as some conservation benefits or it might be because they recognize that the biophysical properties of the environment have changed (or will change in future as climates change) so that it is simply not feasible to attempt to restore the original ecosystem. For these reasons the forests they develop may include some native species but may also contain exotic species as well. This is because these have a higher commercial value, because they are able to tolerate the new environmental conditions better than the original native species or because they can facilitate the establishment of some of these native species (Brockerhoff et al. 2008; Carnus et al. 2006; Lamb

Fig. 1.2 Ecological restoration in Thailand. The multi-species forest is now 15 years old. It was established by planting seedlings but the site has been enriched by natural colonists from nearby undisturbed natural forest

Fig. 1.3 A Norway spruce forest in Sweden in which canopy openings have been created to allow the forest to be enriched with broadleaved species (beech, *Fagus sylvatica*)

1998; Parrotta et al. 1997; Stanturf et al. 2009). Hobbs et al. (2009) have referred to such forests as 'novel' ecosystems because of the new species assemblages present. These forests, too, might qualify as restored forests under the SER definition if they contained most of the main species that occur in a reference ecosystem and provide an appropriate community structure. On the other hand they might never develop to resemble the original ecosystems because of environmental changes that have occurred at the sites and because of the management methods being used.

These three options simplify a much richer and more diverse range of options available for landholders to restore forest cover at particular sites. However, within the present context, and for the sake of simplicity, all may be thought of as different forms of forest restoration (accepting the limitations of this terminology e.g. Hobbs et al. 2011). More than one approach may be used within a particular landscape with the actual methods used at a particular site being tailored to the landholder's objectives and the ecological conditions at that site. Note also that the methods used can change over time as ecological and economic circumstances change meaning that option 1 – simple monocultures producing easily saleable timbers – may initially be the only realistic choice for many forest managers but, over time it may become possible to introduce a wider variety of species (involving option 3 or even option 2) as environmental conditions improve and the emphasis changes from the production of goods to the production of services (e.g. Lee and Suh 2005; Madsen et al. 2005; Tak et al. 2007).

One further word on terminology is that we do recognize that while forest restoration shares many of its terms and techniques with traditional forest management, they are not always synonymous. For example we reserve the term "reforestation" for the artificial regeneration of a forest almost or completely clear-felled by harvesting, wildfire, or wind storm. Similarly we regard deforestation as the process of removing forest cover along with conversion to another land use or abandoned from management; it does not refer to periodic removal of forest cover like clear-cutting and stand regeneration in normal forest management.

1.3 The Landscape Mosaic

The opportunities for restoration and the type of restoration that is carried out depend upon the landscape in which it is being done. There is some confusion about the meaning of 'landscape' with some users of the term indicating a spatial extent while others do not. Lindenmayer and Fisher (2006) argue the definition depends on the context in which the term is being used; from a human perspective it may cover areas of hundreds or thousands of hectares but from a conservation biology perspective it depends more on the scales over which a particular species moves. Nassauer and Opdam (2008) define a landscape as a heterogeneous mosaic of ecosystems that is constantly being adapted by humans to increase its perceived value. Boedhihartono and Sayer (Chap. 16, this volume) suggest it is best thought of as the scale at which it is necessary to intervene if one is to balance trade-offs

Table 1.1 Some components of the landscape mosaic

Biophysical components	Socio-economic components
Topography: Hills and flatlands	*Population density*: areas where human population is concentrated (urban areas) and areas where it is sparse
Soils: Areas with fertile and less fertile soils supporting productive and less-productive agricultural lands	*Land ownership*: large and small farms; resident and absentee landowners, state ownership, communal ownership
Vegetation: annual and perennial crops, large and small patches of disturbed and undisturbed forests, regrowth forest, shrublands and grasslands; areas with invasive exotics	*Landholder status*: rich and poor landowners; traditional users
Biodiversity: areas with residual populations of endemic, endangered or vulnerable species (within forests, on forest margins and outside forests); areas with over-abundant populations of native or exotic species that may have become weeds or pests	*Infra-structure*: roads and railways which provide access and affect transport costs
Erosion: areas with severe erosion and others with none	*Commercial value or productivity*: tend to be greater in fertile lands close to transport and densely populated areas
Hydrological: rivers, wetlands and areas with high run-off; areas with and without severe erosion; areas with modified infiltration or drainage	
Wildfire: fire-prone areas with high fire frequencies and areas that are only rarely burned	

and optimize conservation and livelihood benefits. Clearly landscapes have both structural and functional components, which are influenced by the scale at which one approaches defining a landscape (Aylward 2005; Bruijnzeel 2004; Lindenmayer et al. 2008; Omernik and Bailey 1997).

Landscapes are not uniform and nor is it useful to think of them simply as containing 'forest' and 'non-forest' (Lindenmayer et al. 2008). Most landscapes are represented by spatially diverse mosaics of different types of vegetation and land use practices and it is useful to recognize some of the features that are responsible for this heterogeneity. Table 1.1 shows some of the biophysical and socio-economic attributes of landscapes. Biophysical attributes such as topography and soil fertility influence where deforestation is likely to have occurred in the past but will also influence where there may be opportunities for restoration in the future. For example, flat lands with fertile soils are likely to have been cleared at an early stage and are less likely to have been degraded and abandoned than sites with less fertile soils on steep slopes. It is the latter that are more likely to be available for restoration. These patterns will also determine where undisturbed forest persists, where regrowth is more likely to develop and where most of the original biodiversity will be retained. Some areas are also more likely to be burned by wildfires making regeneration more difficult (e.g. areas near roads and railroads or human habitations).

The socio-economic mosaic will be equally variable. Some areas will have high human population densities while others will have low densities. Farms in some places will be large because they are owned by wealthy landholders while others will be small and held by poorer landholders. Some larger farms that are close to transport may be farmed intensively while others may be farmed only episodically such as those with marginal soils or owned by absentee landholders. These various landowners are likely to have differing perceptions about whether or not to undertake restoration on their land and, if so, what sort of tree-planting to carry out. Some will be motivated by commercial considerations and driven by perceptions of the opportunity costs of reforestation. Others, perhaps those better-off farmers who have larger land holdings may be more interested in restoration on less productive parts of their landholdings to generate environmental services and protect crops.

None of these patterns are fixed. Populations of plants and animals move about landscapes depending on the types and spatial patterns of residual forests and crops. Land uses and vegetation patterns change as markets and market prices change and human populations may increase in some areas and decrease in others. Historical events such as wars, disease, fires and other natural disasters also influence settlement patterns and shape landscapes (Chazdon 2003; Foster et al. 1998). Some of these issues are explored further in subsequent chapters (See Convery and Dutson, Chap. 12; Crow, Chap. 2; Han and Oliver, Chapter 5 in Stanturf et al. 2012; Hughes et al. Chapter 15 in Stanturf et al. 2012; Nagy and Lockaby, Chap. 4; Oliver et al. Chap. 3; Wimberly et al., Chap. 6 in this volume).

As far as restoration is concerned these spatial patterns have several consequences. One is that some areas are more likely to be available for restoration than others. For example, marginal agricultural lands may be available but fertile croplands will not. A second is that natural recolonization and regeneration will be more likely in some parts of the landscape mosaic than others (e.g. regrowth is more likely at sites that have not been too heavily disturbed and are close to residual forests). And, thirdly, the areas available for restoration or able to regenerate naturally are not necessarily those parts of the landscape that are most in need of restoration to conserve biodiversity or to protect watersheds.

This last issue is critical. Most watersheds will be more effectively protected by continued forest cover on steep slopes and along riparian strips. Biodiversity conservation will be most effectively maintained by ensuring connectivity between forest remnants, by enlarging small forest patches and by creating protective buffer areas around forest patches subject to disturbances such as fires or continued clearing. Areas critical for watershed protection and for biodiversity conservation will sometimes overlap but in other cases will not.

However the location of any actual restoration will largely depend on land ownership patterns and landscape with many small farmers being unlikely to completely restore their land even when the market for forest products (or environmental services) is strong. Instead most will be inclined to use only some of their land for trees and the remainder for other purposes unless they are able to obtain most of their income from off-farm sources. And the location of any tree planting that does take place will depend upon farmers' perceptions of their opportunity costs.

While it may make ecological sense to restore steep hills or to form a link between two patches of remnant forest, individual landowners may have different perceptions of the value of such undertakings and prefer to use commercially attractive exotic species grown in a plantation monoculture.

1.4 What Is Forest Landscape Restoration and How Is It Different from Site-Level Restoration?

Forest Landscape Restoration (FLR) differs from site-level restoration because it seeks to restore ecological processes that operate at a larger landscape scale such as those maintaining the populations of species requiring large habitat areas or those responsible for hydrological flows.

But, because complete restoration of a landscape is usually unrealistic, choices must be made about where in the landscape mosaic this restoration is undertaken. FLR seeks to do this by using a strategic approach that targets key locations rather than relying on the individual decisions of separate landholders. At the same time it also seeks to improve the livelihoods of these landholders so that restoration is not carried out at their expense. According to Maginnis and Jackson (2007) FLR is defined as "a process that aims to regain ecological integrity and enhance human well-being in deforested or degraded forest landscape". The definition implies that FLR is a considered process and not simply a series of *ad hoc* treatments that eventually cover large areas.

Decisions about restoration will always depend on the resources available and on patterns of land tenure. But they will also depend on the aspiration and goals of individual landholders. This means some form of landscape-wide planning process will be needed to ensure key areas are restored and that incentives or compensation is provided to individual landholders to achieve this and the costs as well as the benefits are shared between landholders and the broader community.

1.5 How Is Forest Landscape Restoration Carried Out?

There are a large variety of approaches that have been used to address FLR and many of these are considered in the chapters that follow. Some involve relatively informal techniques while others require considerable planning. However the process of implementing FLR usually involves the consideration of four quite explicit questions:

(i) How much restoration should be carried out in a particular landscape?
(ii) Where should this be carried out?
(iii) What type of restoration should be done at each location?
(iv) How should the FLR process be managed?

1.5.1 How Much Restoration Is Needed?

There is no simple answer to this question. One obvious determining factor is the amount of forest that still remains in the landscape. But, more to the point, it depends on whether the previous loss of forests has given rise to particular problems. Other things being equal, a landscape with only 10 % forest cover remaining is likely to have acquired more problems and be more in need of some restoration treatment than one where 90 % of the original forest cover still remains. But, in principle, is there some kind of minimum forest cover threshold that one should aim for?

There has been considerable debate over the idea of ecological thresholds but the general conclusion is that these are hard to define (Groffman et al. 2006). There are several reasons for this. In the case of wildlife it appears that the populations of some species are adversely affected by only small proportional losses in habitat (or even simply declines in habitat quality) while others are more tolerant. For example, some woodland birds are only affected when habitat is reduced to less than 30% of the original cover (Radford et al. 2005). Species such as beetles may tolerate even greater levels of deforestation. But, even when the habitat requirements of a particular animal species are understood, it can be difficult to specify how the overall collective species richness is affected by decreases – or increases - in forest cover. The same uncertainty is likely to be true of plants with the added twist that the regional loss of plant biodiversity caused by deforestation appears to be slower than for wildlife biodiversity (probably due, in part at least, simply because species such as trees can be long-lived so that the impact of forest loss on species diversity is less obvious). Increasing the populations of residual tree species may assist the survival of what otherwise might be seen as the 'living dead' (Janzen 1988). But, in this case, the spatial location of any plantings may be more critical than the amount of restoration undertaken.

The relationship between the amount of forest cover and the extent of soil erosion is also difficult to define. Deforestation is known to affect erosion but it is difficult to specify a threshold cover below which accelerated erosion occurs and above which it ceases or returns to natural background levels (Lal 2001). Erosion is affected by rainfall intensity and soil type. It is also affected by slope: small areas of deforestation on steep slopes will sponsor more erosion than large areas on flatter lands. In short, increasing levels of forest cover is likely to improve many ecological outcomes but there is unlikely to be a simple threshold for forest cover that applies to all the environmental variables likely to be of concern to stakeholders contemplating restoration.

In practice a key issue, of course, is that if forest cover is to increase, even modestly, then other land uses (and habitat types) must decrease. Whether or not landholders are attracted to restoration depends on its opportunity costs. Farmers may be willing to undertake some afforestation on marginal agricultural lands where the opportunity costs are low but most will be reluctant to do so on fertile cropping lands where the opportunity costs are high. Some forms of restoration can generate financial returns but these often occur some years after planting takes place. Even when such restoration is financially rewarding many landowners will still need additional livelihood support if they are to participate in a broad landscape restoration program. Under these conditions many will only commit a small portion of their land to tree-planting.

There are some broader financial perspectives that are also relevant to this question. The financial returns from small forest plantations scattered across a landscape will only be realised when they are collectively large enough to provide a market with a regular supply of timber. In most cases a few, small plantations will not be enough. The type of products produced is also influential. Higher-value timber can be transported to more distant markets but more modestly priced utility timbers cannot. This means there may, in fact, be a distinct economic threshold that must be exceeded. It is difficult to specify this threshold area because it will depend so much on biological and financial conditions specific to a particular location. This topic is referred to in subsequent chapters in Stanturf et al. 2012 including Booth et al., Chap. 13, Rosengren Chap. 17 and Wilson et al., Chap. 11.

1.5.2 Where Should Restoration Be Carried Out?

The functional consequence of any new forest area depends on where in the landscape it is established. As noted earlier, afforestation of a steep hillside will reduce erosion more than afforesting the same area on flat land. Likewise, a new forest that provides a link between two patches of remnant natural forest will probably help species move across the landscape and conserve biodiversity more than the same area of planted forest isolated in the midst of an otherwise homogenous agricultural landscape (Llewellyn et al. 1996). The task for those planning FLR is to strategically distribute new forest areas across the landscape in a way that maximises their ecological impact. Landscape ecologists have generated a number of recommendations for priority locations for restoration. These are shown in Table 1.2.

Economic considerations may suggest other priority locations. These might include agriculturally marginal lands (where the opportunity costs are lower), sites distant from roads (where, again, opportunity costs are lower) or, alternatively, sites near roads (where log transport costs are lower). Needless to say, certain locations will achieve some outcomes but not others and this is where trade-offs will be needed. This question is referred to in subsequent chapters in this volume including Bentrup et al., (Chap. 5), Booth et al. (Chap. 13), Gobster (Chap. 8), Harper et al. (Chap. 14) and Larsen et al. (Chap. 9).

1.5.3 What Type of Restoration Should Be Done at Each Location?

There can be large differences in the composition and structure of undisturbed natural forests, regrowth forests and plantation forests. Part of the difference is because of the relative youth of the newly created forests relative to the undisturbed forests.

Table 1.2 Priority areas for restoration in degraded landscapes to improve functional outcomes

Location of new forests	Advantage of new forests at this location
Areas able to regenerate naturally	The cost of restoration is low (although the costs of protecting these areas may be significant)
Buffer strips planted around remnant patches of natural forests	Protect these remnants from further disturbances, enlarge their effective areas and soften edge effects (highest priority being given to remnants with endangered or vulnerable species)
Corridors planted between remnant patches of natural forests	Facilitate movement of species and genetic exchange between isolated populations
Corridors or 'stepping stones' planted along altitudinal and longitudinal gradients	Facilitate movement of species in response to environmental stresses such as climate change
Steep slopes	Protect erosion-prone soils
Riparian strips	Protect erosion-prone soils and act as filters to limit sediments reaching waterways. Act as corridors for species movement
Areas subject to sheet erosion and with compacted soils	Protect erosion prone soils and increase infiltration capacity
Groundwater recharge areas in salinity-prone areas	Increase evapo-transpiration thereby increasing depth of water table and decreasing salinity problems
Coastal protection zones	Decrease storm impacts
Urban areas	To improve recreational opportunities

However, there can also be large differences in the composition and structure of many newly established forests even when these are of the same age. This is usually the consequence of a deliberate choice by the landholder. Plantation monocultures are obviously the simplest type of new forest (Fig. 1.1). Many are managed on short rotations, especially those established in the tropics. These represent a high proportion of most newly established forests because landowners perceive them as being more profitable and easier to manage. These plantations can restore productivity and some ecological functions but will not provide the habitats needed by many forest-dependent species. This means they will not be as useful in improving connectivity between natural forest fragments to allow species or genes to move across a landscape. On the other hand, some monoculture plantations and especially older ones with shrubby understories, can provide good watershed protection.

Mixed-species plantings are better at providing habitats for a wider variety of other species and are also able to protect watersheds. There are a variety of these including even-aged plantings and forests where an existing monoculture is enriched with additional species (Lamb 2011). But these types of plantings can have other advantages as well including improving productivity, improving nutrition or reducing damage from insect pests or disease (Dey et al. 2010; Lockhart et al. 2008; Stanturf et al. 2009; Lamb 2011). One particular advantage of mixtures that make them attractive to some landholders is that they diversify the goods and services provided and thereby reduce economic risks. This may not be an especially attractive advantage for large industrial plantation owners for whom the added management complexity is a disadvantage but could be for smallholders. This advantage may increase as markets for environmental services develop and there is a need to develop robust and resilient new forests able to supply these services over the longer term.

The best type of new forest from a conservation viewpoint would be a species-rich forest established to restore the habitats of forest-dependent species and that was not subject to future disturbances such as those caused by tree-felling (e.g. Fig. 1.2). These forests might be established using seedlings or seeds, enriching existing monocultures by manipulating canopy covers to enhance the establishment of additional species or by protecting natural regrowth. Such forests may not produce commercially attractive goods but may be attractive where there are markets for ecosystem services.

There is often an interaction between the 'what type' question and the 'where' question. A species-rich and structurally complex new forest would be preferable when developing a corridor to provide improved landscape connectivity but a simple monoculture might be quite sufficient if there is a need to simply increase evapo-transpiration in order to lower water tables and reduce the risk of salinization (see Chap. 14 by Harper et al., this volume). Likewise, a monoculture grown for pulpwood might be commercially attractive if grown next to a road but could be worthless if grown by a smallholder in a remote mountain area without good roads and where the cost of transport was high. This question is discussed by Bentrup et al. (Chap.5), Davis et al., (Chap. 15), Gobster (Chap. 8), Harper et al., (Chap.14), Jim (Chapter 6 in Stanturf et al. 2012), Han and Oliver (Chapter 5 in Stanturf et al. 2012), Larsen et al., (Chap. 9 this volume), Rosengren, (Chap. 17 in Stanturf et al. 2012) and Wilson et al., (Chap. 11) later in this volume.

1.5.4 How Should This Process of Restoration Be Managed?

A single private landholder or public agency might be able to resolve each of these questions without too much trouble but it is rather more difficult to do across a broader landscape mosaic where a variety of landholders are present. Under these circumstances several different approaches have been used. One is a largely top-down approach in which a government land use planning agency sets objectives and decides where and how to restore forests across the landscape. This approach appeals to many ecologists because it allows them to apply their hard won scientific knowledge in a way they believe will generate widespread benefits.

Some sophisticated modelling techniques have been developed that facilitate this approach (e.g. see Booth et al., Chap. 13 Stanturf et al. 2012 and Wimberly et al., Chap. 6, this volume). The planners can then use incentives and compensation to try to persuade landholders in key locations to adopt their restoration plans. The advantage of this approach is that treatments can be targeted to overcome specific conservation problems such as creating more habitat in particular areas for an endangered species or to solve erosion or watershed protection problems at certain sites. The disadvantage, however, is that it can be politically difficult because different stakeholders may have contrasting views about the need for any restoration on their land or about priorities for restoration (e.g. should the priority be given to species conservation or watershed protection?). And even where a need for restoration is accepted, landholders in priority locations might find that the compensation on offer does not match what they believe are their opportunity costs. Top-down approaches can be seemingly efficient but may be politically and economically contentious.

An alternative approach is rather more bottom-up and involves having the stakeholders themselves identify priority areas for restoration and thus where compensation might be needed. The advantage of this is that disagreements can be identified at an early stage, trade-offs can be discussed and the stakeholders as a group can decide on treatment priorities. The process is likely to be rather more complex than top-down planning but perhaps more politically acceptable and, because of this ultimately prove to be more sustainable. The disadvantage of leaving decisions entirely to local landholders can be that national priorities (e.g. including watershed protection or conservation issues) may be ignored unless external facilitators include them in the discussion. An example of a largely bottom-up approach but where external facilitators were involved is given by Boedhihatono and Sayer (Chap. 16).

There is, of course, a third possibility and that is a combination of top-down and bottom-up. There are a variety of ways this might be achieved but most involve a regional planning group identifying a series of alternative restoration scenarios and taking these to a meeting of stakeholders (or stakeholder representatives) who then discuss them and choose one (or develop a new alternative of their own based on the original proposals). Examples of this approach have been given by Bouroin and Castell (2011) as well as by Bentrup et al. (Chapter 5), Convery and Dutson (Chapter 12), Siregar et al. (Chapter 3 in Stanturf et al. 2012) and Pullar and Lamb (Chapter 1 in Stanturf et al. 2012).

1.6 Issues Deserving Further Study

Explicit (or at least implicit) answers to these four questions are needed for the implementation of all FLR projects. But finding answers to these questions often raises other more fundamental questions that must be resolved first.

1.6.1 Who Are the Stakeholders and How Can They Participate in FLR?

The most obvious stakeholders are those owning or managing land found within the landscape. But others with a legitimate interest are those who might be affected by restoration. They could include other farmers, national or regional water supply managers, consumers of forest products, those with an interest in conservation and the broader community as a whole. These various stakeholders differ in the extent of their interest, their financial resources and their political power or influence.

It is likely to be difficult to identify all of these people and devise an equitable method by which their voices can be heard and their interests represented. An even more difficult task will be to do this in a way that ensures this participation over the longer term that most FLR will normally take to implement.

1.6.2 How to Make Collaborative Decisions?

It is often difficult to get a large number of people to agree on a particular course of action, especially when it involves changing the way in which they use their land. Part of the problem is they may not all have an equal understanding of the facts of the case or of the implications of certain choices. But even when they do, some will bring quite different sets of cultural attitudes to the discussion than others. For example, some may be inclined to support 'conservation' especially if others, such as neighbours whom they respect, do so. But some people may be hostile to the notion of outsiders having any say at all in how they manage their land.

There are also difficulties in the decision-making process with the political elite tending to dominate proceedings and loud, self-confident speakers likely to overshadow quieter, less assertive speakers (and men over women). Traditional societies often have long-standing institutions or methodologies to make collaborative decisions. But communities made up of more recent arrivals are less likely to have these. It can be difficult to arrive at mutually satisfactory decisions in situations where a government agency seeks to generate a change that benefits the broader community at the (perceived) expense of a private landholder. In such cases it is usually necessary to bring in a facilitator acceptable to all parties to initiate the process. Ideally, this should be institutionalised and a new collective decision-making body established to continue to manage the process over the longer time period that is usually necessary for FLR.

1.6.3 How Can Reforestation/Restoration Be Made Economically Attractive to Landholders and Especially Those in Key Locations?

Some landholders will occupy especially important locations within the landscape such as areas that might be used to create corridors between existing forest remnants or steep lands that are currently eroding. Some of these landholders may be uninterested in planting trees or may only want to plant simple monocultures using exotic species that have limited value for conservation or watershed protection. Those with existing monocultures may not wish to enlarge the diversity of tree species present even when this is clearly necessary to improve biodiversity conservation. Some of these landholders may be persuaded to engage in restoration once the practice is more fully explained. But restoration may be seen by many landholders as an unconventional land use activity and they may be unwilling to change unless they perceive that they will benefit (Convery and Dutson, Chap. 12). Compensation or the development of a market for ecosystem services may tip the balance in favour of some kind of restoration but then the task becomes one of coordinating activity and payments across a large number of landholders.

1.6.4 How to Accommodate Disagreements?

Disagreements can sometimes be resolved, especially if a facilitator is available. But sometimes they cannot. Where only a few individuals are involved it may be possible to offer some kind of compensation (e.g. a cash payment or alternative land elsewhere). But more fundamental disagreements between, say, the resident and non-resident stakeholders may be more difficult to resolve and could mean that certain restoration goals may take many years to achieve.

1.6.5 How to Integrate Biophysical and Socio-economic Constraints/Imperatives?

Forests are often managed to generate a variety of benefits including timber and various ecosystem services. The same is true, in theory, of restoration. But maximum financial gains to landholders may come at the expense of certain services. For example, it may be more attractive for landholders near a busy timber market to grow fast growing trees in monocultures than to grow species-rich forest for biodiversity conservation. The landholder's task of judging how to restore a forest is made more difficult because of the uncertainty over future markets for forest products or ecosystem services. In some places it seems the latter (in the form of markets for carbon sequestration) could become more important than the former. Likewise, for government agencies seeking to manage the balance between national and private benefits it is difficult to decide whether to spend a large amount of resources restoring a highly degraded resource or to use the same amount of resources improving a larger but less degraded area.

1.6.6 How to Ensure That Populations of Threatened or Endangered Species Can Persist Across Landscapes?

Restorationists generally assume that wildlife will respond to increased amounts of forest and this is probably true in many cases. But wildlife species differ in their habitat requirements. Some are generalists and are able to use pretty well any form of forest cover. But others are rather more specific in their requirements and need particular structural features or food species to be restored before they will begin to use a site. The most difficult species to cater for are those with large home ranges and needing extensive areas of particular types of habitat. These species are often those that are most affected by deforestation and which become classed as 'threatened' or 'endangered'. A key task for restorationists, therefore, is not so much one of restoring 'biodiversity' but to protect or restore the populations of these most vulnerable species.

Han and Oliver (Chapter 7 in Stanturf et al. 2012) provides an interesting case on forest restoration and management to protect and conserve a tiger population in north-eastern China, where the authors identify key requirements for the forest habitat that are quite different from the virgin forest habitat that traditionally has been described as key tiger habitat. This case strongly demonstrates how important it is to clearly identify the mechanisms behind a target species population decline and properly link this to informed management now and in the future. This includes proper interpretation of land-use history and its impact on the target species to reach a sound explanation of how the species became threatened; and it paves the road for efficient and reliable FLR-efforts to support such a vulnerable species.

1.6.7 Monitoring Outcomes

Restoring degraded landscapes is difficult and success is not assured (Hobbs et al. 2003). The problem of predicting successional outcomes and changes in the population of wildlife as forests are restored is complex enough but is made even more so when economic and social factors must be taken into account. Both ecological and economic circumstances can alter, unexpected events may occur and people can change their minds. The most common approach to dealing with such problems is to develop some form of monitoring and adaptive management (Walters and Holling 1990). Suggestions about how this might be approached are given in Danielson et al. (2005), Lindenmayer and Likens (2010) and Lamb (2011) but there are few examples where these have been used to monitor large landscape-scale restoration over the long time periods that are necessary to establish appropriate successional trajectories. The issue is considered further by Allan et al. (Chap.10).

1.7 Conclusion

The extent of global deforestation and of land needing some form of restoration means that ways must be found to increase the scale at which restoration is carried out. The task is not simply to scale-up using an existing set of established silvicultural techniques but to find ways of intervening within already complex landscape mosaics in order to improve ecological functioning and also improve the livelihoods of people now living within that landscape as well of those of the future.

There may be a variety of approaches actually used to restore forests within a particular landscape mosaic. Landowners at some sites may wish to maximise the production of goods such as timber while those at other sites may wish to maximise the provision of ecosystem services. Some landowners may want to achieve both.

The advantage of FLR is that it is easier to make trade-offs involving these contrasting options at a landscape scale than at an individual landholding. Future challenges such as those imposed by climate change mean that the nature of these trade-offs will vary over time. To date there are few examples of where Forest Landscape Restoration, as defined here, has been successfully achieved and where the process has been in place for any substantial length of time. This means the subsequent chapters in this book provide only a first indication of the processes involved, the variety of methods that might be employed and the problems we are still to overcome.

References

Aronson J, Dhillion S, Floc'h L (1995) On the need to select an ecosystem of reference, however imperfect: a reply to Pickett and Parker. Restor Ecol 3:1–3

Asner GP, Loarie SR, Heyder U (2010) Combined effects of climate and land-use change on the future of humid tropical forests. Conserv Lett 3:395–403

Aylward B (2005) Towards watershed science that matters. Hydrol Process 19(13):2643–2647

Beaumont L, Pitman A, Poulsen M, Hughes L (2007) Where will species go? Incorporating new advances in climate modelling into projections of species distributions. Glob Chang Biol 13(7):1368–1385

Berry PM, Rounsevell MDA, Harrison PA, Audsley E (2006) Assessing the vulnerability of agricultural land use and species to climate change and the role of policy in facilitating adaptation. Environ Sci Policy 9(2):189–204

Biot Y, Blaikie PM, Jackson C, Palmer-Jones R (1995) Rethinking research on land degradation in developing countries. World Bank, Washington, DC

Bolte A et al (2009) Adaptive forest management in central Europe: climate change impacts, strategies and integrative concept. Scand J For Res 24(6):473–482

Bourgoin J, Castella J-C (2011) PLUP Fiction: land use simulation for participatory land use planing in northern Lao PDR. Mt Res Dev 31:78–88

Brockerhoff E, Jactel H, Parrotta J, Quine C, Sayer J (2008) Plantation forests and biodiversity: oxymoron or opportunity? Biodivers Conserv 17(5):925–951

Bruijnzeel LA (2004) Hydrological functions of tropical forests: not seeing the soil for the trees? Agric Ecosyst Environ 104(1):185–228

Carnus JM et al (2006) Planted forests and biodiversity. J For 104(2):65–77

Chazdon RL (2003) Tropical forest recovery: legacies of human impact and natural disturbances. Perspect Plant Ecol Evol Syst 6(1–2):51–71

Chomitz K (2007) At loggerheads? Agricultural expansion, poverty reduction and environment in the tropical forest. World Bank, Washington, DC

Clewell A, Rieger JP (1997) What practioners need from restoration ecologists. Restor Ecol 5:350–354

Danielson F, Burgess ND, Balmford A (2005) Monitoring matters: examining the potential of locally based approaches. Biodivers Conserv 14:2507–2542

Dey DC, Gardiner ES, Kabrick JM, Stanturf JA, Jacobs DF (2010) Innovations in afforestation of agricultural bottomlands to restore native forests in the eastern USA. Scand J For Res 25(1 supp 8):31–42

Elliott S, Blakesley D, Maxwell JF, Doust S, Suwannaratana S (2006) How to plant a forest: the principles and practice of restoring tropical forests. Biology Department, University of Chiang Mai, Chiang Mai

FAO (2011) State of the world's forests 2011. Food and Agricultural Organization of the United Nations, Rome

Foley JA et al (2005) Global consequences of land use. Science 309(5734):570–574

Foster DR, Knight DH, Franklin JF (1998) Landscape patterns and legacies resulting from large, infrequent forest disturbances. Ecosystems 1(6):497–510

Fox J, Vogler J (2005) Land-use and land-cover change in montane mainland southeast Asia. Environ Manag 36:394–403

Gerber J-F (2011) Conflicts over industrial tree plantations in the South: who, how and why? Glob Environ Chang 21:165–176

Groffman P, Baron J, Blett T, Gold A, Goodman I, Gunderson L, Levinson B, Palmer M, Paerl H, Peterson G, Poff N, Rejeski D, Reynolds J, Turner M, Weathers K, Wiens J (2006) Ecological thresholds: the key to successful environmental management or an important concept with no practical application? Ecosystems 9:1–13

Hahn K, Emborg J, Larsen JB, Madsen P (2005) Forest rehabilitation in Denmark using nature-based forestry. In: Stanturf J, Madsen P (eds) Restoration of boreal and temperate forests. CRC Press, Boca Raton

Hobbs RJ, Cramer VA, Kristjanson LJ (2003) What happens if we can't fix it? Triage, palliative care and setting priorities in salinising landscapes. Aust J Bot 51:647–653

Hobbs RJ, Higgs E, Harris JA (2009) Novel ecosystems: implications for conservation and restoration. Trends Ecol Evol 24:599–605

Hobbs RJ, Hallett LM, Ehrlich PR, Mooney HA (2011) Intervention ecology: applying ecological science in the twenty-first century. BioScience 61(6):442–445

Hudson PF, Alcántara-Ayala I (2006) Ancient and modern perspectives on land degradation. Catena 65:102–106

IPPC (Intergovernmental Panel on Climate Change) (2007) Synthesis report. Summary for Policy Makers. IPCC Secretariat, Geneva

Iverson LR, Schwartz MW, Prasad AM (2004) Potential colonization of newly available tree-species habitat under climate change: an analysis for five eastern US species. Landsc Ecol 19(7):787–799

Janzen D (1988) Tropical ecological and biocultural restoration. Science 239:233–234

Kolstad I, Søreide T (2009) Corruption in natural resource management: implications for policy makers. Resour Policy 34:214–226

Lal R (2001) Soil degradation by erosion. Land Degrad Dev 12(6):519–539

Lamb D (1998) Large-scale ecological restoration of degraded tropical forest lands: the potential role of timber plantations. Restor Ecol 6(3):271–279

Lamb D (2011) Regreening the bare hills: tropical forest conservation in the Asia-Pacific region. Springer, Dordrecht

Lee DK, Suh SJ (2005) Forest restoration and rehabilitation in Republic of Korea. In: Stanturf JA, Madsen P (eds) Restoration of Boreal and Temperate forests. CRC Press, Boca Raton, pp 383–396

Lindenmayer DB, Fischer J (2006) Habitat fragmentation and landscape change. CSIRO Publishing, Collingwood

Lindenmayer DB, Likens G (2010) The science and application of ecological monitoring. Biol Conserv 143:1317–1328

Lindenmayer D et al (2008) A checklist for ecological management of landscapes for conservation. Ecol Lett 11:78–91

Lindner M et al (2002) Integrated forestry assessments for climate change impacts. For Ecol Manag 162(1):117–136

Liu Y, Stanturf J, Goodrick S (2010) Trends in global wildfire potential in a changing climate. For Ecol Manag 259(4):685–697

Llewellyn DW et al (1996) A decision-support system for prioritizing restoration sites on the Mississippi River Alluvial Plain. Conserv Biol 10(5):1446–1455

Lockhart BR, Gardiner E, Leininger T, Stanturf J (2008) A stand-development approach to oak afforestation in the Lower Mississippi Alluvial Valley. South J Appl For 32(3):120–129

Madsen P, Jensen FA, Fodgaard S (2005) Afforestation in Denmark. In: Stanturf J, Madsen P (eds) Restoration of Boreal and Temperate forests. CRC Press, Boca Raton

Maginnis S, Jackson W (2007) What is FLR and how does it differ from current approaches? In: Rietbergen- McCracken J, Maginnis S, Sarre A (eds) The forest landscape restoration handbook. Earthscan, London, pp 5–20

MEA (2005) Ecosystem and human well-being: synthesis. Island Press, Washington, DC, www.millenniumassessment.org/en/Synthesis.aspx

Milad M, Schaich H, Bürgi M, Konold W (2011) Climate change and nature conservation in Central European forests: a review of consequences, concepts and challenges. For Ecol Manag 261:829–843

Nassauer JI, Opdam P (2008) Design in science: extending the landscape ecology paradigm. Landsc Ecol 23:633–644

Omernik JM, Bailey RG (1997) Distinguishing between watersheds and ecoregions. J Am Water Resour Assoc 33(5):935–949

Parrotta JA, Knowles OH (2001) Restoring tropical forest on lands mined for bauxite: examples from the Brazilian Amazon. Ecol Eng 17:219–239

Parrotta J, Turnbull J, Jones N (1997) Catalyzing native forest regeneration on degraded tropical lands. For Ecol Manag 99(1–2):1–7

Radford J, Bennett A, Cheers G (2005) Landscape-level thresholds of habitat cover for woodland-dependent birds. Biol Conserv 124:317–337

Ramankutty N, Foley JA (1999) Estimating global changes in land cover: croplands from 1700 to 1992. Global Biogeochem Cycles 13:997–1027

Saxon E, Baker B, Hargrove W, Hoffman F, Zganjar C (2005) Mapping environments at risk under different global climate change scenarios. Ecol Lett 8(1):53–60

SER (2004) The SER International primer on ecological restoration. Science and Policy Working Group, Society for Ecological Restoration International.

Stanturf JA (2005) What is forest restoration? In: Stanturf JA, Madsen P (eds) Restoration of boreal and temperate forests. CRC Press, Boca Raton, pp 3–11

Stanturf JA et al (2009) Restoration of bottomland hardwood forests across a treatment intensity gradient. For Ecol Manag 257(8):1803–1814

Stanturf JA, Madsen P, Lamb D (2012) A goal oriented approach to forest landscape restoration. Springer, Dordrecht.

Tak K, Chun Y, Wood PM (2007) The South Korean forest dilemma. Int For Rev 9:548–557

Walters CJ, Holling CS (1990) Large scale management experiments and learning by doing. Ecology 71:2060–2068

Chapter 2
What Can Landscape Ecology Contribute to Forest Landscape Restoration?

Thomas R. Crow

2.1 Introduction

The art and science of restoration is now commonly viewed as integral to managing natural resources (e.g., Hobbs and Harris 2001; Stanturf and Madsen 2004; Falk et al. 2006; Bosworth and Brown 2007). This view is due in large part to an increased awareness that the cumulative impacts of human activities are profoundly affecting the biosphere – the ultimate ecosystem on which we all depend for our life sustaining goods and services. Effective restoration, however, requires confronting some rather perplexing questions. What is to be restored and to what end? How much restoration will be required in order to obtain the desired outcomes? Where and when on the landscape should restoration occur in order to maximize the benefits? And what defines success? These questions require considering subjects ranging from environmental ethics to cultural values in addition to the technical knowledge and professional judgment that tend to dominate the practice of restoration (Light and Higgs 1996; Light 2000; Vining et al. 2000; Davis and Slobodkin 2004).

Landscape ecology, with its emphasis on pattern and process at large spatial scales, when combined with ecological restoration creates the potential for a "big picture" approach to supporting restoration activities and to aiding the related decision-making that occurs. Landscape ecology provides a useful contextual framework for considering the many dimensions of restoration – historic, social, cultural, political, aesthetic, moral, and ecological – that are implicit in the above questions. Landscape ecology also adds an important spatial component to the practice of restoration.

Others have recognized the value of applying a landscape perspective to restoration. Naveh (1994), for example, was an early advocate for applying a landscape perspective to restoration, and in his 1994 paper, he explored the relation between restoration and

T.R. Crow (Retd) (✉)
Rocky Mountain Research Station,
US Forest Service, Fort Collins, Colorado, USA
e-mail: crow.thomasr@gmail.com

J. Stanturf et al. (eds.), *Forest Landscape Restoration: Integrating Natural and Social Sciences*, World Forests 15, DOI 10.1007/978-94-007-5326-6_2,

landscape-level processes. Bell et al. (1997) suggest that a union between restoration and landscape ecology could benefit the science and application in both disciplines. In their paper *A Historical Perspective and Future Outlook on Landscape Scale Restoration in the Northwest Wisconsin Pine Barrens*, Radeloff et al. (2000) state "The concurrent discussions of landscape scale restoration among restoration ecologists, and of historic disturbance pattern as a guideline for forest management among forest scientists, offer a unique opportunity for collaboration between these traditionally separated fields." My purpose is to further explore this union between ecological restoration and landscape ecology to address the question: What can landscape ecology contribute to forest landscape restoration.

2.2 Addressing the Questions

Answers to the questions posed in the Introduction depend largely on the social and political contexts in which restoration is being conducted as well as what is known about the ecosystems being restored (Light and Higgs 1996; Higgs 1997). The common desire to return an ecosystem to some previous state could represent a condition that was once common and has been largely lost or the restored ecosystem has some special quality such as providing critical fish habitat or providing an important ecosystem service such as clean water. As Lackey (2001, 2004) suggests, the very terminology used to characterize the condition of an ecosystem – terms such as healthy or damaged – are more likely to be value-based rather than science-based. Because societal values and preferences are important in the decision-making process, there are many possible "correct choices" as outcomes for restoration. And if we accept this premise, then defining ecological restoration as "the process of restoring one or more *valued processes or attributes* of a landscape" makes good sense (Davis and Slobodkin 2004).

2.3 Restoring Forests and Climate Change

The utility of using historic ecosystem conditions as a goal or even as a reference point for restoration is questionable when global climate change is considered. Given the long time frames necessary for forest restoration and given the likely changes in temperature, precipitation, and the frequency of extreme events, prescribing goals for restoration based solely on historical references will be at best challenging and at worst a recipe for failure (Harris et al. 2006).

As Davis and Slobodkin (2004) suggest in their definition of restoration, a strategy for dealing with this conundrum is to focus on restoring the "valued processes" in addition to emphasizing ecosystem attributes. But by expanding the scope of restoration to include valued processes as well as attributes, restoration becomes a more forward thinking concept (Crow 2008). That is, forest restoration becomes an adaptive practice for dealing with changing conditions. The goal is to create and maintain healthy, productive, sustainable forest ecosystems through restoration practices. Here, restoration involves *active* management, in which a future state is defined and management is applied to create that state. Hence it is not a question of restoring valued processes *or* attributes of a landscape, but instead, one of restoring valued processes *and* attributes because the two are inextricably linked and can not be separated. When restoration is viewed as encompassing both processes and attributes, then a much broader range of viable approaches and acceptable outcomes are available to the forest manager.

A rapidly changing climate not only shifts restoration from a backward to a forward looking concept, it changes the fundamental rationale for managing forests toward resilience and sustainability (Crow 2008). Not only are there challenging scientific and technical issues associated with this shift, but there are important moral and social issues as well. At present, only limited practical guidance is available to managers about how to address climate change, and the uncertainties about the specific conditions that will prevail at local and regional levels are huge. In a report published in 2007 by the United States Government Accountability Office (GAO) entitled "Climate Change: Agencies Should Develop Guidance for Addressing the Effects on Federal Land and Water Resources," the GAO found that land and water management agencies such as the Bureau of Land Management, Forest Service, Fish and Wildlife Service, National Oceanic and Atmospheric Service, and the National Park Service "have not made climate change a priority, and the agencies' strategic plans do not specifically address climate change" (GAO 2007). Instead, the report contends, managers tend to focus on near-term problems and mandated activities, with inadequate attention given to critical longer-term issues such as climate change. Strategies for managing forest, grassland, and aquatic ecosystems for adaptation and mitigation with a large dose of uncertainty are among the major challenges facing resource managers now and for many years to come.

2.4 Restoring Landscape Composition and Structure

At the forest stand level, attributes such as composition and structure are generally expressed as the dominant species and by age- and size-class distributions. At the landscape level, categories of land cover are common attributes, with categories ranging from broad, dichotomous classifications such as forest and non-forest, to specific cover classes that recognize individual species, species groups, or vegetative communities, e.g., communities in the United States such as red pine plantations, oak-hickory forests, red cedar glades or tallgrass prairies. The number, size, shape, and spatial arrangement of these classes define landscape structure. Just as reference stands help guide the compositional and structural goals for restoring forest stands, reference landscapes when available can help define the desired conditions for land cover and landscape structure.

Restoration of landscape structure is becoming more common. Fragmentation in which larger blocks or patches of habitat are converted to smaller patches through land use and ownership is pervasive in landscapes subject to intense human activity

(Turner et al. 1996; Crow et al. 1999). Landscape fragmentation profoundly affects the fluxes of radiation, the movement of air (wind) and nutrients as well as the movement of organisms within and across the landscape (Saunders et al. 1991; Gustafson and Gardner 1996). One consequence of fragmentation is the loss of landscape connectivity that facilitates movement by organisms among habitat patches and the loss of interior environments that provide critical habitat for many species. Restoring connectivity to facilitate the movement of organisms within landscapes will receive increased attention as resource managers and planners explore ways to adapt to a changing climate.

In natural environments, variability in the physical environment coupled with natural disturbance creates structural and compositional complexity that is expressed as landscape heterogeneity. This means that disturbance regimes and the resulting patterns of stand density, species composition, and age-class distribution differ significantly across the landscape and throughout a region. This variation enhances ecological processes and biological diversity. But human-dominated landscape tend to become homogenized as measured by their landscape composition and structure because many forest management practices, and more generally, land-use practices tend to simplify the variation that naturally occurs. An example can be found in the northern Great Lakes region of the United States where Schulte et al. (2007) quantified the consequences of a century of Euro-American land use based on historical public records, current forest inventories and other land cover data. Their analyses show "a distinct and rapid trajectory of vegetation change toward historically unprecedented and simplified conditions." In addition to overall loss of forestland, current regional forests have lower species diversity and less structural complexity compared to forests prior to Euro-American settlement (Schulte et al. 2007). They recommend a coordinated effort at the regional level among land management agencies to restore forest diversity and complexity.

Ponderosa pine (*Pinus ponderosa*) in the southwestern United States offers another good example of restoring attributes. The composition and structure of these forests are shaped through time by fire, episodic regeneration, insect infestation, and regional climate events such as drought (Covington et al. 1997; Allen et al. 2002). These processes created complex forest patterns at both the stand and landscape levels, with patterns shifting through time within a natural range of variability; but with Euro-American settlement came logging, livestock grazing, and fire suppression. Logging greatly reduced the number of large trees, and livestock grazing and fire suppression promoted the development of unnaturally dense stands of understory trees. Covington et al. (1997) provide a striking example of the changes that have occurred in forest structure in their study area located in the Fort Valley Experimental Forest near Flagstaff, Arizona. Small diameter trees (<40 cm dbh) increased dramatically from 1.4 per ha in 1876 to 513 in 1992, while large trees decreased from 19 per acre in 1876 to 13 in 1992. Decreases in the abundance and diversity of grasses and forbs occurred as well due to increased overstory competition, to increased competition for moisture, and to the formation of thick needle mats on the forest floor. Changes in composition and structure in these ponderosa forests, in turn, created changes in ecological processes such as the hydrologic cycle. More densely forested watersheds are likely to decrease total streamflows, peak flows, as well as base flows (Allen et al. 2002).

Managing the age-class structure of forests on the landscape can help create structural complexity. All age-classes, young and old, should occur somewhere on the landscape. When all age-classes are represented, greater compositional and structural complexity exists at the landscape and regional scales. This model for forest landscape management emphasizes the dynamic nature of stand development in which forest age-classes are viewed as a shifting mosaic of stands or patches on a landscape that includes young, mature, and old forests, and in which young forests become mature forests, mature forests become old, and harvesting or natural disturbances change mature and old forests into regenerating forests. Using a negative exponential model, Johnson et al. (1995) predicted the rate at which parts of the landscape will survive disturbance, and consequently predicts the percentage of the landscape that will survive to become old forest.

How do we fit the dynamic, shifting-mosaic model into a working landscape? Timber management generally truncates the process of stand development at rotation age. If management is uniformly applied across the landscape, i.e., all stands are managed to rotation age, the landscape structure becomes homogenized. That is, the landscape is dominated by forests that are rotation age or younger, and thus, the diversity of the forest is diminished at the landscape and region levels. If the landscape is managed, however, so that all age classes are represented, from the very young to forest stands 200, 300, or 400+ year old, then the structural diversity or, more generally, the ecological diversity of the forest is enhanced.

This model requires active management of the age-class structure on the landscape to ensure that all-age classes are present currently, or more likely, in the future. Good planning and landscape management create a 'working landscape' that is producing multiple values and benefits including forest products. A long-term institutional commitment is obviously required for creating a "working landscape." Such a commitment is more likely to occur on public lands such as National Forests than on private lands. Heterogeneous landscapes as defined by genetic structure, compositional diversity, and age-class structure are more resistant to disease and insect outbreaks and climate change as compared to homogeneous landscapes.

2.5 Restoring Processes on the Landscape

Although restoration activities are often directed at changing ecosystem composition and structure, the ultimate goal is restoring ecological processes to create healthy forest ecosystems that are sustainable with changing conditions. Again, ponderosa pine in the southwestern United State provides a good example. As the open, parkland forest was transformed into a closed-canopy forest, these changes as noted resulted in lower forest productivity, less understory plant diversity, decreases in stream flows, and increases in fire size and severity. A combination of mechanical thinning and surface fires have been used to restored the structural integrity of these forests. A return to the open parkland is projected to "increase soil moisture and mineralization and uptake of nitrogen, leading to increased photosynthesis, growth efficiency, and resin production" in the residual trees as well as increase the diversity and productivity of the understory vegetation (Covington et al. 1997).

Most restoration activities are local in their application. The challenge is to take these local efforts and make them operational at the landscape level. Large-scale restoration efforts to return functionality to the landscape, such as the Everglades in south Florida or to clean-up the Chesapeake Bay in the mid-Atlantic region in the United States, have proven to be difficult due to the scale at which restoration must occur and because of the large number of people, organizations, and political entities that have a stake in the outcome.

There are, however, efforts directed at maintaining inherent landscape patterns through forestry practices that serve as useful case studies. In response to a commitment in Sweden to maintain biological diversity in forests, the Swedish Forestry Act of 1994 directs harvesting of timber in a way that mimics natural processes and builds more complex structure at the stand and, where applicable, at the landscape level (Lämås and Fries 1995). Further, Bondrup-Nielsen (1995) describes an approach that involves retaining forest reserves and managing harvest patterns in the landscape matrix in order to mimic patterns on the landscape created by the physical environment and through natural disturbance. Other authors have used models to consider harvest patterns in both space and time. Gustafson and Crow (1996), for example, used the model HARVEST to evaluate alternative timber harvesting strategies in actual landscapes in which the size, spatial distribution, and rate of harvesting are varied and the landscape structure as measured by patch size distribution, linear forest edge, amount of forest interior is quantified after an extended period of treatment. The goal of these exercises is not to optimize timber production, but to simulate alternative management strategies by incorporating decisions that are typical of those made by resource managers (Crow and Gustafson 1997). Their results help define some guiding principles for landscape management. Larger harvest units that are aggregated always create diversity in patch sizes, reduce the amount of forest edge, and protect interior habitats better than small, dispersed harvest units. As the size of the harvested units decreases, however, the advantage of aggregation of harvest units is quickly lost. There is also an economic advantage to creating large, aggregated harvest units. In this case, both economics and ecology argue for the same landscape management strategy.

A different landscape, in this case a matrix dominated by agriculture, offers another example of restoring functionality. Landscape simplification in agricultural systems combined with the genetic engineering and intensive use of synthetic pesticides and fertilizer, along with irrigation have produced dramatic increases in food production. These increases have come with significant environmental costs: excessive sedimentation in waterways, contamination of drinking water, seasonal hypoxic zones in coastal areas, increased frequency of severe flooding, loss of critical fish and wildlife habitat to name a few. Perennial systems embedded in agricultural landscapes can provide water purification, help reduce the loss of nutrients and sediments, control pests, reduce water temperatures, and enhance biological diversity (Schulte et al. 2006). Among the options, perennial riparian buffers along streams are best known, but many other options such as vegetative filter strips composed of perennial grasses located along the contour of hillslopes or a variety of agroforestry systems are options as well. A small amount of the right perennial system located in the right places in the landscape can greatly improve ecosystem health in these agricultural landscapes.

The assumption that a small amount of perennial vegetation placed in the right locations in the landscape can greatly improve water quality is based on concepts of thresholds, tipping points, and nonlinear responses. These concepts are implicit in questions such as: What are the cumulative impacts of restoration applied at the local level? How much local restoration is needed to make a difference at a landscape and regional level? Where in the landscape should restoration be practiced to maximize the benefits at the larger spatial scales? Addressing these questions represent some of the fundamental challenges facing forest landscape restoration.

The importance of addressing these questions can be illustrated by considering the action plan for reducing, mitigating, and controlling hypoxia in the northern Gulf of Mexico as developed by the Mississippi River/Gulf of Mexico Watershed Nutrient Task Force (2001). Here, algal blooms resulting from excessive nitrogen, primarily from nonpoint sources within the Basin, reduces the dissolved oxygen in the water column, and correspondingly, results in the loss of aquatic habitat. Although even small reductions in nitrogen loads are desirable, modeling studies suggest that reductions in nitrogen loads into the Gulf of at least 30–40% will be necessary to reach the goal of reducing the size of the hypoxic area to $<5,000\ km^2$. What is not clear, however, is to what extent land use practices will need to change in order to reach this goal. Again, the fundamental questions at the landscape and region level are: where, how much, and what kind of change in land use is needed?

2.6 Managing for Resilience

The concept of resilience relates to the notions of robustness, sustainability, and risk. It applies to both biological and social systems, and at the most fundamental level, resilience deals with the ability of biological and social systems to adapt and even to benefit from change (Holling 1973; Walker et al. 2004). Resilient forests, then, are those that can accommodate a changing climate or changes in natural or anthropogenic disturbance while maintaining their diversity, health, and productivity.

Promoting resilience is commonly suggested as an adaptive management strategy for dealing with a changing climate. Millar et al. (2007) provide a list of forest management practices to enhance resilience (Table 2.1), and many of these are, in fact, restoration activities. Because of climate change, less emphasis is being given to restoring ecosystems to prior states and more emphasis is given to restoring ecological processes that create diverse, productive, and healthy ecosystems. Restoration

Table 2.1 Recommendations from Millar et al. (2007) for promoting resilience to climate change in forests

Assist transitions, population adjustments, range shifts, and other natural adaptations – This recommendation involves managing for conditions anticipated in the future. Many of these involve traditional silviculture treatments such as thinning or the establishment of target species in plantation outside their present ranges
Increase redundancy – Risk is decreased by increasing redundancy. View this as spreading your risk as opposed to concentrating it
Expand genetic diversity guidelines – In the past, guidelines for genetic management stressed the use of local genotypes to avoid contamination of gene pools with poorly adapted genotypes. This assumed that environmental conditions were static. Reformulating these guidelines may be needed to reflect the reality of climate change
Manage for asynchrony and use establishment phase to reset succession – Because climate change is a global phenomena, synchronous changes in biota at the local and landscape levels can be expected. This could provide less diversity and less resilience to future changes. Management to increase the diversity of forest structure and composition, age-classes, as well as geotypic variation might be needed
Establish "neo-native" forests – Paleohistorical records provide information about historic ranges for species and their responses to climate change in the past. Once again, the emphasis is on managing for anticipated conditions, not current conditions
Promote connected landscapes – Landscape level planning is necessary to reduce forest fragmentation and thus enhance forest connectivity and the movement of forest species
Realign significantly disrupted conditions – Realign means restore, but with a forward, not backward, perspective of restoration
Anticipate surprises and threshold effects – There are always surprises, many of which are due to nonlinear responses to changing conditions and to management
Experiment with refugia – The concept of refugia needs rethinking in the light of climate change. For example, many actively managed public lands will be increasingly viewed as important refugia for recolonizing disturbed landscapes

Many of these recommendations relate to landscape restoration

then becomes a proactive and anticipatory strategy with the goal of creating forests that are sustainable under likely future conditions. It becomes a means for promoting adaptability for both terrestrial and aquatic ecosystems. In many cases, intensive management will be needed. These treatments could include modifying harvest schedules, altering thinning prescriptions, replanting with different species or genotypes, or creating landscapes that facilitate the migration of species thought to be at risk. Restoration still deals with transforming the composition, structure, and function of ecosystems. But instead of looking backward, the goal is to transform ecosystems to be better suited, that is, more resilient, to future conditions.

2.7 Dealing with Uncertainty

Enormous uncertainties are associated with restoration activities at any spatial scale and this is especially true at the landscape level. These uncertainties are due in part to a lack of understanding about the responses of complex landscapes, but also to uncertainties related to changing conditions, e.g., climate change, and our limited abilities to predict these changes at local levels. A prudent response to uncertainty is to use an adaptive strategy that involves monitoring with adjustments as necessary through time. The key here is effective monitoring – a subject that is much discussed but often inadequately implemented.

In addition to an adaptive approach, other strategies help resource managers and planners deal with uncertainty. Fuller et al. (2008), for example, assessed the relative uncertainty associated with parameter values and input data used in computer models to project the response restoration in the Everglades by altering the hydrologic regimes. In their test, Fuller et al. (2008) compare the potential impacts of two alternative 30-year hydrologic regimes on wildlife habitat quality in the Florida Everglades. The approach, called relative assessment, can be used with a variety of modeling approaches, and it provides managers as well as stakeholders an objective method for testing the robustness of different management options.

Adaptive management is really about adaptive learning; that is, the willingness and the ability to incorporate new knowledge and information into the decision-making process (Lynch et al. 2008). The concept of "learning as you go" may seem rather lame, but involvement by scientists remains critical throughout the adaptive process, and both practical experience and new scientific information are powerful guides for the future. This interaction between scientist and manager is worth considerable thought. Moreover, the concept of adaptive learning also applies to the three-way relation among scientist, manager, and stakeholder, with the learning flowing in all directions. The goal, however, is to create a sense of ownership among stakeholders for the restoration effort in order to build and maintain support. There needs to be a meaningful way to give stakeholders a voice when addressing the question: to what end do we manage a landscape? As experiences such as the Greater Yellowstone Ecosystem have demonstrated, building this support is an arduous task over large areas such as the Greater Yellowstone Ecosystem with its many stakeholders and diverse interests (see Lynch et al. 2008). But success is impossible without involving those affected by the decisions being made.

2.8 Determining Success

A fair amount has been written about measuring the success of restoration efforts (e.g., Aronson and Le Floc'h 1996; Ruiz-Jaen and Aide 2005); however, little of the published material deals explicitly with assessing success at the landscape level. The material available, however, does provide some useful guidelines for landscape restoration. A good place to start is the primer published by the Society for Ecological Restoration International (SER 2004) that lists attributes that should be considered when assessing restoration projects. These attributes fall within three broad categories: (1) diversity, (2) vegetation composition and structure, and (3) ecological processes. Most efforts to measure success have focused on diversity and some measure of composition or structure. Ecological processes are often viewed with some justification as much more difficult and thus more expensive to measure.

The first point is an obvious one – criteria for success should be established prior to doing anything in the field. Also, select multiple variables and select variables that can be applied at multiple spatial scales. Some measurement of richness of the biota within the landscape is necessary as is some indication whether the composition and structure of the restored landscape are meeting the preset criteria for success. Has forest fragmentation increased or decreased as a result of restoration? Has the amount of edge habitat within the landscape been reduced? Has the amount of interior forest habitat increased or decreased? Are old as well as young forests present in the landscape? Are both large and small patches present in the landscape? Are species of concern reproducing? These are the type of question relating to diversity, composition, and structure that might be appropriate depending on the goals and objectives of the landscape restoration project.

For ecological processes, foresters have much experience in measuring processes, largely indirectly, by measuring attributes that generally reflect forest health. These include levels of productivity, rates of mortality, the abundance of regeneration for targeted species, and the dominance of indigenous (or its inverse, the dominance of invasive) species within the restored landscape. Other ecological processes or interactions, such as herbivory, dispersal, pollination, predation, and parasitism can be measured indirectly through surrogates (Ruiz-Jaen and Aide 2005).

2.9 Conclusion

A broad perspective is useful for several key issues relating to forest restoration. The global trend toward lower levels of heterogeneity in human dominated landscapes; the possibility that current and future conditions will fall outside the natural range of variability due to climate change; the critical and extensive losses in habitat that has caused declines in biological diversity; and the impacts on important ecosystem services that sustains us all; and our inability to deal with the whole while considering the pieces, all argue for a more comprehensive, integrative approach to resource management. Earlier the question was raised 'What can Landscape Ecology contribute to forest landscape restoration'. The answer is that landscape ecology provides managers, planners, scientists, and policymakers the opportunity to move from the individual pieces in which restoration projects are conducted at the local level to the whole in which the landscape that includes mixed ownerships, mixed land uses, and multiple land covers becomes the platform for restoring ecosystem health. It is this latter perspective that will be needed to deal effectively with the big threat – the interaction of climate change with many other anthropogenic and natural disturbances. It is a huge challenge for resource managers and planners, but the stakes are too high and the risks of doing nothing are too great to ignore the challenge.

The critical need for those involved in forest landscape restoration is determining the proper balance between recreating past conditions and attempting to direct landscape ecosystem toward compositional, structural, and functional conditions that are better suited for future environments. Although both goals may be achievable in some case, Harris et al. (2006) caution that achieving both may not always be possible.

There are no universal prescriptions for forest landscape restoration and to a great extent we will need to learn by doing and it will be necessary to tailor the approach to accommodate local conditions and to specific goals and expectations that are expressed by stakeholders. There are, however, a few guiding principles that are widely applicable. First, the strong support and active participation of all involved stakeholders is a prerequisite in order for restoration to be successful (Reitbergen-McCracken et al. 2007). Second, the fundamental goal for restoration is to keep all the biotic pieces and the related ecological processes so that people can continue to benefit from their ecosystem services. And third – an extension of the second – restoration is, above all, about keeping options open for future generations. In the final analysis, it is really very simple – practicing forest landscape restoration is part of being a good steward.

References

Allen CD, Savage M, Falk DA, Suckling KF, Swetnam TW, Schulke T, Stacey PB, Morgan P, Hoffman M, Klingel JT (2002) Ecological restoration of southwestern ponderosa pine ecosystems: a broad perspective. Ecol Appl 12:1418–1433

Aronson J, Le Floc'h E (1996) Vital landscape attributes: missing tools for restoration ecology. Restor Ecol 4:377–387

Bell SS, Fonseca MS, Motten LB (1997) Linking restoration and landscape ecology. Restor Ecol 5:318–323

Bondrup-Nielsen S (1995) Forestry and the boreal forest: maintaining inherent landscape patterns. Water Air Soil Pollut 82:71–76

Bosworth D, Brown H (2007) Investing in the future: ecological restoration and the USDA Forest Service. J For 105:208–211

Covington WW, Fulé PZ, Moore MM, Hart SC, Kolb TE, Mast JN, Sackett SS, Wagner MR (1997) Restoring ecosystem health in ponderosa pine forests of the southwest. J For 95:23–29

Crow TR (2008) A framework for restoration in the national forests. In: Deal RL (Technical editor) Integrated restoration of forested ecosystems to achieve multiresource benefits. In: Proceedings of the 2007 National Silviculture Workshop. USDA Forest Service, General Technical Report. PNW-GTR-733. Portland, OR, pp 13–21

Crow TR, Gustafson EJ (1997) Ecosystem management: managing natural resources in time and space. In: Kohm KA, Franklin JF (eds) Creating a forestry for the 21st century. The science of ecosystem management. Island Press, Washington, DC, pp 215–228

Crow TR, Host GE, Mladenoff DJ (1999) Ownership and ecosystem as sources of spatial heterogeneity in a forested landscape, Wisconsin, USA. Landsc Ecol 14:449–463

Davis MA, Slobodkin LB (2004) The science and values of restoration ecology. Restor Ecol 12:1–3

Falk D, Palmer M, Zedler JB (eds) (2006) Foundations of restoration ecology. Island Press, Washington, DC

Fuller MM, Gross LJ, Duke-Sylvester SM, Palmer M (2008) Testing the robustness of management decisions to uncertainty: Everglades restoration scenarios. Ecol Appl 18:711–723

GAO (2007) Climate change. Agencies should develop guidance for addressing the effects on federal land and water resources. GAO-07-863. Government Accountability Office, Washington, DC, 184 p

Gustafson EJ, Crow TR (1996) Simulating the effects of alternative forest management strategies on landscape structure. J Environ Manage 46:77–94

Gustafson EJ, Gardner RH (1996) The effect of landscape heterogeneity on the probability of patch colonization. Ecology 77:94–107

Harris JA, Hobbs RJ, Higgs E, Aronson J (2006) Ecological restoration and global climate change. Restor Ecol 14:170–176

Higgs ES (1997) What is good ecological restoration? Conserv Biol 11:338–348

Hobbs RJ, Harris JA (2001) Restoration ecology: repairing the earth's ecosystems in the new millennium. Restor Ecol 9:239–246

Holling CS (1973) Resilience and stability of ecological systems. Annu Rev Ecol Syst 4:1–23

Johnson EA, Miyanishi K, Weir JMH (1995) Old-growth, disturbance, and ecosystem management. Can J Bot 73:918–926

Lackey RT (2001) Values, policy, and ecosystem health. BioScience 51:437–443

Lackey RT (2004) Societal values and the proper role of restoration ecologists. Front Ecol Environ 2:45–46

Lämås T, Fries C (1995) Emergence of a biodiversity concept in Swedish forest policy. Water Air Soil Pollut 82:57–66

Light A (2000) Ecological restoration and the culture of nature: a pragmatic perspective. In: Gobster PH, Hull RB (eds) Restoring nature: perspectives from the social sciences and humanities. Island Press, Washington, DC, pp 49–70

Light A, Higgs ES (1996) The politics of ecological restoration. Environ Eth 18:227–247

Lynch HJ, Hodge S, Albert C, Dunham M (2008) The Greater Yellowstone Ecosystem: challenges for regional ecosystem management. Environ Manag 41:820–833

Millar CI, Stephenson NL, Stephens SL (2007) Climate change and forests of the future: managing in the face of uncertainty. Ecol Appl 17:2145–2151

Mississippi River/Gulf of Mexico Watershed Nutrient Task Force (2001) Action plan for reducing, mitigating, and controlling hypoxia in the Northern Gulf of Mexico. Mississippi River/Gulf of Mexico Watershed Nutrient Task Force, Washington, DC

Naveh Z (1994) From biodiversity to ecodiversity: a landscape-ecology approach to conservation and restoration. Restor Ecol 2:180–189

Radeloff VC, Mladenoff DJ, Boyce MS (2000) A historical perspective and future outlook on landscape scale restoration in the northwest Wisconsin pine barrens. Restor Ecol 8:119–126

Reitbergen-McCracken J, Maginnis S, Sarre A (eds) (2007) The forest landscape restoration handbook. Earthscan, London

Ruiz-Jaen MC, Aide TM (2005) Restoration success: how is it being measured? Restor Ecol 13:569–577

Saunders DA, Hobbs RJ, Margules CR (1991) Biological consequences of ecosystem fragmentation: a review. Conserv Biol 5:18–32

Schulte LA, Asbjornsen H, Liebman M, Crow TR (2006) Agroecosystem restoration through strategic integration of perennials. J Soil Water Conserv 61:165A–169A

Schulte LA, Mladenoff DJ, Crow TR, Merrick LC, Cleland DT (2007) Homogenization of northern U.S. Great Lakes forests due to land use. Landsc Ecol 22:1089–1103

SER (Society for Ecological Restoration) International Science & Policy Working Group (2004) The SER International Primer on Ecological Restoration. Available at http://www.ser.org. Society for Ecological Restoration International, Tucson, AZ

Stanturf JA, Madsen P (eds) (2004) Restoration of boreal and temperate forests. CRC Press/Lewis Publishers, Boca Raton, 569 p

Turner MG, Wear DN, Flamm RO (1996) Land ownership and land-cover change in the southern Appalachian highlands and the Olympic Peninsula. Ecol Appl 6:1150–1172

Vining J, Tyler E, Kweon B-S (2000) Public values, opinions, and emotions in restoration controversies. In: Gobster PH, Hull RB (eds) Restoring nature, perspectives from the social sciences and humanities. Island Press, Washington, DC, pp 143–161

Walker BCS, Holling SR Carpenter, Kinzig A (2004) Resilience, adaptability and transformability in social-ecological systems. Ecol. Soc 9(2):5. [Online] URL: http://www.ecology *and society*.org/vol2/iss2/art5

Chapter 3
Landscape Management

Chadwick Dearing Oliver, Kristofer Covey, Aaron Hohl, David Larsen, James B. McCarter, Andrew Niccolai, and Jeremy Wilson

3.1 Introduction

Natural resources are neither uniformly nor randomly distributed across the Earth. Rather, they are commonly grouped within geomorphologic and climatic boundaries. These groups—"Ecological Zones"—are generally large and cross political and socio-economic boundaries. It is cumbersome to coordinate effective management for many values across these large areas. Consequently, we subdivide ecological zones into smaller areas and then further subdivide these, creating a hierarchy of sizes for management (Fig. 3.1; Oliver 2003). The term "ecosystem" refers to an ecological grouping of biotic and abiotic factors at any scale (Chapin et al. 2002; Kimmins 2003).

C.D. Oliver (✉) • K. Covey
School of Forestry and Environmental Studies, Yale University, New Haven, CT, USA
e-mail: chad.oliver@yale.edu; kristofer.covey@yale.edu

A. Hohl
Department of Forestry and Wildland Resources, Humboldt State University, Arcata, CA, USA
e-mail: amhohl@yahoo.com

D. Larsen
School of Natural Resources, University of Missouri, Columbia, MO, USA
e-mail: LarsenDR@missouri.edu

J.B. McCarter
School of Environment and Forest Sciences, College of the Environment, University of Washington, Seattle, WA, USA
e-mail: jmac@uw.edu

A. Niccolai
Research and Development Center, United States Coast Guard, New London, CT, USA
e-mail: Andrew.M.Niccolai@uscg.mil

J. Wilson
Harris Center for Conservation Education, Hancock, NH, USA
e-mail: wilson@harriscenter.org

J. Stanturf et al. (eds.), *Forest Landscape Restoration: Integrating Natural and Social Sciences*, World Forests 15, DOI 10.1007/978-94-007-5326-6_3,

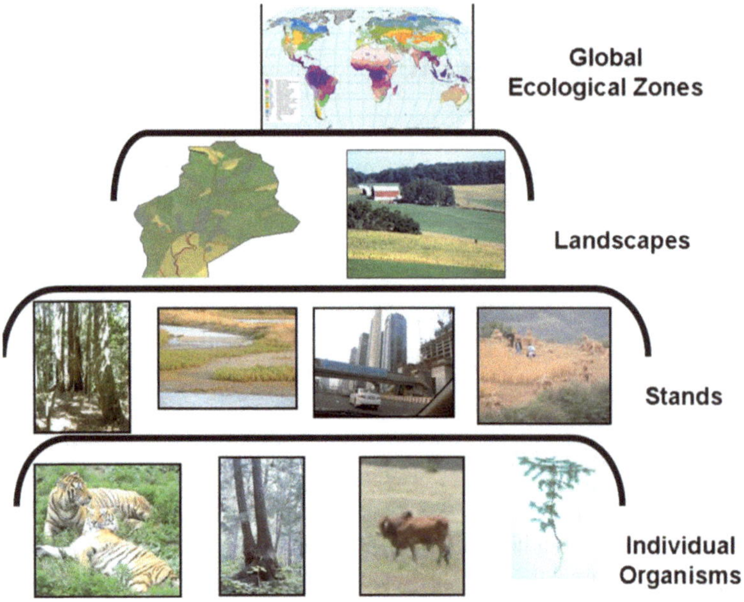

Fig. 3.1 Ecosystems can be viewed as a single whole (the Earth) or it can be subdivided into smaller areas (Ecological Zones) and then further subdivided, creating a hierarchy of different scales. Different management activities are effective at each scale

At a fine scale, we manage individual organisms such as crop plants, weeds, domestic and wild animals, and disease vectors. Groups of individual organisms and their abiotic environment are aggregated into contiguous areas that are relatively homogeneous and managed uniformly. These are generally referred to as 'stands' if forested but may be known by other names if they contain non-forest land covers. Contiguous aggregations of these stands are also the focus of management. These aggregations have many names, but will be termed 'landscapes' here.

In addition to being different in physical size, each scale also manifests different properties significant for management, can provide different values, and so is managed differently. This chapter focuses on the landscape scale of management, briefly comparing and contrasting it with the other scales. The chapter first describes the physical delineation and properties of landscapes. Then, it looks at values that can be provided at the landscape and other scales. Finally, it describes how different temporal and spatial scales can be integrated to achieve objectives in terrestrial landscapes.

3.2 The Landscape Scale

3.2.1 *Physical Delineation*

Each hierarchical level is described—or bounded—at a different scale. Organisms, for example, are readily described at the scale of the individual. Groups of organisms are bounded within areas of relatively uniform species, growth rates, spatial distributions, ages, and past histories. They are generally known as 'stands' or 'woodlots' if forested; but are termed agriculture fields, pastures, bogs, lakes, stream or river reaches, or residential or commercial lots if used for other purposes.

Landscapes—aggregations of contiguous stands, agriculture fields, bogs, stream reaches, and other such features—have been addressed by many disciplines and for many purposes. Consequently, the aggregations vary in size and boundaries at scales of tens to thousands of ha. Chapman (1931), Davis et al. (2001), Boyce (1995), O'Neil et al. (1988), Gosz (1993), Lidicker (1995), and Forman and Gordon (1986) have each defined hierarchies of sizes and parameters for classification; however, these definitions are not widely agreed-upon. Some described names and purposes are:

- Drainage basin or river catchment area: An area bounded by the ridge tops of a stream or river catchment area. This delineation is very useful for water management purposes.
- Watershed: Technically, this area is the reverse of a drainage basin, and includes an entire area that water flows away from—bounded by streams, rivers, or lakes on all sides; however, 'watershed' is often used as a synonym for 'drainage basin.'
- Estate: An estate is generally a large forested area under a single ownership, and management is coordinated to provide values to the landowner.
- Timbershed: An area from which all timber for a wood mill is expected to flow. It is usually of a size to supply the wood needs of a mill, biomass plant, or other production facility.
- Sustained Yield Unit: A timbershed, from which the wood can be supplied in perpetuity; i.e., the wood will regrow at least as fast as it is harvested.
- Home range: The area that an animal needs or uses to provide all of its needs in some or all seasons.
- Irrigation district: The total area that captures, stores, and utilizes water for irrigation.
- Political area: An area that is within a single political jurisdiction.

Managers working under the umbrella of each of these classifications have developed knowledge and techniques that help understanding and management at the multiple-stand level. It is appropriate to amalgamate this knowledge and those technologies. Agreement on a single way to describe 'landscapes' may eventually be reached with time as more management effort is concentrated at this scale.

Nomenclature for areas larger than landscapes, such as Global Ecological Zones (FAO 1997) or ecosystem provinces (Bailey 1983) are also still being refined. Their physical boundaries are commonly mountains, seas or oceans, and/or topographic changes that alter the climate and soils or political boundaries. As with individual landscapes, the similar cultural relations within the political boundary are often a major factor driving land use, water and species distributions, soil structure and nutrient content, and biodiversity (Goudie 2006).

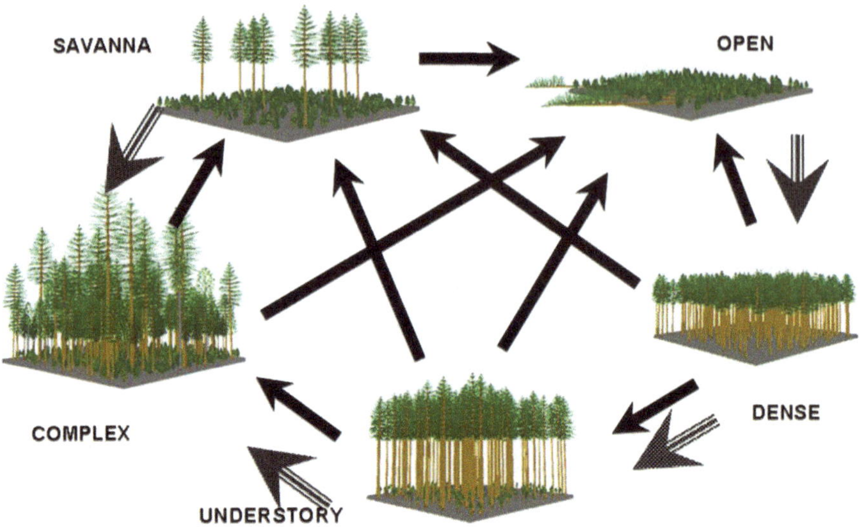

Fig. 3.2 Forests exist in a variety of structures, such as shown here. An individual area changes with growth (*grey arrows*) and with natural or human disturbances (*black arrows*). Each structure provides some, but not all, values. Consequently, a landscape that maintains a variety of stand structures will probably provide the most values that people want

3.2.2 Landscape Properties

Each hierarchical level (Fig. 3.1) has observable properties corresponding to scale. At the organism scale, we observe the size and health of the organism and its parts—for example, the size of seeds or fruit, the size and quality of tree stems, the health of animals.

These emergent properties at integrated scales are usually aggregated (Johnson 2006) properties that were not apparent at finer scales. At the stand scale we discuss volume of fruit or timber per unit area (e.g., ha or acre), plant density such as tree basal area per area, stand structure (Fig. 3.2), wind or fire susceptibility, tree age distribution, or soil structure. Some inherent properties are not readily predictable by simply aggregating measures at finer scales. While a stand's structure, for example, is not easily discernible by measuring individual organisms, at the stand level new properties driven by interactions between individuals become apparent. In forest stands, the differing shade tolerances and relative growth patterns of individuals and nearly universal patterns of canopy stratification and shifting times of dominance (Oliver 1992) are readily observable. These emergent properties determine what values are being provided and how the organism, stand, landscape, or larger ecosystem will change. A tree with a high height/diameter ratio is unlikely to be growing rapidly and will easily fall in a windstorm or be attacked by insects. Similarly, a stand in the open structure will host certain bird and butterfly species, but will not provide timber until it grows to another structure. Emergent properties have not yet been systematically studied at the landscape scale. Some reasonable emergent properties are suggested and described below and in Fig. 3.3.

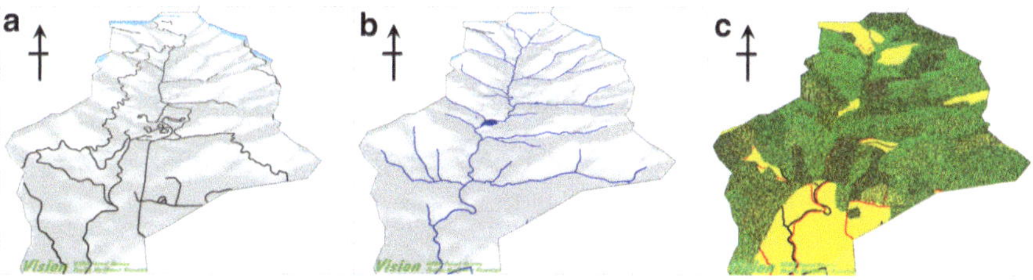

Fig. 3.3 Properties that are expressed at the landscape scale and provide many values include Landforms (**a** and **b**), Land Cover (**c**), Corridors (**a**), and Water Availability (**b**). Other properties are Seasonal Cycles, Disturbance Patterns, and others. (Envision copy of Bent Creek Forest, USDA Forest Service: +/− 2,500 ha)

3.2.2.1 Landform

The geomorphology, topography, and soil condition that characterize a landscape is here referred to as the landform. A landscape may consist of all or part of an alluvial floodplain, karstic topography, loess soils, sandy deposits, or metamorphic or igneous bedrock and resulting soils. It can also contain a combination of these. The landscape may be level, gently sloping, steep, or contain bench-like terrain. It can contain soils of different textures, structures, drainage, fertility, and erosiveness or distributions of these soil properties in different parts of the landscape. Each landform property is reflected in the plant growth potential; its ease of access; and in the minerals, wood, food, recreation, and other societal values expected from the landscape (Swanson et al. 1988).

3.2.2.2 Land Cover

The proportion of different land covers is an emergent property of a landscape. A landscape may be covered with a mixture of forests, grasslands, shrublands, deserts, wetlands, and rivers and lakes. Any of the land covers can form a matrix across the landscape, with other land covers embedded within. The forests within a landscape can contain different mixtures of tree species and varied amounts of each structure (Fig. 3.2); and non-forest areas can also have variations. Landscapes can also contain human modifications of cover—changed amounts of natural vegetation such as grasslands and forests, changed amounts of forest structures, and replacements of natural land covers with agriculture fields, pastures, lakes, or human dwellings. The arrangement of the different cover types may form meaningful patterns. For example, the closed forests or open vegetation covers can be clumped and form few edges with other vegetation covers; or, the closed and open covers can be dispersed in small areas with many edges but few large interiors away from edges. The proportion of edge to interior area has been shown to have an effect on a number of biological processes (Murcia 1995; Dijak and Thompson 2000; Chen et al. 1992). While it is clear that this ratio and other landscape indices like it provide useful measures of both habitat diversity and continuity, their use in predicting habitat suitability for a particular species has been questioned (Schumaker 1996).

As with landform, land cover largely determines what values a landscape is capable of providing. Animal and plant habitats are closely tied to land cover, as are the ability of the landscape to provide food, timber, and other products.

3.2.2.3 Water Availability

The amount and distribution of water within a landscape is another important property. Water can be absent, concentrated in one part of the landscape, or dispersed throughout it; its availability may vary seasonally and annually. Water can be stored and transported by streams and rivers, lakes, aquifers, springs, wells, water tanks, and/or irrigation canals.

Water is essential for most values from wildlife and plant survival to crop production, human habitation, and construction. Its distribution often influences other resources within the landscape, dictating whether the many values will become concentrated and compete vigorously for a few sources of water in a limited area or be dispersed over many parts of the landscape.

3.2.2.4 Corridors

The combination of availability, locations, and distribution of various terrestrial and aquatic corridors for movement of animals and people is an important property of a landscape (Tischendorf and Fahrig 2000; Bennett 1998). Terrestrial corridors are commonly animal and human paths or trails—and more recently highways, railroads, and other structures. Aquatic corridors are commonly rivers and streams. Stand structures can be important for movement of species (Haddad 1999; Fry and Money 1994). The landform can help dictate the amount and distribution of corridors. Sometimes corridors are lacking in parts of a landscape rendering these areas isolated. Corridors can also contain narrow passages where all travel must be concentrated—such as passes or river fords.

People can increase connectivity in a landscape intentionally by removing blocks to potential corridors, or by creating continuous habitat connections between fragmented landscape patches. They can also intentionally increase, decrease, or divert corridors by using roads, fences, and gates. Highways and railroads can inadvertently eliminate animal trails and change animal use of a landscape (Alexander and Waters 2000). The relative benefits and cost of corridors has been the subject of decades of intense debate (Haddad et al. 2000). In a series of papers Daniel Simberloff questioned the scientific basis for the use of corridors in reserve design and argued that corridors may be poor conservation investments (Simberloff and Cox 1987; Simberloff et al. 1992). Other authors have reached similar conclusions (Hobbs 1992). Harrison and Bruna (1999) cautioned that corridors were likely inadequate substitutions for overall habitat loss and one study investigating the efficacy of an existing corridor system found that despite being structurally suitable, corridors failed to provide for genetic exchange between populations in corridor-connected patches (Horskins et al. 2006). Beier and Noss (1998), however, conducted a review of published studies investigating corridors and argued that there was ample evidence supporting their use in landscape management. More recently, large-scale studies have demonstrated that corridors can be effective in aiding the dispersal—and subsequently the richness—of species in forest gaps (Damschen et al. 2006; Haddad 1999; Tewksbury et al. 2002).

Much of the debate surrounding the role of corridors in landscape management may be settled through careful consideration of the many roles corridors play in the landscape. Corridors may act, as they are often intended, as conduits. However, they also increase habitat area; act as barriers that block the movement of species attempting to cross them; as filters that permit the movement of some species while not others; as a source of individuals; or as sinks (Hess and Fischer 2001).

3.2.2.5 Seasonal Cycles

Each landscape has its own seasonal cycles based on an interaction of the climate, landform, and other factors. The cycles can be wet and dry seasons, cold and hot seasons, or a combination of these. Many landscapes are periodically inundated with floods that occur somewhat predictably based on the origin of a river running through the landscape. Other landscapes become inaccessible part of the year because of very deep snows or saturated soils, while frozen soils and rivers can make some landscapes more accessible. Additionally, certain landscapes become dangerous during particular seasons because of potential fires or tornados. Even day length changes dramatically by season in landscapes at higher latitudes.

Native animals and plants have commonly adapted their cycles of reproduction, growth, and dormancy to these seasonal variations. Human activities and domestic animals commonly obtain the most value from a landscape by similarly adjusting to these cycles.

3.2.2.6 Disturbance Patterns

The disturbance pattern—type, frequency, distribution, and intensity—is also a characteristic property of each landscape. Severe floods commonly occur in low lying areas near water courses. Severest windstorms often occur on exposed slopes or in fortuitously placed valleys. Fires often become most intense on slopes with intense sunlight and upper slopes and ridges. Slope failures are most common on steep slopes of certain geomorphic origin. Earthquakes are commonly most severe along fault lines and impact human structures the most on alluvial soils. Artificial reservoir dams can become dangerous if old and not maintained. The patterns of these and other disturbances are strongly influenced by the landform, land cover, and seasonal variations of the landscape (Pickett and White 1985; Wilson et al. 1998).

Disturbances can be disastrous if not prepared for; however, appropriate mitigation practices can minimize disturbances, reduce their ability to destroy values, and even make use of their impacts (Attiwill 1994).

3.2.2.7 Human and Animal Populations

The population of human and animal species is an emergent property of the landscape. Each population is the consequence of the ability of the landscape to sustain it by providing food, water, and shelter; the population's competitiveness with other animals; and the historic presence and size of the population and its competitors. The survival and competitiveness of each population is a result of other emergent properties such as water distribution, annual cycles, disturbance patterns, land cover, and such landform patterns as ruggedness and viewpoints.

3.3 Resources and Management Objectives

Terrestrial ecosystems can provide, or fail to provide, many things that people value. On the one hand, species periodically went extinct, forests burned, drastic erosion occurred and declined, and carbon dioxide increased and abated in the atmosphere long before widespread human influence. On yet another hand, many, diverse species have evolved and found habitats and ecosystems have re-grown after fires and other disturbances long before people evolved.

We manage ecosystems to avoid those consequences that we do not want and to obtain those that we do want. Historically, our understanding and communication was limited to small areas. For example, people first concentrated on individual plants and animals as hunter-gatherers, then on agricultural fields, pastures, and stands as farmers. People rapidly began to coordinate among stands, fields, and pastures to ensure a flow of values—e.g., vegetables, meat, and wood—over time. As more of each landscape became utilized, people began to realize that renewable resources such as wood can become temporarily unavailable and create human hardships in the process. And, techniques were developed for managing landscapes so they provided suitable wood and other values sustainably (Johann 1997; Kirby and Watkins 1998; Perlin 1989).

3.3.1 Objectives of Management

Each landscape can provide many resources in various amounts. Important questions are: What resources should the landscape provide, and how much of each resource? People often manage landscapes for a single, dominant resource or for many resources—agriculture, timber, grazing, biodiversity, and water quality and/or quantity.

Changing technologies often increase the efficiency—and decrease the landscape area needed—for some resources by increasing transportation of resources among landscapes and regions and increasing the amount of the resource that can be produced on each area. These efficiencies of transportation and production sometime outstrip the world population's demand and more of the resource is produced than can be consumed. There is a surplus of food and wood in the world, for example; and these resources are grown and/or extracted from a few landscapes and shipped elsewhere, leaving other landscapes to shift their management emphasis away from agriculture, timber, or grazing (Oliver 1999). This shift has occurred in different landscapes for centuries and is still ongoing.

With global communication and transportation, we are now realizing that both the identifying and providing of resource values needs to be coordinated at a broad scale and over a long time. Concerns have been raised for centuries over impending shortages of timber and food. More recently, we are also becoming concerned about such issues as a human population explosion, species extinctions, air and water pollution, soil degradation, and energy sustainability. Similar concerns over the instability of labor and other technical facets of extraction industries have been raised.

Brundtland (1987) summarized the many environmental and development concerns using the term "sustainable development" to emphasize intergenerational equity. More recently the same concept has been applied to sustaining whole ecosystems (Farcy 2004; Grumbine 1994; Kessler et al. 1992). Oliver (2003) suggested that "sustainable forestry" be extended to spatial equity as well as temporal equity; that is, "people living in one place and time should provide their 'fair share' of values—neither unfairly exploiting nor depriving themselves of certain values to the detriment or benefit of people in another place or time."

Following the Rio Earth Summit in 1992, forestry officials in various countries joined into groups and developed a relatively few criteria to be sustained by forestry (Aplet et al. 1993; Burley 2001; Johnson 1993). "Although intended as measures, these criteria are similar among the different groups and robust enough to be used as a reasonable set of values for a working definition of sustainable forestry" (Oliver 2003). These criteria include:

- productive capacity,
- biodiversity,
- soil and water quality,
- forest health,
- ability to sequester carbon,
- socioeconomic benefits, and
- infrastructure to sustain the other criteria.

3.3.2 *Achieving Objectives at the Landscape Level*

Although intended for forestry, these criteria are robust enough to be extended to providing other resources sustainably from non-forested ecosystems as well. If these values are provided, the ecosystems will probably be managed in ways that can easily accommodate other values as they emerge. Each of these values is best provided if multiple scales are coordinated, as shall be discussed below.

3.3.2.1 Productive Capacity

Productive Capacity refers to the ongoing ability of the landscape to provide forests and other plant and animal crops. Each ecological zone has innate soil and climate characteristics that provide limits of how readily the productive capacity can be enhanced or degraded. Consequently, some plant, animal, and forest crops are concentrated within specific ecological zones.

The productive capacity and its enhancement for each resource can vary within different areas of a landscape. Some areas are so unproductive, inaccessible, prone to disturbances or steep that they are not utilized for production—and may have no such capacity in the future. If enough of a landscape is actually or potentially productive enough to invest in an infrastructure, individual portions within it can be modified to be more productive. Usually landform positions with deeper, well drained soils of sufficient moisture and appropriate textures and nutrients are most productive; however, large areas of gentle slopes and low compaction and erosion are important for agriculture crops and grazing. Access is also important for a part of a landscape to be productive, as are seasonal variations and disturbance regimes that allow the desired food, wood, or other crops to be grown. If conditions are suitable to make the effort worthwhile, the productivity of individual stands or fields within a landscape can be enhanced by terracing, adding water through irrigation, creating levees to avoid flooding, appropriate plowing to enhance the soil productivity, and similar measures.

Improvements to the productive capacity are also made at the individual organism scale. Selective breeding, training, pruning, weeding, and the application of pesticides or herbicides to individual plants or animals, can enhance the ecosystem's productivity.

3.3.2.2 Biodiversity

Biodiversity refers to the genetic diversity of life—the species, variations within species, and the actual and potential changes based on the environments in which the species live. Ecological zones vary dramatically in the total number, concentration, and degree of endangerment of plant and animal species. Small islands and tropical areas, for example, generally contain the greatest biodiversity and greatest danger of plant and animal extinctions (Myers et al. 2000). Consequently, global biodiversity can first be protected by identifying and investing in greatest habitat protection in these most at-risk ecological zones. Potentially endangered species need to be addressed at this broad level to ensure a viable population will be maintained, with some or all landscapes within the ecological zone contributing.

Species' habitats can be protected at the landscape level, although there is disagreement over the best means of protection. Some argue that large areas should be isolated as reserves and left to 'non-human' processes (Groves et al. 2002; Margules and Pressey 2000). This practice assumes native species will best flourish in this environment. Others contend that wilderness preservation and biodiversity conservation are distinctly separate goals (Bengtsson et al. 2003; Sarkar 1999). A compromise has been proposed by Seymour and Hunter (1999) in which part of each ecosystem is allocated to reserves; part to intensive production of plant, animal, or tree crops; and a third part to integrated management in which the landscape is managed for multiple objectives, including the enhancement of habitats. Human activities can enhance native species habitats by ensuring that all stand structures are maintained (Fig. 3.2), since some species depend on each structure. For example, in the Pacific Northwestern United States the endangered grizzly bear and spotted owl survive in very different habitat structures; the grizzly thrives in open habitats, while the owl requires intact complex old-growth habitat structures (Forsman et al. 1984; Servheen 1983).

Landscapes can be allocated to these different uses or a landscape can contain different areas of reserves, integrated management, and intensive cropping. Topographic, water, corridor, and disturbance properties can help dictate where each use is best located. And, management can enhance the water supply and distribution, create or block corridors, and change the land cover distribution to affect the biodiversity. At the individual stand level, biodiversity can be enhanced by providing specific host species, stand structures, and/or structural features (e.g., snags; Franklin 1993). Features of these individual stands need to be coordinated across the landscape to ensure an appropriate distribution. If a species is still endangered after its habitat is protected at the stand, landscape, and ecological zone levels, the individuals are then nurtured to ensure their population increases to viable levels.

3.3.2.3 Soil and Water Quality

The need for protection of soil and water quality and quantity varies among ecological zones. For example, unlike many other ecosystems, boreal ecosystems generally contain soils of inherently low productivity, but are resistant to degradation and generally contain abundant water and relatively little physical or chemical contamination (Bonan and Shugart 2003; Haila 1994; Setala et al. 2000).

Soil and water quality are generally protected at the landscape level by minimizing overland flow of water. Overland flow creates erosion, incises the streams, creates floods followed by times of low streamflow, and reduces the water flowing to aquifers. This overland flow is best avoided by maintaining good soil structures, minimizing the amount of ditches, and minimizing the water flow rate in temporary or permanent channels. Overland flow and erosion can be especially harmful in roads and trails that are not appropriately protected. Soil structures are maintained by retaining forest land cover—and to a lesser extent other land covers (Osborne and Kovacic 1993).

On the other hand, the amount of water available as runoff is reduced by forests since they evapotranspire water. This loss occurs especially when forest canopies are dense—in the dense, understory, and complex structures (Fig. 3.2; Zhang et al. 2001). The tradeoff between maintaining the soil structure but reducing the canopy can be accomplished by keeping the forest in a variety of structures through growth and natural or human disturbances—as well as by strategically placing land uses that harm soil structures where their impact will be least harmful. Consequently, a varied landscape cover may be most helpful in maintaining the combination of water quality and quantity.

Individual stands can be addressed where appropriate to reduce erosion and increase soil productivity. And, individual animals and plants can be managed to avoid problems of compaction and/or erosion, even in sensitive soils such as those found on China's Loess Plateau (Qiang-guo 2001).

3.3.2.4 Ecosystem Health

Ecosystem Health refers to the vigor, resilience, and complexity of the interactions found in an ecosystem (Rapport et al. 1998). When an ecosystem becomes dominated by a pest—a single insect, disease, plant, or animal species that can exclude other values from the ecosystem—often these keys to health decline (Fischer et al. 2006; Pimentel et al. 2005; Simberloff 2005; Zavaleta et al. 2001). These pests include insect outbreaks, plant diseases, an excessively large population of a plant or animal, or an exotic plant or animal that excludes native ones. Whereas small insect and disease outbreaks and species dominations are common—for example in an individual stand—they can become problematic if an entire landscape is dominated.

Ecosystem health concerns are generally avoided by maintaining a diversity of land covers over an area. Even in forested landscapes, a diversity of species and structures will help keep pests from expanding to such large areas that they threaten other values (Jactel et al. 2005). Too much uniformly closed forest can also allow fires to become catastrophic, whereas they are much more likely to be benign where there is a diversity of structures and/or other land covers (Picket and White 1985; Pollet and Omi 2002).

Individual stands are usually addressed in a coordinated way to prevent a pest from moving to an adjacent stand and its population building excessively. And, in some cases, individual pests are eradicated—or eradication is attempted—because the pest is so harmful to other plants and animals.

3.3.2.5 Carbon Sequestration

Forests have the potential to sequester carbon in three ways: in the growing and standing forest; in non-decomposed forest products; and by using wood in energy production or construction to replace products that emit more carbon dioxide in their production and use—i.e., steel, concrete, brick, and aluminum (Perez-Garcia et al. 2007).

The amount of forests grown and harvested for carbon sequestration needs to be balanced across the landscape so that the carbon sequestration is maximized and other values are provided. Too little harvest can lead to a loss of diversity of stand structures, overcrowding of forests and resulting forest health and catastrophic fire issues (Covington and Moore 1994). These changes have the potential to increase the export of carbon dioxide to the atmosphere as forests burn and carbon dioxide-producing materials are used in construction. Too much harvest can lead to loss of carbon dioxide intake and carbon storage in forests, loss of forest habitats, reduction of soil structure and water quality, and loss of the productive capacity of the landscape.

3.3.2.6 Socioeconomic Benefits

Socioeconomic Benefits refer to the human wellbeing in terms of employment and economic wellbeing, recreation opportunities, and aesthetic enjoyment of ecosystems beyond the specific values described above.

Employment, recreation, and aesthetics are enjoyed by the most people in landscapes that have accessible corridors, a diversity of land covers, disturbances of low impact, and few forest health issues (de Groot 2006). Maintenance of corridors, diverse land covers, soil and water quality, forest health, and products requires active management. This active management requires direct employment in labor and indirect employment in producing machinery to perform the management. As shall be discussed, the diverse nature of resource management commonly provides a variety of jobs throughout the year. And, the diverse conditions across the landscape mean that many different activities can be performed.

3.3.2.7 Infrastructure to Sustain the Other Criteria

This refers to the laws, economic incentives and capacity, skilled labor and machinery, and many improvements to the landscape. Such improvements include roads, bridges, fences, irrigation systems, dams, wells, terraces, buildings, and other human structures that change the landscape properties. The laws and economic incentives are usually instituted at broad, political levels—such as ecological zones. Skilled labor, machinery, and improvements are generally specific to each landscape and are part of the landscape management process described in the rest of this paper.

3.4 The Ecological, Technical, and Social Facets of Management

Oliver and Deal (2007) showed that global data could be used to approximate what and how much resources the different ecological zones could sustainably provide as their 'fair share.' Using an extension of such analyses, it is technically possible to estimate what resources could be appropriately provided by each landscape to be most sustainable.

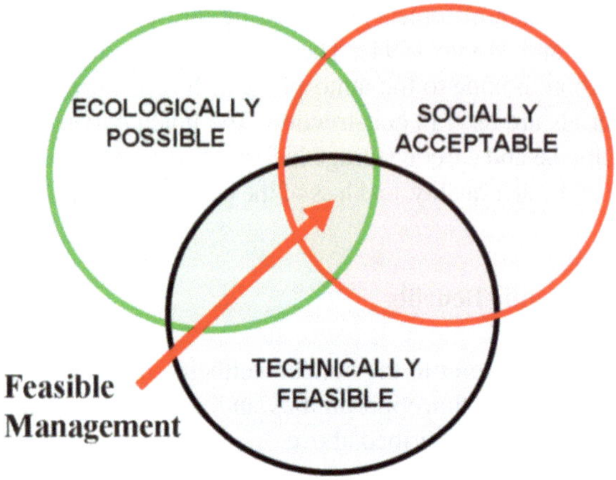

Fig. 3.4 Resource management will only be successful if it is ecologically possible, socially acceptable, and technically feasible. This diagram is similar to the "Sustainable Forestry" *Venn diagram* and the "Triple Bottom Line" of sustainable management; however, this diagram refers to "technically feasible," instead of "economically feasible" and considers economic feasibility as just one component of technical feasibility

The World Trade Organization is attempting to ensure resources are provided in efficient locations rather than in places that are heavily subsidized (Shell 1995). By contrast, the World Bank is beginning to emphasize local production and markets (Swinburn et al. 2006). Efficient provision is not just based on appropriate soils and climates; it is also based on an in-place technological infrastructure of transportation, irrigation, processing, energy efficiency, and skilled labor. Such an infrastructure may have been established decades, or even centuries, before and can continue to make the landscape efficient in providing the resource.

Although conceptually ideal, coordinating management at broad, ecological zone scales and then to lower hierarchies is difficult for many reasons. For example, if it is being superimposed upon traditional practices of managing landscapes independently and according to local cultures. In addition, people commonly want different things from landscapes. Some things people want are compatible, sometimes they are in conflict, and some things are impossible to achieve. Sometimes several decades are needed to provide a desired result from the landscape. Meanwhile, people's desires, their technical abilities, and the landscape itself may all change—often dramatically. The shifting values are also difficult for technological facets of management—long term investments such as roads, skilled labor, or dams—since such investments can be lost with changes in values.

Landscape management is the task of merging:

- what an area can possibly provide,
- how the area can be technically changed, and
- what people want.

These are similar to the "triple bottom line" of sustainable management, and can be shown as a Venn diagram (Fig. 3.4).

3.4.1 *The Natural Facet of Management: What an Area Can Possibly Provide*

As described above a landscape can only provide resource objectives within its natural capabilities, based on the soils, climate, and other factors.

3.4.2 *The Technical Facet of Management: How the Area Can Be Technically Changed*

The range of values that are possible to achieve from a landscape are partly limited by the technical abilities of people to manage. This ability includes a variety of factors, such as skilled people, appropriate tools, and knowledge. Skilled people and knowledge are constantly changing and the needed technologies also change as the landscape changes. But, tools such as bridges, dams, levees, factories, equipment, and machinery remain constant for many years or decades. A combination of dynamic needs and the periodic, static nature of tools create an interesting need for periodic stability in the management of very dynamic landscapes. Knowledge of outside influences on the landscape is needed to ensure that the values to be managed for cannot be more easily provided elsewhere. Appropriate tools are needed to manage many resources; and these tools need to be installed, operated, and serviced by knowledgeable personnel. Appropriate incentives also need to be present for people to expend their time, energy, and equipment in performing tasks.

Appropriate knowledge, tools, and incentives can potentially change the landscape in many ways that range from short term, unsustainable extraction to longer term sustaining resources and enhancing the landscapes capabilities. For example, timber, grasses, and minerals can be unsustainably extracted from a landscape. On the other hand, technologies can coordinate changes in land cover or water flow sustainably so that some parts of the landscape provide exploitable conditions at all times in the future. And, the forest area can sustain all habitats by sustaining some area in each structure at all times (Fig. 3.2). Similarly, appropriate grazing rotations can provide a healthy rangeland and other resources. Technologies can also change the total capacity of a landscape to provide values by affecting the landscape properties. Dams and irrigation systems can change the seasonal water flow and water availability to uplands, while fertilizers and various plowing practices can increase the total productivity of soils for crop and forest growth.

3.4.3 *The Social Facet of Management: What People Want*

Landscapes are managed for many values; are highly visible; and involve large sums of money, government agencies, and many people with different values. Access to landscapes cannot be easily controlled. Consequently, it is important that the many people interested in the landscape acquiesce—and preferably agree—to management decisions. Those with an interest in the landscape—stakeholders—can include landowners within and near the landscape, workers or other technology providers, users of the landscape's resources or competitive resources, and others.

Acceptable management depends on social values and varies among cultures and subcultures within an area. Even people with similar values may differ in the scientific theory they use to achieve desired values (Clark 2002). For example, some people who value biodiversity feel it is best achieved by excluding active management from forests, while others feel it is best achieved by actively managing for a diversity of structures (Han et al. 2012).

Cultures and social values are dynamic, and what is socially acceptable management practices at one time may not be at another as local populations change, communication draws in more distant stakeholders, and different values gain or lose appreciation.

3.4.4 Addressing the Dynamic Nature of Landscape Management

All three facets of management are not only dynamic, but also somewhat unpredictable. That is, we do not always get the same results for a given management activity because of the natural, technical, and social changes described above (Botkin and Sobel 1975).

Resource management is a social investment that requires several years or decades of relatively predictable results to be worthwhile because of the static nature of the technologies. Resource management has avoided famines, plagues, and water and timber shortages. On the other hand, many of our resource management plans that were intended to span many decades have been abandoned or dramatically modified after one or a few decades. Agriculture lands and accompanying processing plants and railroads have been abandoned (Moreira and Russo 2007; Preiss et al. 1997; Whitney 1994), levee systems intended to provide more agriculture area are being considered for modification (Galloway 1995), forests intended for harvest have been redirected toward preservation (Kessler et al. 1992; Rubin et al. 1991), unanticipated water shortages have arisen (Barten et al. 2008), and large energy projects have been abandoned before completion (Pope 2008). Plans for sustainable harvest of timber in British Columbia are being modified because insects have destroyed many trees planned for future harvest (Fettig et al. 2007; Robbins 2008). Even planned 'sustainable yield' units that were designed to ensure timber production perpetually in the United States have been abandoned after a few decades (Henderson and Krahl 1996).

What society desires as outputs may also change over time—from wanting timber that is harvested from forests to wanting unharvested forests (Kessler et al. 1992; Mather 1992). Since shifts in societal values often occur much faster than landscapes develop, well intentioned management inputs based on today's needs may not always provide desirable future outputs. The previous, partially successful techniques of resource management need to be modified to recognize explicitly that ecological, social, and economic inputs and desired outputs are not completely stable.

Balancing the needs for stability in order to invest in management technologies with the inherent instability of natural, innovation, and social systems can be done in two ways:

- We assume the natural, technical, and social facets are stable for short "management cycles" of 5 or 10 years (or longer for investing in such things as dams, roads, and bridges). During the cycles, the work units, markets, and other investments are assumed to be stable.
- We design the various technical aspects with flexibility, so they can be resilient to variations (Oliver et al. 2008).

3.5 Managing the Landscape

Landscape management is the task of merging what an area can possibly provide, what people want, and how the area can be technically changed to achieve what people want. The skill of management is to know what values can be provided; how to assess and decide among conflicting values that can be provided; how to plan over time and space to achieve these values; and how to implement the plan.

Even where certain outputs are desirable and can be provided from a landscape, they may not be immediately achievable for several reasons:

- Resource change is not instantaneous. It may take decades to provide timber or to develop forests in the complex structure if the landscape presently contains only forests in the open structure.
- The appropriate financing, skilled labor, machinery, markets, and roads may not be readily available for the needed management operations.
- Even when the management infrastructure is in place, time—and the correct season--is needed to accomplish each operation.

The range of values potentially provided by a landscape often increases with time as the above factors are overcome. Consequently, it is appropriate first to determine what values will be desired in the long term—and how to achieve them. Then, increasingly shorter management time horizons are planned with the ultimate goal of reaching the long term values (Bare 1996). The allocation of different parts of the landscape to different land uses—forestry, agriculture, grazing, buildings, and others—is done first because this allocation is expected to endure for the longest time. Following allocation, managing over a large area such as one or several landscapes and at a long time horizon of many decades into the future is commonly referred to as "strategic management." "Tactical management" focuses within a single landscape and often on specific stands and on an intermediate time horizon of 5 or

10 years, and "operational management" generally focusing on smaller areas and an annual cycle (Kangas et al. 2008).

3.5.1 Land Use Allocation

Land use allocation entails the determination, implementation, and enforcement of what land uses such as housing, agriculture, and forestry occur—or are prioritized--on each specific place within a landscape (Roberts 1979). Such allocation can occur by zoning or free market mechanisms. Housing and urban construction are generally the most profitable allocations of land, with agriculture being second if the soil is productive. Forests have commonly been confined to the least productive or otherwise unprofitable lands; however, this allocation is beginning to be modified to ensure that forest areas are interspersed among other uses so the forest can address recreation; habitat protection; hazard reduction; and water quality, flow, and quantity concerns.

Land uses can be competitive in some respects and complementary in others. They are competitive since more land in one use means less available to another; however, the roads used—and partly paid for—by farming are also used and paid for by forestry. And, the seasonal labor and equipment of farming can be used by forestry during other seasons. In some areas of the world, endangered wild animals are killed as a source of meat—bushmeat. If some of the forest is converted to farms for intensive husbanding of domestic meat, the hunting pressure on such wild bushmeat could decline (Wilkie and Carpenter 1999).

Land use allocations change because of deliberate planning and/or unexpected economic and social factors (Lambin et al. 2001). The area of forest land has stabilized or increased as technology has allowed food production to concentrate on the best soils and marginal agriculture land to be abandoned in much of the United States, Europe, and Japan. The trend appears to be happening elsewhere as well, as other countries transition from agrarian-dominated to technological cultures (FAO 2001; Mather 1992).

3.5.2 Strategic Management

Landscape-scale management of forests has been applied and studied for many centuries in Europe (Johann 1997; Kirby and Watkins 1998; Rackham 1986). Much of the knowledge and skills gained from managing forests can be applied to other resources. These skills include operations analysis, decision analysis, social skills, silviculture, various engineering techniques, and economics (Patton 1971; Dykstra 1984; Davis et al. 2001; Oliver and Twery 1999; Clark 2002).

Strategic management involves determining what the desirable condition of the various landscape properties is in the long term for each land use allocation. For example, of the possible forest conditions,

- What are the desired management intensities of the various soils?
- What is the desired land cover distribution?

- What are the desired distribution and amount of water?
- What is the desired distribution of corridors?
- How should the seasonal cycles be adjusted to?
- How can the disturbance patterns be best addressed?

These questions require information in the form of soil, hydrologic, topographic, road, and land cover maps as well as inventories of existing vegetation conditions.

Strategic management necessitates understanding the natural system and knowing how to emulate or change its behavior where feasible so it can provide the desired outputs—such as maintaining a dynamic balance of forest structures within the landscape (Boyce 1995; Fig. 3.2). It can entail replacing disturbances (Fig. 3.2) with targeted operations such as removal of trees through thinnings, selection harvesting, clearcutting, and others (Smith et al. 1997; Waring and Schlesinger 1997).

One challenge of management is to identify explicitly as many positive and negative consequences of management as possible, so that unintended consequences are minimized.

Another challenge is to decide tradeoffs among incompatible values since all outputs cannot be produced at the highest level simultaneously at all times (Heilig 2003). For example, a decision must be made of the amount of each stand structure to provide for different habitats. Such decisions are commonly made through a 'tradeoff' decision analysis process using a matrix (Oliver and Twery 1999). The potentially important consequences of management are listed in the array, and various management alternatives are listed and compared against each of the consequences. Such a process allows the decision maker to understand all of the consequences that could be affected by management as well as the effects and tradeoffs among values of each management alternative.

Another challenge is to identify the equipment, prices, labor, materials, and investment costs needed to accomplish each alternative strategic plan. These needs are collectively termed work units. Identification of these work units is sometimes known as gap analysis (Scott et al. 1993). Broad estimates of these work units for different management cycles are commonly done during strategic planning and refined during tactical planning.

A strategic plan generally includes long term goals for the different landscape properties and equipment units as well as intermediate goals and equipment units for each tactical management cycle. There are strengths and drawbacks to strategic management. As strengths, it anticipates long term influences on the landscape that may emerge from within or outside; it can set long term management trajectories to ensure that the landscape provides desired values such as a sustainable harvest of timber and/or sustained habitats; and it can give specificity to short term actions so they can be accomplished. The drawback is that strategic management cannot anticipate all intervening changes such as transforming markets, natural catastrophic disturbances, and changing labor and machinery inputs. To balance between the needs for specificity and flexibility, only broad management goals are set at a long time horizon and a large scale. More specific, short term goals are then set at shorter time horizons.

3.5.3 Tactical Management

Tactical management builds on strategic planning by selecting specific projects to be accomplished to specific stands or other locations for each year during the management cycle to meet the goals determined at the strategic level. Tactical management also specifies and fills the work unit needs broadly identified by strategic management for the current management cycle. For example, tactical management must match specific silvicultural operations to specific stands within the forest to achieve the movement toward the strategic forest cover goals identified for the current management cycle. Variations and spatial patterns across the landscape that are often ignored at the strategic level are addressed during tactical management. Various computer systems have been developed to assess and help design appropriate management at the tactical level that accounts for variations and spatial patterns (Millspaugh and Thompson 2001; Wilson and Baker 1998; Walters et al. 1997).

Once the stands and their specific treatments are identified, a gap analysis can determine what work units need to be acquired to best accomplish the treatments. For example, if much timber harvesting is to be done on steep slopes, then cable harvesting systems or other steep-terrain work units would be needed. A decision can also be made between investing in a highly mechanized, laborsaving machine for thinning and investing in large labor crews that use less costly machines (Krick 1962).

Financing of machinery and recruiting of skilled people usually require a commitment of five or more years; consequently, acquisition of work units is done with an extended time horizon. Some work units are commonly still operational from the previous management cycle, and ones being newly acquired often remain functional through the next cycle. Although planned during specific management cycles, the actual changing of work units is a continuous and gradual process rather than one punctuated by management cycles.

Each stand to be treated requires scheduling of operations so they are coordinated to accomplish a given task. For example (Fig. 3.5), to harvest and regenerate a stand requires the operations of road construction, harvesting, and site preparation be done sequentially, with each step being completed before the other can begin. Parallel to this sequence are the operations of collecting seeds, growing seedlings, and lifting and transporting them to the prepared stand. The second sequence must be coordinated with the first sequence so that the stand is ready when the seedlings are delivered. And a third input, labor to plant the seedlings, must arrive at the prepared stand at the same time as the seedlings. All of these schedules must account for seasons of easy, difficult, and impossible logging and road building; the time needed for seeds to germinate and grow; the possibilities of equipment breakdown and labor unavailability; and other variables (Davis et al. 2001). The scheduling of operations recognizes the instability of natural systems and constantly makes adjustments to account for changes in weather and equipment, seedling, and labor availability.

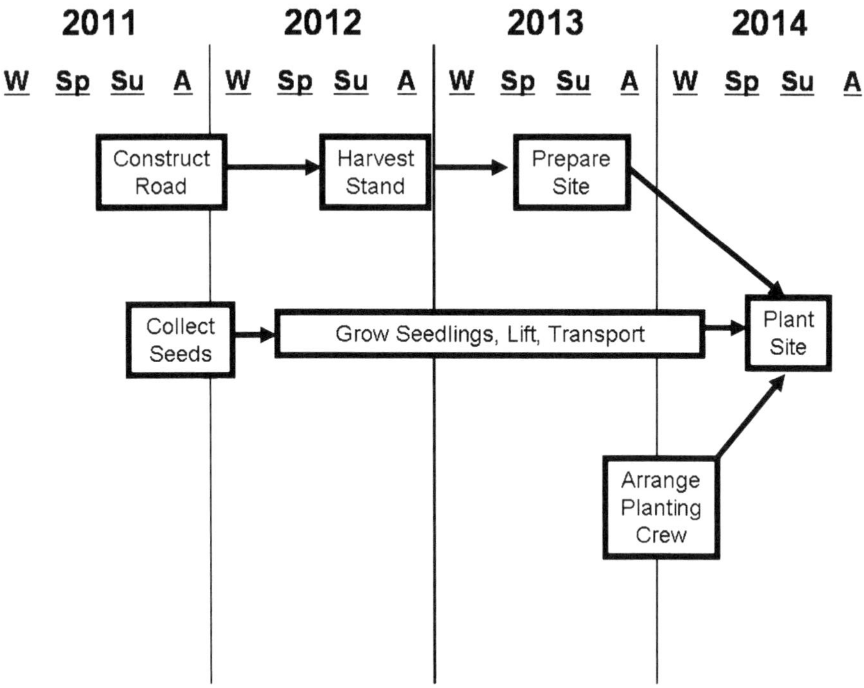

Fig. 3.5 Tactical management supplements strategic management by specifying specific operations (*in boxes*) that must be accomplished sequentially (*on a single row*) or in parallel (*each row*) at specific times and seasons in order to accomplish a task. The accomplishment of many such tasks leads to the goals specified in strategic management. Operational management involves the day-to-day accomplishment of each operation

For implementation purposes, the identification of stands to treat and equipment to purchase is considered stable during a management cycle. This stability gives assurance to investors and laborers alike. Dramatic changes such as wildfires, labor shortages, market turmoil, or sudden and unforeseen changes in public perception or in legal regulations can destroy the infrastructure, the ability to implement the plan, and the willingness of local people to trust future plans. For a landscape to be managed sustainably in the long term, it is best that the infrastructure be allowed to change slowly as the equipment is depreciated and the labor skills and occupations also change. At the same time, it is best that the equipment units not be so specialized that they cannot adjust to within-cycle changes.

Near the end of each management cycle, the strategic and tactical plans are revisited to determine if both the allocations among land uses are still appropriate and the previously targeted stands for treatment are still desirable. By periodically adjusting the strategic and tactical plans, the current condition of the landscape remains on a trajectory to meet the shifting goals of society.

3.5.4 *Operational Management*

One output of tactical management is an annual plan for each year of the management cycle. This plan designates what stand operations will be carried out—and to which stands. Operational management matches the equipment units to stands for each day to accomplish the targets. Operational management must continuously make adjustments to the scheduling to account for local variations in weather, machinery breakdowns, human errors, and market fluctuations.

The actual deployment of an operation on a stand requires much skill and experience. It entails designating where each person, piece of equipment, material (e.g., the seedling storage) will be initially located within each stand, how each will move, and how fast they can be expected to move.

3.5.5 *Feedback and Management Adjustments*

Feedback is continuously needed from the operations to the tactical to the strategic management levels to ensure the expectations are adjusted to coincide with reality (Oliver et al. 1999). When managing at all levels, there is a tremendous need for flexibility while focusing on the long term goals. Unexpected natural or human actions can cause dramatic shifts in the actual or desired outputs. Strategic, tactical, and annual management try to anticipate changes and then provide short time intervals when a degree of stability is presumed. Tactical management adjusts the annual plans each year to account for the variation in operations actually accomplished while trying to meet the management cycle goals provided by strategic management. Similarly, strategic management adjusts the management cycle plans at the end of each cycle.

Where necessary, annual plans and management cycle plans may be adjusted in the middle of the year or cycle; however, the benefits of this adjustment needs to be balanced against the resulting disruptions to equipment investments, operations, and other inputs.

Literature Cited

Alexander SM, Waters NM (2000) The effects of highway transportation corridors on wildlife: a case study of Banff National Park. Transp Res Part C Emerg Technol 8:307–320

Aplet GH, Johnson N, Olson JT, Sample VA (eds) (1993) Defining sustainable forestry. The Wilderness Society, Washington, DC

Attiwill PM (1994) The disturbance of forest ecosystems: the ecological basis for conservative management. Forest Ecol Manag 63:247–300

Bailey RG (1983) Delineation of ecosystem regions. Environ Manag 7:365–373

Bare BB (1996) Hierarchical forest planning: some general observations. In: Workshop on hierarchical approaches to forest management in public and private organizations. Candian Forestry Service, University of Toronto, Petawawa National Forestry Insititute

Barten PK, Jones JA, Achterman GL, Brooks KN, Creed IF, Folliott PF, Hariston-Strang A, Kavanaugh MC, Macdonald L, Smith RC, Tinker DB, Walder SB, Wemple BC, Weyerhaeuser GH, Alexander L, de Guzman E (2008) Hydrologic effects of a changing forest landscape. National Academic Press, Washington, DC
Beier P, Noss RF (1998) Do habitat corridors provide connectivity? Conserv Biol 12:1241–1252
Bengtsson J, Angelstam P, Elmqvist T, Emanuelsson U, Folke C, Ihse M, Moberg F, Nyström M (2003) Reserves, resilience and dynamic landscapes. Ambio 32:389–396
Bennett AF (1998) Linkages in the landscape: the role of corridors and connectivity in wildlife conservation. IUCN, Gland
Bonan GB, Shugart HH (2003) Environmental factors and ecological processes in boreal forests. Annu Rev Ecol Syst 20:1–28
Botkin DB, Sobel MJ (1975) Stability in time-varying ecosystems. Am Nat 109:625–646
Boyce SG (1995) Landscape forestry. Wiley, New York
Brundtland GH (1987) Our common future: the report of the World Commission on Environment and Development. Oxford University Press, New York
Burley J (2001) International initiatives for the sustainable management of forests. In: Shepard SRJ, Harshaw IW (eds) Forests and landscapes linking ecology, sustainablility, and aesthetics. CABI Publishing, New York
Chapin FS, Matson PA, Mooney HA (2002) Principles of terrestrial ecosystem ecology. Springer Science, New York
Chapman HH (1931) Forest management. J.B. Lyon Co., Albany
Chen J, Franklin JF, Spies TA (1992) Vegetation responses to edge environments in old-growth Douglas-fir forests. Ecol Appl 2:387–396
Clark TW (2002) The policy process: a practical guide for natural resource professionals. Yale University Press, New Haven
Covington WW, Moore MM (1994) Postsettlement changes in natural fire regimes and forest structure – ecological restoration of old-growth ponderosa pine forests. J Sustain For 2:153–181
Damschen EI, Haddad NM, Orrock JL, Tewksbury JJ, Levey DJ (2006) Corridors increase plant species richness at large scales. Science 313:1284–1286
Davis LS, Johnson KN, Bettinger P, Howard TE (2001) Forest management to sustain ecological, economic, and social values, 4th edn. Waveland Press, Long Grove
de Groot R (2006) Function-analysis and valuation as a tool to assess land use conflicts in planning for sustainable, multi-functional landscapes. Landsc Urb Plan 75:175–186
Dijak WD, Thompson FR III (2000) Landscape and edge effects on the distribution of mammalian predators in Missouri. J Wildl Manag 64:209–216
Dykstra DP (1984) Mathematical programming for natural resource management. MacGraw Hill, New York
FAO (1997) State of the world's forests. FAO, Rome
FAO (2001) State of the world's forests. FAO, Rome
Farcy C (2004) Forest planning in Europe: state of the art, international debate and emerging tools. In: Andersen F, Birot Y, Paivinen R (eds) Towards the sustainable use of Europe's forests-forest ecosystem and landscape research: scientific challanges and opportunities. European Forest Institute, Joensuu
Fettig CJ, Klepzig KD, Billings RF, Munson AS, Nebeker TE, Negrón JF, Nowak JT (2007) The effectiveness of vegetation management practices for prevention and control of bark beetle infestations in coniferous forests of the western and southern United States. Forest Ecol Manag 238:24–53
Fischer J, Lindenmayer DB, Manning AD (2006) Biodiversity, ecosystem function, and resilience: ten guiding principles for commodity production landscapes. Front Ecol Environ 4:80–86
Forman RTT, Godron M (1986) Landscape ecology. Wiley, New York
Forsman ED, Meslow EC, Wight HM (1984) Distribution and biology of the spotted owl in Oregon. Wildl Monogr 87:1–64
Franklin JF (1993) Preserving biodiversity: species, ecosystems, or landscapes? Ecol Appl 3:202–205

Fry ME, Money NR (1994) Biodiversity conservation in the managemtent of utility rights-of-way. In: Proceedings of the 15th Annual Forest Vegetation Management Conference, Redding, CA
Galloway GE (1995) New directions in floodplain management. J Am Water Res Assoc 31:351–357
Gosz JR (1993) Ecotone hierarchies. Ecol Appl 3:370–376
Goudie AS (2006) The human impact on the natural environment: past present and future, 6th edn. Blackwell Publishing, Malden
Groves CR, Jensen DB, Valutis LL, Redford KH, Shaffer ML, Scott JM, Baumgartner JV, Higgins JV, Beck MW, Anderson MG (2002) Planning for biodiversity conservation: putting conservation science into practice. BioScience 52:499–512
Grumbine RE (1994) What is ecosystem management? Conserv Biol 8:27–38
Haddad NM (1999) Corridor and distance effects on interpatch movements: a landscape experiment with butterflies. Ecol Appl 9:612–622
Haddad NM, Rosenberg DK, Noon BR (2000) On experimentation and the study of corridors: response to Beier and Noss. Conserv Biol 14:1543–1545
Haila Y (1994) Preserving ecological diversity in boreal forests: ecological background, research, and management. Annales Zoologici Fennici 31:203–217
Han X, Oliver CD, Ge J, Guo Q, Kou X (2012) Managing forest stand structures to enhance conservation of the Amur Tiger (*Panthera tigris altaica*). In: J. Stanturf et al. (eds.), A Goal-Oriented Approach to Forest Landscape Restoration, Springer
Harrison S, Bruna E (1999) Habitat fragmentation and large-scale conservation: what do we know for sure? Ecography 22:225–232
Heilig GK (2003) Multifunctionality of landscapes and ecosystem services with respect to rural development. In: Helming K, Wiggering H (eds) Sustainable development of multifunctional landscapes. Springer, Berlin
Henderson D, Krahl L (1996) Public management of federal forest land in the United States. Unasylva 47:55–61
Hess GR, Fischer RA (2001) Communicating clearly about conservation corridors. Lands Urb Plan 55:195–208
Hobbs RJ (1992) The role of corridors in conservation: solution or bandwagon? Trends Ecol Evol 7:389–392
Horskins K, Mather P, Wilson J (2006) Corridors and connectivity: when use and function do not equate. Landsc Ecol 21:641–655
Jactel H, Brockerhoff E, Duelli P (2005) A test of the biodiversity-stability theory: meta-analysis of tree species diversity effects on insect pest infestations, and re-examination of responsible factors. In: Scherer-Lorenzen M, Körner C, Schulze E-D (eds) Forest diversity and function temperate and boreal systems. Springer, Heidelberg
Johann E (1997) From overutilization to sustainable management regulations; a case study of long-term historical changes in the forest resources of Austria. Summaries of Voluntary Papers; XI World Forestry Congress, Antalya, Turkey
Johnson SP (1993) The earth summit: The United Nations conference on Environment and Development (UNCED). Graham and Trotman, London
Johnson CW (2006) What are emergent properties and how do they affect the engineering of complex systems? Reliab Eng Syst Saf 91:1475–1481
Kangas A, Kangas J, Kurttila M (2008) Decision support for forest management. Springer, Dordrecht
Kessler WB, Salwasser H, Cartwright CW Jr, Caplan JA (1992) New perspectives for sustainable natural resources management. Ecol Appl 2:221–225
Kimmins JP (2003) Forest ecology, 3rd edn. Benjamin Cummings, Redwood City
Kirby KJ, Watkins C (1998) The ecological history of European forests. CAB International, Wallingford, Oxon
Krick EV (1962) Methods engineering: design and measurement of work methods. Wiley, New York

Lambin EF, Turner BL, Geist HJ, Agbola SB, Angelsen A, Bruce JW, Coomes OT, Dirzo R, Fischer G, Folke C, George PS, Homewood K, Imbernon J, Leemans R, Li X, Moran EF, Mortimore M, Ramakrishnan PS, Richards JF, Skånes H, Steffen W, Stone GD, Svedin U, Veldkamp TA, Vogel C, Xu J (2001) The causes of land-use and land-cover change: moving beyond the myths. Glob Environ Chang 11:261–269
Lidicker WZ (1995) Landscape aproaches in mammalian ecology and conservation. University of Minnesota Press, Minneapolis
Margules CR, Pressey RL (2000) Systematic conservation planning. Nature 405:243–253
Mather AS (1992) The forest transition. Area 24:367–379
Millspaugh JJ, Thompson AM (eds) (2001) Models for planning wildlife conservation in large landscapes. Academic, New York
Moreira F, Russo D (2007) Modelling the impact of agricultural abandonment and wildfires on vertebrate diversity in Mediterranean Europe. Landsc Ecol 22:1461–1476
Murcia C (1995) Edge effects in fragmented forests: implications for conservation. Trends Ecol Evol 10:58–62
Myers N, Mittermeier RA, Mittermeier CG, da Fonseca GAB, Kent J (2000) Biodiversity hotspots for conservation priorities. Nature 403:853–858
Oliver CD (1992) Similarities of stand structure patterns based on uniformities of stand development processes throughout the world–some evidence and the application to silviculture through adaptive management. In: Kelty MJ, Larson BC, Oliver CD (eds) The ecology and silviculture of mixed-species forests: a festschrift for David M Smith. Kluwer Academic Publishers, Boston
Oliver CD (1999) The future of the forest management industry: highly mechanized plantations and reserves or a knowledge-intensive integrated approach? For Chron 75:229–245
Oliver CD (2003) Sustainable forestry: what is it? How do we achieve it? J For 101:8–14
Oliver CD, Deal RL (2007) A working definition of sustainable forestry and means of achieving it at different scales. J Sustain For 24:141–263
Oliver CD, Twery M (1999) Decision support systems/models and analyses. In: Sexton T, Malk AJ, Szaro RC, Johnson NC (eds) Ecological Stewardship: a common reference for ecosystem management. Elsevier Science Ltd, Oxford
Oliver CD, Boydak M, Segura G, Bare BB (1999) Forest organization, management, and policy. In: Hunter ML Jr (ed) Maintaining biodiversity in forest ecosystems. Oxford University Press, Oxford
Oliver CD, McCarter JB, Comnick JM, Ceder K, Nelson CS (2008) Simulating landscape change using the landscape management system. In: Millspaugh JJ, Thompson FR III (eds) Models for planning wildlife conservation in large landscapes. Academic Press, Burrlington
O'Neill RV, Krummel JR, Gardner RH, Sugihara G, Jackson B, DeAngelis DL, Milne BT, Turner MG, Zygmunt B, Christensen SW, Dale VH, Graham RL (1988) Indices of landscape pattern. Landsc Ecol 1:153–162
Osborne LL, Kovacic DA (1993) Riparian vegetated buffer strips in water-quality restoration and stream management. Freshw Biol 29:243–258
Patten BC (ed) (1971) Systems analysis and simulation in ecology. Academic, New York
Perez-Garcia J, Lippke B, Comnick J, Manriquez C (2007) An assessment of carbon pools, storage and wood products market substitution using life-cycle analysis results. Wood Fiber Sci 37:140–148
Perlin J (1989) A forest journey: the role of wood in the development of civilization. Harvard University Press, Cambridge
Pickett STA, White PS (1985) The ecology of natural disturbance and patch dynamics. Academic, San Diego
Pimentel D, Zuniga R, Morrison D (2005) Update on the environmental and economic costs associated with alien-invasive species in the United States. Ecol Econ 52:273–288
Pollet J, Omi PN (2002) Effect of thinning and prescribed burning on crown fire severity in ponderosa pine forests. Int J Wildland Fire 11:1–10

Pope D (2008) Nuclear implosions: the rise and fall of the Washington public power supply system. Cambridge University Press, New York
Preiss E, Martin J-L, Debussche M (1997) Rural depopulation and recent landscape changes in a Mediterranean region:consequences to the breeding avifauna. Landsc Ecol 12:51–61
Qiang-gu C (2001) Soil erosion and management on the Loess Plateau. J Geogr Sci 11:53–70
Rackham O (1986) The history of the countryside: the full fascinating story of Britain's landscape. J. M Dent & Sons, London
Rapport DJ, Costanza R, McMichael AJ (1998) Assessing ecosystem health. Trends Ecol Evol 13:397–402
Robbins J (2008) Bark beetle kills millions of acres of trees in the West. New York Times, New York
Roberts JC (1979) Principles of land-use planning. Planning the uses and management of land. ASA-CSSA-SSSA, Madison
Rubin J, Helfand G, Loomis J (1991) A benefit-cost analysis of the northern spotted owl. J For 89:25–30
Sarkar S (1999) Wilderness preservation and biodiversity conservation: keeping divergent goals distinct. BioScience 49:405–412
Schumacher N H (1996) Using landscape indices to predict habitat connectivity. Ecology 77: 1210–1225
Scott JM, Davis F, Csuti B, Noss R, Butterfield B, Groves C, Anderson H, Caicco S, D'Erchia F, Edwards TC Jr., Ulliman J, Wright RG (1993) Gap analysis: a geographic approach to protection of biological diversity. Wildlife Monographs No. 123
Servheen C (1983) Grizzly bear food habits, movements, and habitat selection in the Mission Mountains, Montana. J Wildl Manag 47:1026–1035
Setala H, Haimi J, Siira-Pietikainen A (2000) Sensitivity of soil processes in northern forest soils: are management practices a threat? Forest Ecol Manage 133:5–11
Seymour RS, Hunter ML (1999) Principles of ecological forestry. In: Hunter ML (ed) Maintaining biodiversity in forest ecosystems. Cambridge University Press, Cambridge
Shell GR (1995) Trade legalism and international relations theory: an analysis of the World Trade Organization. Duke Law J 44:829–927
Simberloff D (2005) Non-native species DO threaten the natural environment! J Agric Environ Ethics 18:595–607
Simberloff D, Cox J (1987) Consequences and costs of conservation corridors. Conserv Biol 1:63–71
Simberloff D, Farr JA, Cox J, Mehlman DW (1992) Movement corridors: conservation bargains or poor investments? Conserv Biol 6:493–504
Smith DM, Larson BC, Kelty MJ, Ashton PMS (1997) The practice of silviculture applied forest ecology, 9th edn. Wiley, Hoboken
Swanson FJ, Kratz TK, Caine N, Woodmansee RG (1988) Landform effects on ecosystem patterns and processes. BioScience 38:92–98
Swinburn G, Gogo S, Murphy F (2006) Local economic development: a primer developing and implementing local economic development strategies and action plans. The World Bank, Washington, DC
Tewksbury JJ, Levey DJ, Haddad NM, Sargent S, Orrock JL, Weldon A, Danielson BJ, Brinkerhoff J, Damschen EI, Townsend P (2002) Corridors affect plants, animals, and their interactions in fragmented landscapes. Proc Natl Acad Sci U S A 99:12923–12926
Tischendorf L, Fahrig L (2000) On the usage and measurement of landscape connectivity. Oikos 90:7–19
Walters C, Christensen V, Pauly D (1997) Structuring dynamic models of exploited ecosystems from trophic mass-balance assessments. Rev Fish Biol Fish 7:139–172
Walters BE, Sugerman DF (1992) WPPSS and the Pacific Northwest: a mere possibility of unraveling a hopelessly twisted debt. Environ Law 16:91–161
Waring RH, Schlesinger WH (1997) Forest ecosystems concepts and management. Academic, New York

Whitney GG (1994) From coastal wilderness to fruited plain: a history of environmental change in temperate North America from 1500 to the present. Cambridge University Press, Cambridge

Wilkie DS, Carpenter JF (1999) Bushmeat hunting in the Congo Basin: an assessment of impacts and options for mitigation. Biodivers Conserv 8:927–955

Wilson JS, Baker PJ (1998) Mitigating fire risk to late-successional forest reserves on the east slope of the Washington Cascade Range, USA. Forest Ecol Manage 110:59–75

Wilson JS, Isaac ES, Gara RI (1998) Impacts of mountain pine beetle (*Dendroctonus ponderosae*) (Col., Scolytidae) infestation on future landscape susceptibility to the western spruce budworm (*Choristoneura occidentalis*) (Lep., Tortricidae) in north central Washington. J Appl Entomology-Zeitschrift Fur Angewandte Entomologie 122:239–245

Zavaleta ES, Hobbs RJ, Mooney HA (2001) Viewing invasive species removal in a whole-ecosystem context. Trends Ecol Evol 16:454–459

Zhang L, Dawes WR, Walker GR (2001) Response of mean annual evapotranspiration to vegetation changes at catchment scale. Water Resour Res 37:701–708

Part II
Natural Science Perspectives

Chapter 4
Hydrologic Connectivity of Landscapes and Implications for Forest Restoration

R. Chelsea Nagy and B. Graeme Lockaby

Abbreviations

OM	organic matter
DOC	dissolved organic carbon
POC	particulate organic carbon
NPP	net primary productivity
ANPP	aboveground net primary productivity
CRNWR	Cache River National Wildlife Refuge
NTU	nephelometric turbidity units
N	nitrogen
P	phosphorus
NO_x	nitrate plus nitrite
TN	total nitrogen
OC	organic carbon

R.C. Nagy (✉)
Department of Ecology and Evolutionary Biology, Brown University,
80 Waterman St., Providence, RI 02912, USA
e-mail: nagyrc@gmail.com

B.G. Lockaby
School of Forestry and Wildlife Sciences, Auburn University,
602 Duncan Dr., Auburn, AL 36849, USA
e-mail: lockabg@auburn.edu

J. Stanturf et al. (eds.), *Forest Landscape Restoration: Integrating Natural and Social Sciences*, World Forests 15, DOI 10.1007/978-94-007-5326-6_4,

4.1 Influence of Hydrologic Processes on Landscape Functions

The hydrological cycle is intricately linked with landscape patterns and processes. An understanding of interactions between terrestrial and aquatic systems is increasingly important as conservationists promote restoration at the landscape scale. Accounting for these interactions in restoration planning will move the field towards more comprehensive, successful efforts. In addition, with a better understanding of the complexities and linkages of aquatic and terrestrial systems, realistic expectations of restoration potential will help to build the support of the local community, thus bridging the gap between scientists and the general public. This chapter describes stream and floodplain hydrology in the landscape, as well as the effects of landscape activities and forest restoration on the hydrological cycle.

4.1.1 Floodplain Dynamics and Utilization

A river functions in four dimensions including three spatial dimensions (longitudinal, lateral, and vertical) as well as temporally. Longitudinal connectivity refers to the upstream-downstream transport. The lateral connection is the interaction of the channel with the floodplain. Vertical connectivity is the interface of the surface water with the groundwater. The focus of this chapter is the lateral connection, but for a description of the other dimensions, readers are referred to Amoros and Bornette (2002), Nilsson and Svedmark (2002), and Kondolf et al. (2006).

Periodic inundation of the floodplain is a naturally occurring process. Known as the flood-pulse, these phases of inundation are influenced by the hydrology and geomorphology of the system and can vary in duration, frequency, and intensity (Junk et al. 1989). As floodwaters advance, the littoral zone, or the aquatic-terrestrial interface, moves away from the channel with the flood and then back toward the channel as the flood retreats (Junk et al. 1989; Bayley 1995). Periodic floods create a dynamic environment with a high degree of habitat heterogeneity and thus promote a range of biodiversity and productivity among systems (Naiman and Décamps 1997).

In order to inhabit and utilize floodplains, humans have altered the natural flooding regimes of rivers and streams. Channelization and the construction of dams, levees, and dikes are common methods of flow regulation. Gergel et al. (2002) reported that levees and similar structures (e.g. floodwalls, embankments, and dikes) cover more than 40,000 km in the US alone. Levees are designed to contain the river and reduce overbank flow. This disconnect of the floodplain from the channel has detrimental effects on the function of both the riparian zone and the stream. For example, modification of floodplains and channels disrupts sediment deposition and erosion processes, as well as alters nutrient cycling, stream and riparian species composition, and habitat condition (Poff et al. 1997; Hupp et al. 2009).

Floodplains have long been recognized as fertile lands that are generally good for farming. For example, the people of Mesopotamia, attracted by fertile soils and the potential to ship goods, began farming the Tigris and Euphrates River floodplains in the early Holocene period (Morozova 2005). In more recent years, technological advancements have greatly increased the degree of floodplain alteration and utilization (Lockaby 2009). While natural flood subsidies created these fertile landscapes, these areas have been altered to such a degree that they now require anthropogenic subsidies (e.g. irrigation, fertilization, and tillage) to maintain agricultural production.

4.1.2 Flooding Provisions

Natural flood events are critical for the maintenance of floodplain ecosystems. "The principal driving force responsible for the existence, productivity, and interactions of the major biota in river-floodplain systems is the flood pulse" (Junk et al. 1989). As flood waters spread out laterally over the floodplain, they provide sediment, nutrients, and organic matter (OM) (Nilsson and Svedmark 2002; Thoms 2003). In return, an inundated floodplain releases dissolved organic carbon (DOC), nutrients, and algal biomass which are transported back to the stream channel (Junk et al. 1989; Bayley 1995; Tockner et al. 1999; Thoms 2003; Lockaby et al. 2008). However, not all aspects of flooding and hydrological connectivity are beneficial to the system. For example, floods also transport contaminants to the floodplains (Jackson and Pringle 2010) and create low-oxygen environments (Nilsson and Svedmark 2002).

The amount and spatial distribution of sediment deposition relates to the floodplain topography, flow velocity, period of inundation, stream load, and sediment particle size (Lambert and Walling 1987; Walling and He 1998; Rayburg et al. 2006). The sediment load in the stream depends on the watershed geology, as well as current and historical land use practices. Legacy effects of historical agricultural practices have been linked to high rates of sediment deposition in floodplains (Binkley and Brown 1993; de la Crétaz and Barten 2007; Lockaby 2009). Smaller floods often deposit sediment, while floods of larger magnitude with greater energy, can scour sediment (Toda et al. 2005). In terms of spatial distribution, there is generally a higher rate of sediment deposition near the channel and the rate decreases when moving away from the channel into the floodplain (Walling and He 1998). This pattern coincides with coarse particle deposition closer to the channel while fine particles travel further from the channel (Walling and He 1998; Rayburg et al. 2006).

The wetting and drying phases that occur in the floodplain may cause many changes in nutrient cycling simultaneously. Floods deliver nutrients in particulate or dissolved forms (Nilsson and Svedmark 2002; Tockner et al. 1999; Junk et al. 1989) and may increase nutrient availability in floodplain soils (Bayley 1995; Baldwin and Mitchell 2000), providing riparian vegetation greater access to the nutrients required for growth. Partial wetting of the floodplain (e.g. with precipitation) initially increases the rate of nitrification, but denitrification dominates when completely inundated by flooding (Baldwin and Mitchell 2000). Drying of wet sediments may create a zone of nitrification and denitrification in close spatial proximity at the oxic-anoxic boundary, while complete desiccation terminates denitrification in the floodplain (Baldwin and Mitchell 2000; Austin and Strauss 2011). Flooding also alters rates of decomposition. In some cases, decomposition may be enhanced with flooding (Glazebrook and Roberston 1999; Naiman et al. 2005) although typically, flooded soils are expected to have reduced decomposition (Schlesinger 1991). The difference lies in the duration of inundation: permanently flooded conditions inhibit or slow decomposition and shorter or intermittent flooding enhances decomposition (Lockaby et al. 1996). Further, Neckles and Neill (1994) found that seasonal flooding may increase aboveground decomposition but decrease belowground decomposition.

Plants and animals in the riparian zone depend on the flooding subsidies of OM as sources of nutrients and energy (Toda et al. 2005; Naiman et al. 2005). For example, aquatic insects provide energy to riparian arthropods (Naiman et al. 2005). In natural flood events, particulate organic carbon (POC) may be deposited in the floodplain and DOC may be exported (Tockner et al. 1999). Junk et al. (1989) suggest that the magnitude of OM passing from the channel to the floodplain is much smaller than the DOC that travels from the floodplain to the channel. When flood regimes are altered, inputs of DOC from the floodplain to the channel may be reduced (Thoms 2003).

Periodic inundation of floodplains, differing levels of connectivity, and processes of erosion and deposition create habitat heterogeneity and thus can sustain a diverse assemblage of organisms (Salo et al. 1986; Bayley 1995; Poff et al. 1997; Amoros and Bornette 2002; Nilsson and Svedmark 2002; Richards et al. 2002; Rayburg et al. 2006). Habitat characteristics such as water temperature, suspended solids, nutrients, and substrata composition influence biodiversity in floodplain water bodies (Amoros and Bornette 2002). High diversity of riparian zones is maintained by flooding regimes and thus a first step in floodplain restoration is often the reestablishment of hydrologic connectivity (Steiger et al. 2005).

4.1.3 Flooding: Subsidy or Stress on Productivity?

The subsidy-stress gradient (Odum et al. 1979; Megonigal et al. 1997) has similar underpinnings as the intermediate disturbance hypothesis which predicts maximum function at moderate levels of disturbance. Moderate flooding may be a subsidy, while high intensity flooding (in either frequency or duration) may be a stress (Odum et al. 1979). Past the point of maximum performance (here maximum productivity with moderate flooding), the variability may increase and reduce the stability of the system (Fig. 4.2). In theory, systems can be stable if subsidies are pulsed. Some riparian forest species depend on the floods for regeneration, as seedling supply and production can be tightly linked to flooding regimes (Hughes 1997; Hughes

Fig. 4.1 Buttress trunks and knees in a bald cypress-tupelo gum forest (Photo by Mark A. Musselman, Audubon, South Carolina. Used by permission. Photo accessed from: http://www.namethatplant.net/community_coastal_bottom.shtml)

et al. 2001). It can also be detrimental to eliminate flood pulses in a system acclimated to their occurrence (Odum et al. 1979). For example, following the damming of a river in southern Alberta, Canada, seedling mortality of poplar trees increased due to drought (Rood and Heinz-Milne 1989).

Inundation is stressful to many plants, but some plants adapt more easily than others to flood regimes (Bayley 1995). Low order streams or heavily-modified systems have short, unpredictable pulses and organisms have limited adaptations for utilizing the floodplain in those conditions (Junk et al. 1989). In contrast, long, predictable pulses of rivers sustain organisms with adaptations to utilize the floodplain (Junk et al. 1989). Bottomland hardwood species such as bald cypress and water tupelo displayed extended buoyancy of seeds and fruits to overcome inundation (Schneider and Sharitz 1988). Other adaptations to flooding include buttressed roots, stems, and knees (Lockaby et al. 2008), examples of which can be seen in the bald cypress-tupelo gum forest in Fig. 4.1.

Several studies support the idea that seasonal flooding or flowing water of low to moderate intensity can increase productivity (Fig. 4.2) by providing an energy subsidy to the community (Conner and Day 1976; Odum et al. 1995; Anderson and Mitsch 2008). In the riparian zone of Rush Creek in California, tree growth of *Populus trichocarpa* was linearly related to streamflow and diversion of streamflow reduced tree growth of *Pinus jeffreyi* (Stromberg and Patten 1990). Clawson et al. (2001) expected to find the highest rates of net primary productivity (NPP) in seasonally flooded sites that occupied the 'intermediate' range of flooding provisions; however, the highest NPP was found in the wettest sites which were flooded for most or all of the year. This suggests that the threshold of flooding stress was not exceeded at the wettest sites (Clawson et al. 2001). Flooding subsidies did not

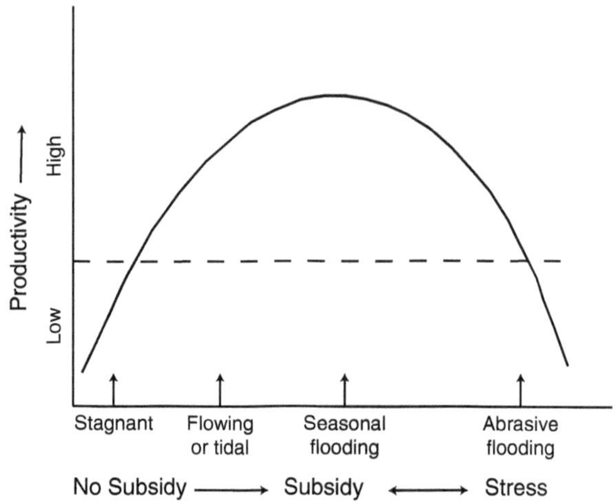

Fig. 4.2 The influence of flood events on productivity (Redrawn from Springer Science+Business Media: Odum et al. 1995, p. 548, figure 3)

significantly enhance productivity on plots of intermediate wetness compared to dry plots in South Carolina and Louisiana (Megonigal et al. 1997). Differences in the productivity response of trees to flooding may be a function of how the flooding categories are defined. For example, Odum et al. (1995) used categories of uplands, bottomland hardwoods, and swamps, Megonigal et al. (1997) and Clawson et al. (2001) used mean water depth to create a gradient of wetness, and Anderson and Mitch (2008) used the number of flooded days as an indicator of the flooding regime. On the high end of the flooding spectrum, low-oxygen environments from extended periods of innundation can slow plant growth (Junk et al. 1989). Megonigal et al. (1997) found significantly lower aboveground net primary productivity (ANPP) in wet plots than intermediate or dry plots. Here, the flood subsidies were outweighed by the stresses of anaerobic soil conditions. Additionally, too much sediment accumulation from flooding may decrease photosynthesis or cause mortality in riparian plants (Nilsson and Svedmark 2002).

4.1.4 Formation and Migration of River Channels

The interactions of rivers with the landscape influence the evolution of the river channel. As summarized in Alabyan and Chalov (1998), three influential factors on river pattern and flow are: (1) discharge which depends on soils and climate, (2) slope or gradient, and (3) erodibility of the bed which depends on sediment properties. Basic classifications distinguish between straight, meandering, and braided or branching channels (Fig. 4.3). These categories can be further divided based on structural levels of fluvial relief including valley bottom, flood channel, or low water channel (Alabyan and Chalov 1998). In reality, there are many intermediate conditions which are not captured by these discrete classes.

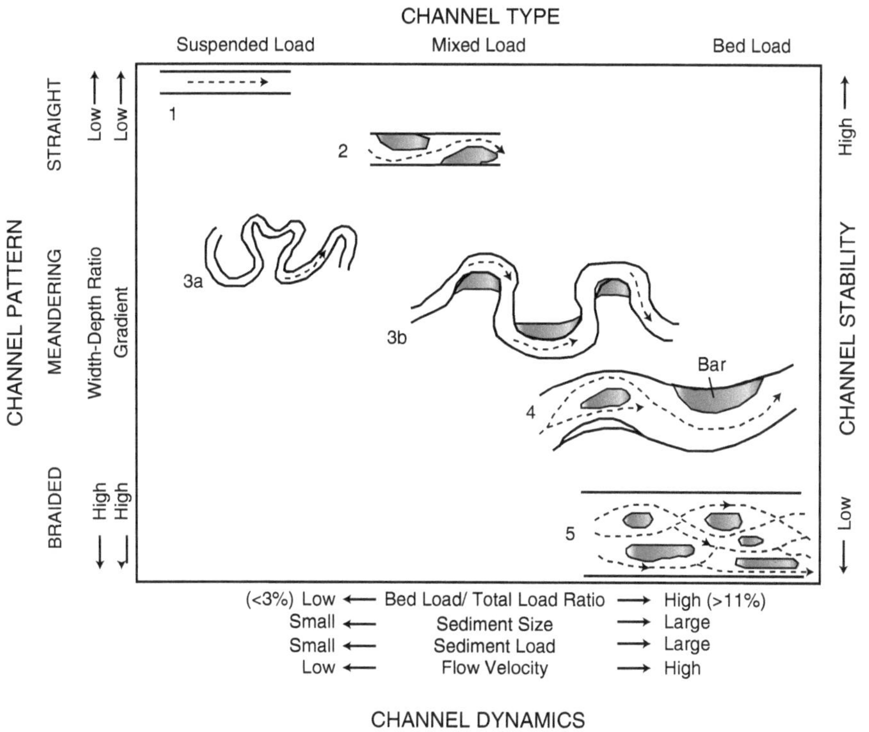

Fig. 4.3 Stream pattern classification (Redrawn from Newson 1994)

Floodplains are primarily formed by two types of deposition: point bar or lateral accretion and overbank deposition or vertical accretion (Wolman and Leopold 1957). Point bars form where sediment is deposited on the inside (convex side) of a river bend (Nanson and Croke 1992). Historically, it has been thought that the proportion of overbank deposits is small (10–20%) compared to channel deposits (80–90%) due to uniform overbank flooding recurrence intervals (Wolman and Leopold 1957). If this were not the case, floodplains would build up very high and consequently overbank flows would be rare. However, more recent studies indicate that lateral accretion may not dominate in all systems; for example in some low-gradient non-braided channels, vertical accretion may dominate (Nanson and Croke 1992). A third fairly common floodplain formation pattern is braid-channel accretion, which results in a mosaic of sediment deposition through dynamic migration and new formation of braid channels (Nanson and Croke 1992).

Lateral movement of a stream channel can be quite rapid. As an example, a river bend in India moved over 180 m laterally in two years (Wolman and Leopold 1957). In general, there is little change in channel width with migration of the channel; the amount of sediment deposited is roughly equal to the amount eroded (Wolman and Leopold 1957; Steiger et al. 2005). Channel migration may vary among channel bends and within individual bends. For example, Meitzen (2009) found the rate of migration was higher at the lower part of bends (downstream) and lower at the higher part of bends (upstream). Riparian vegetation helps to stabilize banks and thus restricts channel movement and can influence whether a single channel or braided channel persists (Murray and Paola 2003). However, high rates of sediment deposition can damage riparian vegetation, clog channels, and thus promote the formation of braided streams (Murray and Paola 2003; Trimble 2008).

Hydrological processes and channel migration influence the structure and composition of riparian forests (Richards et al. 2002). At point bars, differences have been observed in species composition in the edge vs. interior forests. Due to the growth of new vegetation on accreting surfaces, as the distance from the channel increases, early successional species transition to later successional species (Salo et al. 1986; Hughes 1997; Meitzen 2009). In cut banks, or areas of sediment erosion, channel migration may influence forest structure but not composition. Forests on the edges of cut banks have higher structural complexity due to high light conditions and new habitat for colonization created by the removal of mature forests with erosion. The edges of cut banks may also have greater tree density, species richness, and basal area than the interior forests of cut banks (Meitzen 2009).

4.2 Cumulative Effects of Disturbances Within Watersheds

The concept of cumulative effects or 'the combined environmental effects of activities in a watershed' (Brooks et al. 2003) is familiar and implies that the influence of activities may accumulate in time and space (Reid 1993). While it is generally understood that water related impacts observed at the outlet of a watershed reflect the integration of many influences within that catchment, there is often a gap between that knowledge and management or protection strategies. As an example, the Cache River National Wildlife Refuge (CRNWR) in Arkansas is known as an old growth bottomland hardwood forest that provides habitat for many species, possibly including the ivory bill woodpecker. The CRNWR ecosystem has justifiably been the focus of many protection efforts which are predominantly oriented at a local scale. However, the integrity of the system is threatened by high levels of sedimentation resulting from the large proportion of the Cache River basin in agriculture (Lockaby 2009). Upstream disturbances may impact vegetation productivity, composition, and habitat in the CRNWR system. Similarly, Brooks et al. (2003) have suggested that the downstream effects of the 1993 Mississippi River flood (the largest flood in the US during the twentieth century) may have been exacerbated by levee construction, channelization, and urbanization in the headwaters. Cumulative watershed disturbance effects, as illustrated by the Cache River and 1993 Mississippi River flood examples, are very complex from a socioeconomic as well as hydrologic standpoint, and these situations are not easily resolved.

The topographic context of a site is a key consideration in floodplain forest restoration as it influences the hydrological regime and integrates the cumulative effects of watershed disturbance (Lockaby et al. 2008). If all other factors are equal, sites located near low order streams in the upper portions of watersheds will be less subject to catchment level influences than those lower in the drainage basin. Sites lower in watersheds that are associated with higher order streams are more subject to cumulative hydrologic and biogeochemical effects resulting from the integration of processes at higher elevations.

Most research on the cumulative effects at the watershed scale has been devoted to responses of spatial accumulation (Reid 1993) such as the influence of land use. It is well established that land use, particularly agriculture and urbanization, may dramatically alter water quality and stream hydrology (de la Crétaz and Barten 2007). There are many examples including that of Bolstad and Swank (1997) who reported baseflow increases in turbidity and total coliforms from 2.86–5.52 nephelometric turbidity units (NTU) and 9,470–52,140 colonies 100 ml^{-1} respectively moving downstream in Coweeta Creek in North Carolina. These ranges were associated with downstream increases in the proportion of the watershed in agricultural (0.24–4.3%) and urban land uses (0.06–3.4%).

In another example, Fisher et al. (2000) studied the influence of agriculture and urbanization in the Oconee River watershed in Georgia. While poultry production in the headwaters was associated with elevated nitrogen (N), phosphorus (P), and fecal coliform concentrations, those levels diminished downstream as a result of dilution. However, the presence of a mid-sized city (Athens, Georgia) farther downstream doubled N and P concentrations above those observed near the poultry production (Fisher et al. 2000). Consequently, the cumulative effect of different land uses in the Oconee basin was complex and reflected a hierarchy of impacts.

In terms of temporal effects, Weston et al. (2009), working in the Altamaha River floodplain of central Georgia, reported that concentrations and export of nitrate plus nitrite (NO_x) and total nitrogen (TN) increased over time while NH_4^+, P, and organic carbon (OC) declined over the previous 30 years. These trends may be attributed to increased population densities, decreased extent of agriculture, and specifically for P, detergent bans coupled with enhanced P removal at wastewater treatment plants.

A second temporal example is the Piedmont region of the southeastern US which experienced abusive farming practices from the early nineteenth to twentieth centuries. These practices caused the export and accumulation of large volumes of sediment on floodplains and in stream channels (Trimble 2008). This sediment accumulation is still evident today. As examples, Murder Creek, Bonham Creek, and Sally Branch in the Georgia Piedmont currently show sediment accumulation on floodplains of 1.5–2.0 m (Lockaby 2009), the equivalent of 6,000–10,000 years of natural sediment accumulation (Jackson and Pringle 2010). Heavy sediment accumulation is also evident within stream channels. Following forest clearing and construction of unpaved roads in the Bonham Creek and Sally Branch watersheds between 2006 and 2007, large increases in sediment concentrations were observed (Sharif and BalBach 2008). Given the legacy of past land use in the watershed, it is unclear whether the elevated loads originated from unimproved roads or re-suspension of legacy sediment in channels (or some combination of the two) since forest clearing generally stimulates stream flow.

Consideration of cumulative effects and the position of a candidate site within a watershed is critical for floodplain forest restoration. The potential of onsite restoration efforts may be limited if the location is subject to major physiochemical and hydrologic impacts from an assortment of upstream activities. Although some activities within a catchment can be uncontrollable or unpredictable, e.g. urbanization in headwaters, an understanding of the cumulative influence of offsite factors to downstream sites is essential. Furthermore, differences in the extent of riparian zones and associated hydroperiods, or duration of inundation, will be major determinants of restoration trajectories and target conditions.

4.2.1 Land Use/Cover Effects

Previous research has shown that land use/cover change can substantially alter hydrology and water quality (Allan 2004; de la Crétaz and Barten 2007). As a result, it is possible to describe some general influences of land use on water resources with a high degree of confidence. In terms of implications for forest restoration, alterations of stream hydrographs and floodplain hydroperiods are particularly important.

A high proportion of forest cover within a watershed provides a stabilizing effect on the magnitude and rate of change in hydrographs (de la Crétaz and Barten 2007; Nagy et al. 2011). This is due to low overland flow that results from high infiltration capacities in forest soils. High infiltration, in turn, reduces the energy and volume of streamflow. Conversely, when significant proportions of watersheds are devoted to agriculture or urban land uses, the volume and velocity of streamflow increase and rapid changes in stream hydrographs occur (Schoonover et al. 2006; de la Crétaz and Barten 2007; Nagy et al. 2011). The increases in volume stem from increased runoff as well as decreased evapotranspiration. As an example, conversion of forest to agriculture increased streamflow by 10–21% in western North Carolina (Hibbert 1967). These effects are typically more pronounced in urban land use than agriculture due to the sharp reduction in infiltration afforded by paved (impervious) surfaces (Fig. 4.4). Conversion of 10% of watershed area from forest to impervious surfaces was responsible for 32% of the total 159% increase in peak flows in Houston, Texas (Olivera and Defee 2007). In contrast to peak flows, base flows may decrease following urbanization due to reduced infiltration (Paul and Meyer 2001). For example, Rose and Peters (2001) observed baseflow decreases of 25–35% in urbanized vs. non-urbanized watersheds near Atlanta, Georgia.

Whether or not stream hydrograph changes are biologically significant depends on the degree to which similar flooding events (of frequency and duration) occur within riparian areas. Hydroperiod characteristics drive biogeochemical exchange between the stream and floodplain as well as many other floodplain processes (Lockaby et al. 2008). As hydrographs become 'flashier', i.e. rates of change in rising and falling limbs and velocity increase, following substantial land use changes, biotic and abiotic environments of floodplains may change accordingly. These may be reflected by changes in sedimentation and scouring patterns, periodicities of inundation, and

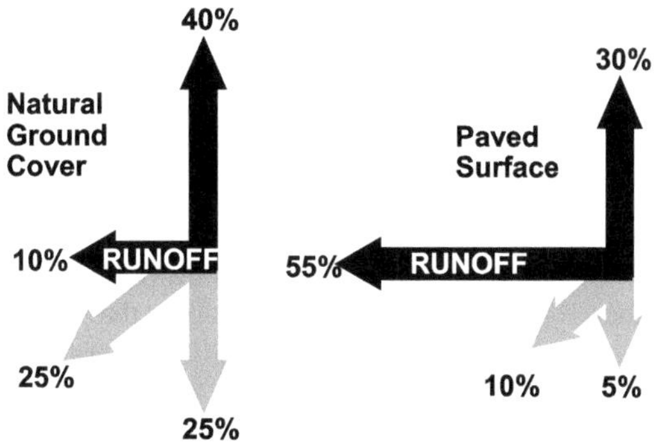

Fig. 4.4 Changes in water components with urbanization (With permission from Groves and Hondorp 2007): runoff-horizontal black arrow, infiltration-gray arrows, and evapotranspiration-vertical black arrow

soil chemical and physical properties (Hupp et al. 2009). Consequently, the vegetation community and ecosystem functions associated with particular sites may become substantially different than those prevalent when the watershed was more heavily forested.

The influences of converting forest to agriculture or urban uses are persistent and stream channels are continually subjected to higher velocities and volumes of streamflow. As a result, channel incisement often occurs (Schoonover et al. 2006; de la Crétaz and Barten 2007; Hupp et al. 2009). Deeper channels are less prone to overbank flooding and, consequently, riparian zones become disconnected hydrologically from adjacent steams (Groffman et al. 2003; Hupp et al. 2009). Consequently, floodplains become drier, a condition which leads to the loss of wetland plant communities and their replacement by species adapted to more mesic conditions. In addition, the disconnect between streams and floodplains leads to degradation of the water filtration function since opportunities for physical and chemical interactions between floodplains and sheetflow become more limited (Lockaby et al. 2008; Hupp et al. 2009). The drying of riparian areas may also lead to increases in nitrogen mineralization, a change that has the potential to affect species succession as well (Groffman et al. 2003).

4.3 Linkages Between Watershed Forest Cover and Water Resources

4.3.1 Perceptions of Forests, Floodplains, and Water Resources

People's perceptions of forests and floodplains have changed considerably over time. Common attitudes towards forest resources span a range from awe and gratitude to irritation and repulsion to apathy. In different periods, forests have been viewed as incredible features of nature, obstructions to be managed or controlled, and more recently precious resources to be preserved and carefully utilized. There may even be a newer phase of disconnect between people and nature as society becomes increasingly urbanized (Lockaby 2009).

The interactions between forests and water have been described for centuries or even millennia. People noted relationships between forests and water resources as early as the first century A.D. (Andréassian 2004). Similarly, people's perceptions of the hydrologic functioning of forests have changed over time. Some ideas portrayed the forest as a siphon, sponge, or a pump. Forests reportedly attracted clouds and thereby enhanced rainfall, although the mechanisms were not specified for this siphoning action (Andréassian 2004). Forests were believed to act like sponges by soaking up water during periods of high precipitation and then releasing it later in times of lower precipitation (Bruijnzeel 2004). Historically, droughts were often attributed to deforestation practices. As divergent theories on the relationships between forests and water emerged, watershed research too emerged in an attempt to reconcile differences. A recent theory, which identifies the mechanisms associated with the siphon metaphor, suggests that with high evapotranspiration in forests, moist air is drawn in from the coast which rises and condenses, forms clouds, and falls as precipitation, allowing continental interiors to maintain high rainfall regimes (Sheil and Murdiyarso 2009).

There may be differences in the forest-water perceptions of the general public vs. scientific community. It is from this disparity that unproven or even flawed theories may perpetuate among local populations. In case studies in South Africa, India, and Costa Rica, there was a general perception linking forests to 'better' hydrological resources (Calder et al. 2004). However, as will be discussed below, not all influences of forests on water resources are 'positive'. Calder et al. (2004) challenged us to: (1) better understand the origin and evolution of local perceptions and how these affect forest and water policies, (2) develop tools for management support, and (3) understand the impacts of forest and water policies on the poor. It is important to emphasize that the effects of forests on hydrological processes can be variable and are highly site-specific. Integration of scientists, conservationists, and local populations will help to spread realistic expectations of what forests can and cannot do for water resources.

4.3.2 Controls on Water Yield

Paired watershed approaches have been used to experimentally determine the effects of afforestation/reforestation and deforestation. In general, with forest clearing, an increase in water yield, or the runoff from the drainage basin, is observed and conversely, with reforestation/afforestation, water yield and/or runoff are reduced (Douglass and Swank 1972; Bosch and Hewlett 1982; Zhang et al. 1999; Andréassian 2004; Bruijnzeel 2004; Farley et al. 2005; Jackson et al. 2005; Sun et al. 2006; de la Crétaz and Barten 2007; Trabucco et al. 2008). Decreased evapotranspiration and increased soil water content and subsurface flow are likely responsible for the increased water yield following forest cutting (de la Crétaz and Barten 2007).

In general, small changes in forest cover, either afforested or deforested, are not sufficient to observe changes in water yield. The greatest responses occur when the basin is completely afforested or deforested.

While the effects of deforestation or afforestation are often realized throughout the watershed, the location of forest cutting/planting within the watershed and the size of the area converted can also influence the effects on water resources. For example, clearing forests in the headwaters may be problematic because that region is dominated by erosional processes. Furthermore, clearing of riparian forest directly along the stream channel, or near the source of streamflow, would intuitively have a greater influence on streams than clearing forests further away from the stream (Bosch and Hewlett 1982). The water requirements of trees along the channel will be eliminated upon clearing, thus increasing the streamflow. Additionally, the percent of basal area reduction, a surrogate for patch size, is roughly proportional to the change in water yield, with minimal effects observed with less than 20% change in forest cover (Bosch and Hewlett 1982; de la Crétaz and Barten 2007).

Consideration of the temporal scale is also important to the interpretation of forest-water study results. The rates of change in water yield will be different for afforestation/reforestation vs. deforestation (Douglass and Swank 1972). The effects of deforestation should be readily apparent following cutting, while the effects of reestablishing vegetation may be a more gradual process as the forest develops (Douglass and Swank 1972; Andréassian 2004). This inherently depends on the nature of the activity (deforestation is rapid vs. afforestation which occurs slowly over time). Therefore, it is important to consider the timeframe of these studies and determine if the effects were likely captured in the time recorded. This advocates for the study of the transition from non-forest to forest as opposed to merely the endpoints (Andréassian 2004). Furthermore, Scott and Prinsloo (2008) found that the decreased streamflow following afforestation may not be permanent. After a forest has matured, streamflow may approach its pre-afforestation level. Therefore, this study suggests that with longer rotations, the negative effects of afforestation on water yield can be minimized.

Forest composition plays a significant role in how water resources are affected by changes in forest cover due to physiological and structural differences. When grouped by genus, there was a significant difference in streamflow changes; specifically, eucalypts were found to reduce runoff more than pines (Farley et al. 2005). Eucalypts and pines generally have greater influence on water resources than deciduous hardwood trees (Douglass and Swank 1972; Bosch and Hewlett 1982). As an example, with a 10% change in forest cover, a 40 mm change in water yield is predicted for pine and eucalypt forests, a 25 mm change in water yield is predicted for deciduous hardwoods, and a 10 mm change in water yield is predicted for scrub vegetation (Bosch and Hewlett 1982). Pines and eucalypts have higher interception and transpiration than deciduous species which may account for the greater water use following afforestation with these species (Jackson et al. 2005).

Climate influences the relative response of water yield to changes in forest cover. The greatest proportional effects are often observed in arid or semi-arid environments, while the greatest absolute effects may occur in more humid environments with greater rainfall (Bosch and Hewlett 1982; Farley et al. 2005; Sun et al. 2006; Trabucco et al. 2008). This suggests that land management in arid and semi-arid regions carries greater risk for altering the water cycle and thus would benefit from additional pretreatment assessments and application of site-specific practices. The climatic regime must be considered when implementing changes in forest cover in order to anticipate the effects on the hydrological cycle.

There are indications that forest management within watersheds has the potential to exacerbate or buffer the effects of climate change on water yield depending on the nature of the management technique (Ford et al. 2011). As examples, conversion of deciduous cover to evergreen tends to reduce peak flows during wet periods but may also reduce water yields during drought. Similarly, a conversion of evergreen species to eucalypts could reduce yields further (Ford et al. 2011).

Other influential factors in determining the effect of changes in forest cover include topographical, geological, and local and regional conditions. Soil properties, such as moisture holding capacity, can be important factors (Zhang et al. 1999). Physical barriers and nutrient deficiencies in soils will limit plant root growth and thus influence the plant water use (Zhang et al. 1999). Regarding soil texture, one study found that intermediate-textured soils may be more inclined to salinization following plantation establishment (Jackson et al. 2005). Soil depth is also important at the watershed scale. "Watershed soils must be deep enough to allow deep-rooted trees to gain a definite advantage over shallow-rooted grass species. Otherwise, the difference between forest and grass will be reduced to the impact of their different interception capacities" (Andréassian 2004). The four case studies in Trabucco et al. (2008) in Ecuador and Bolivia displayed very different responses and the authors attributed these differences not only to changes in forest cover, but also to "climate, soil types, topography, land uses, population densities, existing infrastructures, and tradeoffs with coexisting demands for water". Consequently, local and regional conditions are important for predicting changes in water yield in response to changes in forest cover.

4.3.3 Forest Cover and Peak Flows

Deforestation typically increases flood peaks and volumes (Douglass and Swank 1972; Robinson et al. 2003; Andréassian 2004) but afforestation/reforestation may have little influence on flood peaks (Andréassian 2004), may decrease peak flows (Bruijnzeel 2004), or may increase peak flows (Robinson et al. 2003). Implementation of reforestation/afforestation on account of reducing flooding may therefore be misguided in some cases. However, with careful planning and implementation according to site characteristics, reforestation will result in enhanced ecosystem functioning and may result in decreased peak flows without concurrent decreases in water yield. In addition to the reduction of flooding events, many other benefits of reforestation will be discussed below.

4.3.4 *Forest Cover and Low Flows*

Effects on low flow may be more pronounced than the effects on water yield (Farley et al. 2005) with great reductions following afforestation (Andréassian 2004; Farley et al. 2005) and increases after deforestation (Andréassian 2004). However, low flow may be reduced after forest clearing if soil compaction and consequently reduced infiltration accompany deforestation (Bruijnzeel 2004). In other cases, forest clearing may have little effect on low flows. For example, very small responses were observed in low flows in Europe at the regional scale and the authors speculated that the low flows of these systems were more influenced by physiography and geology (Robinson et al. 2003). In summary, the effects of changes in forest cover on low flows seem to be more variable than the effects on peak flows.

4.3.5 *Other Benefits of Reforestation/Afforestation*

Reforestation/afforestation practices can be beneficial to enhance ecosystem function. For example, forests increase infiltration, decrease erosion, and reduce water pollution (Pizarro et al. 2006; Sun et al. 2006). Forests are able to sequester and store more carbon than many other ecosystems and thus may be used as a climate regulation strategy (Farley et al. 2005; Jackson et al. 2005; Sun et al. 2006). Riparian forests regulate OC export to estuaries (Naiman et al. 2005). Forests also provide habitat for organisms and in the case of riparian forests, supply rivers with organic material that is critical in the aquatic food web (Naiman et al. 2005; Sun et al. 2006; Lockaby et al. 2008). Lastly, forests are valued as systems of high biodiversity, which is one driving factor for forest restoration efforts, especially in the tropics (Lamb et al. 2005).

4.4 Conclusions for Forest Landscape Restoration

4.4.1 *Reestablishing Hydrologic Connectivity*

Since hydrology controls the nature of all functions associated with forested wetlands (Mitsch and Gosselink 2007), reparation of hydrologic connectivity is critical to restoring the function of forested floodplains (Steiger et al. 2005). However, this is often not practical since restoration of original hydrologic regimes (i.e. periodic flooding) would preclude many human activities such as agriculture or urban development in some cases (Brinson and Verhoeven 1999). An example is the restoration of forests in the southern Mississippi River floodplain which consists of regeneration of bottomland hardwood species on land behind levees that was previously in agriculture (Stanturf et al. 2003). Although the floodplain will become forested, the connectivity between the river and floodplain will not be restored and, consequently, some key functions associated with floodplains will not be reestablished.

Many restoration efforts to date have focused on reestablishing the connectivity in order to repair the riparian and stream function. Removal of dikes, levees, culverts, and dams are methods to allow the resumption of flood pulses. These practices enable flooding provisions (e.g. nutrients, sediment, organic matter) to be reinstated and help to create a more natural flooding regime, indicated by a less flashy hydrograph. Examples of hydrologic restoration projects by the Natural Resources Conservation Service (NRCS) in the state of Rhode Island, including the removal of a culvert and a dike, are shown in Fig. 4.5. Much progress has also been made with hydrological restoration efforts to reconnect the channel and the floodplain of the Kissimmee River in Florida (Naiman et al. 2005). However, as Kondolf et al. (2006) pointed out, establishing connectivity may not be appropriate in all cases. Some systems do not have high connectivity naturally, and establishing such conditions may exclude native riparian species (Kondolf et al. 2006). Furthermore, it may be more important to restore the processes and flows than just the connections between a river and its floodplain (Kondolf et al. 2006). Once the hydrologic connections are repaired, fluvial dynamics, such as rates of channel migration and sedimentation, can be used to predict habitat diversity in riparian zones (Richards et al. 2002).

4.4.2 Forest Restoration

Reforestation of key areas such as riparian zones, buffers around protected areas, corridors connecting forest patches, and erosion-prone areas are common starting points for landscape restoration (Lamb et al. 2005). Forests have the potential to meet a variety of needs through multiple uses and provide services for the environment and humans simultaneously. Realistic expectations for forest restoration projects stem from a basic understanding of ecosystem processes, composition, structure, and function, as well as variability in these systems. For example, the potential for nutrient and pollutant filtration and flood regulation that forests provide must be weighed against the short-term reduction in water yield that occurs with reforestation. A major shortcoming of restoration projects to date is the lack of monitoring and evaluation of restoration efforts after implementation. This component is often neglected due to budgetary limitations (DellaSala et al. 2003). Without these assessments, the field of ecological restoration cannot advance. Failures or shortcomings of past restoration efforts should be documented along with detailed descriptions of restoration protocols to avoid encountering the same difficulties in future projects.

Scale, both spatial and temporal, is an important aspect in the processes that govern riparian systems, as well as their restoration (Hughes et al. 2001; Steiger et al. 2005). Considering variation from the species or habitat scale to the community or landscape scale will help establish a range of acceptable restoration outcomes.

Fig. 4.5 Rhode Island NRCS floodplain restoration projects. (**a**) Meshanticut River Headwater (Janet Drive) floodplain in West Warwick- removal of a culvert and creation of a new, open stream channel. (**b**) Pawtuxet and Pocasset River confluence floodplain in Cranston- removal of a dike and floodplain fill to restore connectivity (Photos courtesy of USDA Natural Resources Conservation Service, Rhode Island (2011))

Also, the short- and long-term restoration goals and expectations must be defined during project conception. Short-term objectives may be able to better address restoration at the species or habitat scale, while long-term plans may be better suited for restoration at the landscape scale (Kuuluvainen et al. 2002). Restoring function and processes to the landscape will likely take a long time (Steiger et al. 2005).

Table 4.1 Successful forest restoration efforts

Location	Restoration practice	Benefits of restoration	References
Southern India	Afforestation	Reduced runoff, reduced erosion	UNEP (2009)
Australia	Afforestation	Reduced groundwater recharge and thus decreased salinization	UNEP (2009)
Malaysia	Riparian reforestation	Decreased flood impacts	Maginnis and Jackson (2008)
North Carolina-Duke Forest	Stream restoration and reforestation	Reduced stream NO_3 and P	Richardson and Pahl (2005)
Chesapeake Bay	Protection and riparian restoration	Reduced stream N, P, and sediment	Chesapeake Bay Program (2009)
Ethiopia	Assisted natural regeneration	Increased plant diversity and density	Mengistu et al. (2005)
Southern China	Reforestation-plantations and natural regeneration	Increased carbon storage/climate regulation	Zheng et al. (2008)
Northeast Asia; New Mexico	Afforestation	Improved local economy/created jobs	Moon and Park (2004), Forest Guild (2009)

One of the main ideas emerging from restoration studies is the utilization of multiple methodologies during the planning stage. For example, cross-site comparisons, large-scale manipulations, and predictive modeling all have their own limitations (Holl et al. 2003). However, when these methodologies are used in conjunction with each other, they provide a powerful toolset for evaluating and implementing restoration projects. Further, incorporating adaptive management with research and experimentation will allow a range of restoration outcomes and will help make restoration efforts more successful in meeting the needs of humans and the ecosystem (Poff et al. 1997; Stromberg 2001; Downs and Kondolf 2002; Kuuluvainen et al. 2002; DellaSala et al. 2003; Hughes et al. 2005; Naiman et al. 2005; Chazdon 2008).

Forest landscape restoration efforts are underway globally and aim to enhance both the ecological and socioeconomic integrity of landscapes (UNEP 2009). Successful restoration examples can be found in Table 4.1. Restoration is an inherently interdisciplinary endeavor which must consider economic and social perspectives in the aim to repair and maintain ecosystem function. DellaSala et al. (2003) proposed three core principles in the 'Citizens Call for Ecological Forest Restoration': (1) ecological forest restoration, (2) ecological economics, and (3) community and work force. The interaction of humans with forests and water resources including current and historical contexts must be considered in forest restoration plans. Coupling scientific research with community participation and the involvement of stakeholders, interest groups, and agencies may increase the likelihood of success for restoration efforts (DellaSala et al. 2003; Lamb et al. 2005; Naiman et al. 2005; Chazdon 2008; UNEP 2009). Further, local populations can benefit through incentives provided to landowners for restoration practices (Rodrigues et al. 2009), as well as restoration programs that select plant species with known utility to local populations (Allen et al. 2010).

4.4.3 Management Implications

Specific management recommendations include:

- reestablish the hydrologic connection and attempt to recreate the 'natural' flooding regime
- restore native riparian vegetation species that are well suited to that flooding regime
- be flexible and creative in your management approach and use management actions as experiments (adaptive management)
- learn from other successful restoration efforts (Table 4.1) and set realistic expectations for restoration outcomes with explicit criteria for success based on the site
- involve local citizens in restoration efforts in order to enhance the probability of success.

Acknowledgments Funding for this research was provided by Auburn University's Center for Forest Sustainability.

References

Alabyan AM, Chalov RS (1998) Types of river channel patterns and their natural controls. Earth Surf Process Landf 23:467–474

Allan JD (2004) Landscapes and riverscapes: the influence of land use on stream ecosystems. Annu Rev Ecol Syst 35:257–284

Allen AE, Santana-Michel SJ, Arrona CO et al (2010) Integrating ecological and ethnobotanical priorities into riparian restoration. Ecol Restor 28(3):377–388

Amoros C, Bornette G (2002) Connectivity and biocomplexity in waterbodies of riverine floodplains. Freshw Biol 477:61–776

Anderson CJ, Mitsch WJ (2008) Influence of flood connectivity on bottomland hardwood forest productivity in central Ohio. Ohio J Sci 108(2):2–8

Andréassian V (2004) Waters and forests: from historical controversy to scientific debate. J Hydrol 291:1–27

Austin BJ, Strauss EA (2011) Nitrification and denitrification response to varying periods of desiccation and inundation in a western Kansas stream. Hydrobiology 658:183–195

Baldwin DS, Mitchell AM (2000) The effects of drying and re-flooding on the sediment and soil nutrient dynamics of lowland river-floodplain systems: a synthesis. Regul Rivers Res Manag 16:457–467

Bayley PB (1995) Understanding large river-floodplain ecosystems. BioScience 45(3):153–158

Binkley D, Brown TC (1993) Management impacts on water quality of forests and rangelands. General Technical Report RM-239. U.S. Department of Agriculture Forest Service, Rocky Mountain Forest and Range Experiment Station, Fort Collins, CO

Bolstad PV, Swank WT (1997) Cumulative impacts of land use on water quality in a southern Appalachian watershed. J Am Water Resour 33:519–533

Bosch JM, Hewlett JD (1982) A review of catchment experiments to determine the effect of vegetation changes on water yield and evapotranspiration. J Hydrol 55:3–23

Brinson M, Verhoeven J (1999) Riparian forests. In: Hunter ML (ed) Maintaining biodiversity in forest ecosystems. Cambridge University Press, Cambridge

Brooks KN, Ffolliott PF, Gregersen HM et al (2003) Hydrology and the management of watersheds, 3rd edn. Blackwell Publishing, Ames

Bruijnzeel LA (2004) Hydrological functions of tropical forests: not seeing the soil for the trees? Agric Ecosyst Environ 104:185–228
Calder I, Amezaga J, Aylward B et al (2004) Forests and water policies. The need to reconcile public and science perspectives. Geol Acta 2(2):157–166
Chazdon RL (2008) Beyond deforestation: restoring forests and ecosystem services on degraded lands. Science 320:1458–1460
Chesapeake Bay Program (2009) Bay restoration. http://www.chesapeakebay.net/bayrestoration.aspx?menuitem=13989. Accessed 13 July 2009
Clawson RG, Lockaby BG, Rummer B (2001) Changes in production and nutrient cycling across a wetness gradient within a floodplain forest. Ecosystem 4:126–138
Conner WH, Day JW Jr (1976) Productivity and composition of a baldcypress-water tupelo site and a bottomland hardwood site in a Louisiana swamp. Am J Bot 63:1354–1364
de la Crétaz AL, Barten PK (2007) Land use effects on streamflow and water quality in the Northeastern United States. CRC Press, Boca Raton
DellaSala DA, Martin A, Spivak R et al (2003) A citizen's call for ecological forest restoration: forest restoration principles and criteria. Ecol Restor 21(1):14–23
Douglass JE, Swank WT (1972) Streamflow modification through management of eastern forests. USDA Forest Service Research Paper: SE-94, Asheville, NC
Downs PW, Kondolf GM (2002) Post-project appraisals in adaptive management of river channel restoration. Environ Manag 29(4):477–496
Farley KA, Jobbágy EG, Jackson RB (2005) Effects of afforestation on water yield: a global synthesis with implications for policy. Glob Change Biol 11:1565–1576
Fisher DS, Steiner JL, Endale DM et al (2000) The relationship of land use practices to surface water quality in the Upper Oconee watershed of Georgia. For Ecol Manag 128:39–48
Ford CR, Laseter SH, Swank WT et al (2011) Can forest management be used to sustain water-based ecosystem services in the face of climate change? Ecol Appl 21(6):2049–2067
Forest Guild (2009) Forest Guild success stories. http://www.forestguild.org/success.html. Accessed 7 July 2009
Gergel SE, Dixon MD, Turner MG (2002) Consequences of human-altered floods: levees, floods, and floodplain forests along the Wisconsin River. Ecol Appl 12(6):1755–1770
Glazebrook HS, Robertson AI (1999) The effect of flooding and flood timing on leaf litter breakdown rates and nutrient dynamics in a river red gum (*Eucalyptus camaldulensis*) forest. Austr J Ecol 24:625–635
Groffman PM, Blain DJ, Band LE et al (2003) Down by the riverside: urban riparian ecology. Front Ecol Environ 1(6):315–321
Groves T, Hondorp C (2007) Low impact development: concepts and considerations. http://www.michiganlakeinfo.com/PermaLink,guid,6a215927-af04-4263-8484-21eb4f3810be.aspx. Accessed 15 July 2009
Hibbert AR (1967) Forest treatment effects on water yield. In: Sopper WE, Hull HW (eds) Forest hydrology: proceedings of a National Science Foundation advanced science. Pergamon Press, New York
Holl KD, Crone EE, Schultz CB (2003) Landscape restoration: moving from generalities to methodologies. BioScience 53(5):491–502
Hughes FMR (1997) Floodplain biogeomorphology. Prog Phys Geogr 21(4):501–529
Hughes FMR, Adams WM, Muller E et al (2001) The importance of different scale processes for the restoration of floodplain woodlands. Regul Rivers Res Manag 17:325–345
Hughes FMR, Colston A, Mountford JO (2005) Restoring riparian ecosystems: the challenge of accommodating variability and designing restoration trajectories. Ecol Soc 10(1):12
Hupp CR, Pierce AR, Noe GB (2009) Floodplain geomorphic processes and environmental impacts of human alteration along coastal plain river. Wetlands 29(2):413–429
Jackson CR, Pringle CM (2010) Hydrologic connectivity in intensively developed landscapes. BioScience 60:37–46
Jackson RB, Jobbágy EG, Avissar R et al (2005) Trading water for carbon with biological carbon sequestration. Science 310:1944–1947

Junk WJ, Bayley PB, Sparks RE (1989) The flood pulse concept in river-floodplain systems. In: Dodge DP (ed) Proceedings of the international large river symposium. Canadian Special Publications of Fisheries and Aquatic Sciences, Fisheries and Oceans Canada, Ottawa
Kondolf GM, Boulton AJ, O'Daniel S et al (2006) Process-based ecological river restoration: visualizing three-dimensional connectivity and dynamic vectors to discover lost linkages. Ecol Soc 11(2):5
Kuuluvainen T, Aapala K, Ahlroth P et al (2002) Principles of ecological restoration of boreal forested ecosystems: Finland as an example. Silva Fennica 36(1):409–422
Lamb D, Erskine PD, Parrotta JA (2005) Restoration of degraded tropical forest landscapes. Science 310(5754):1628–1632
Lambert CP, Walling DE (1987) Floodplain sedimentation: a preliminary investigation of contemporary deposition within the lower reaches of the River Culm, Devon, UK. Geogr Ann 69A(3–4):393–404
Lockaby BG (2009) Floodplain ecosystems of the Southeast: linkages between forests and people. Wetl 29(2):407–412
Lockaby BG, Murphy AL, Somers GL (1996) Hydroperiod influences on nutrient dynamics in decomposing litter of a floodplain forest. Soil Sci Soc Am J 60(4):1267–1272
Lockaby BG, Conner WH, Mitchell J (2008) Floodplains. In: Jørgensen SE, Faith BD (eds) Ecosystems. Vol 2 of Encyclopedia of ecology. Elsevier, Oxford
Maginnis S, Jackson W (2008) Restoring forest landscapes. http://cmsdata.iucn.org/downloads/restoring_forest_landscapes.pdf. Accessed 13 July 2009
Megonigal JP, Conner WH, Kroeger S et al (1997) Aboveground production in southeastern floodplain forests: a test of the subsidy-stress hypothesis. Ecology 78(2):370–384
Meitzen KM (2009) Lateral channel migration effects on riparian forest structure and composition, Congaree River, South Carolina, USA. Wetl 29(2):465–475
Mengistu T, Teketay D, Hulten H et al (2005) The role of enclosures in the recovery of woody vegetation in degraded dryland hillsides of central and northern Ethiopia. J Arid Environ 60:259–281
Mitsch W, Gosselink J (2007) Wetlands, 4th edn. Wiley, Hoboken
Moon KH, Park DK (2004) The role and activities of NGOs of reforestation in the northeast Asian region. For Ecol Manag 201:75–81
Morozova GS (2005) A review of Holocene avulsions of the Tigris and Euphrates Rivers and possible effects on the evolution of civilizations in lower Mesopotamia. Geoarchaeology 20(4):401–423
Murray AB, Paola C (2003) Modelling the effect of vegetation on channel pattern in bedload rivers. Earth Surf Process Landf 28(2):131–143
Nagy RC, Lockaby BG, Helms B et al (2011) Water resources and land use and cover in a humid region: the southeastern United States. J Environ Qual 40:867–878
Naiman RJ, Décamps H (1997) The ecology of interfaces: riparian zones. Annu Rev Ecol Syst 28:621–658
Naiman RJ, Décamps H, McClain ME (2005) Riparia: ecology, conservation, and management of streamside communities. Elsevier Academic Press, Burlington
Nanson GC, Croke JC (1992) A genetic classification of floodplains. Geomorphology 4:459–486
Neckles HA, Neill C (1994) Hydrologic control of litter decomposition in seasonally flooded prairie marshes. Hydrobiology 286(3):155–165
Newson M (1994) Hydrology and the river environment. Oxford University Press Inc., New York
Nilsson C, Svedmark M (2002) Basic principles and ecological consequences of changing water regimes: riparian plant communities. Environ Manag 30(4):468–480
Odum EP, Finn JT, Franz EH (1979) Perturbation theory and the subsidy-stress gradient. BioScience 29(6):349–352
Odum WE, Odum EP, Odum HT (1995) Nature's pulsing paradigm. Estuaries 18(4):547–555
Olivera F, DeFee BB (2007) Urbanization and its effect on runoff in the Whiteoak Bayou watershed, Texas. J Am Water Resour Assoc 43:170–182
Paul MJ, Meyer JL (2001) Streams in the urban landscape. Annu Rev Ecol Syst 32:333–365

Pizarro R, Araya S, Jordán C et al (2006) The effects of changes in vegetative cover on river flows in the Purapel river basin of central Chile. J Hydrol 327:249–257
Poff NL, Allan JD, Bain MB et al (1997) The natural flow regime: a paradigm for river conservation and restoration. BioScience 47(11):769–784
Rayburg S, Thoms M, Lenon E (2006) Unraveling the physical template of a terminal flood-plain-wetland sediment storage system. Sediment dynamics and the hydromorphology of fluvial systems (Proceedings of a symposium held in Dundee, UK, July 2006). IAHS Publication 306
Reid LM (1993) Research and cumulative watershed effects. General Technical Report PSW-GTR-141-WWW. Albany, CA. Pacific Southwest Research Station, U.S. Department of Agriculture Forest Service
Richards K, Brasington J, Hughes F (2002) Geomorphic dynamics of floodplains: ecological implications and a potential modeling strategy. Freshw Biol 47:559–579
Richardson CJ, Pahl JW (2005) The Duke Forest stormwater improvement and wetlands restoration project: final report to the North Carolina Clean Water Management Trust Fund and the North Carolina Ecosystem Enhancement Program. http://www.nicholas.duke.edu/wetland/CWMTF05.pdf. Accessed 20 June 2009
Robinson M, Cognard-Plancq A-L, Cosandey C et al (2003) Studies of the impact of forests on peak flows and baseflows: a European perspective. For Ecol Manag 186:85–97
Rodrigues RR, Lima RAF, Gandolfi S et al (2009) On the restoration of high diversity forests: 30 years of experience in the Brazilian Atlantic Forest. Biol Conserv 142:1242–1251
Rood SB, Heinz-Milne S (1989) Abrupt downstream forest decline following river damming in southern Alberta. Can J Bot 67:1744–1749
Rose S, Peters NE (2001) Effects of urbanization on streamflow in the Atlanta area (Georgia, USA): a comparative approach. Hydrol Process 15:1441–1457
Salo J, Kalliola R, Häkkinen I et al (1986) River dynamics and the diversity of the Amazon lowland forest. Nature 322:254–258
Schlesinger WH (1991) Biogeochemistry: an analysis of global change. Academic Press Inc., San Diego
Schneider RL, Sharitz RR (1988) Hydrochory and regeneration in a bald cypress-water tupelo swamp forest. Ecology 69(4):1055–1063
Schoonover JE, Lockaby BG, Helms B (2006) Impacts of land cover on stream hydrology in the west Georgia piedmont, USA. J Environ Qual 35:2123–2131
Scott DF, Prinsloo FW (2008) Longer-term effects of pine and eucalypt plantations on streamflow. Water Resour Res 44:W00A08. doi:10.1029/2007WR006781
Sharif M, Balbach H (2008) Water quality and sediment monitoring program at Ft. Benning, Ga. IPR Presentation to Natural Resources Branch, USCOE. Ft. Benning, GA
Sheil D, Murdiyarso D (2009) How forests attract rain: an examination of a new hypothesis. BioScience 59(4):341–347
Stanturf JA, Gardiner E, Warren M (2003) Restoring forested wetland ecosystems. In: Holland M, Blood E (eds) Achieving sustainable freshwater systems. Island Press, Washington, DC
Steiger J, Tabacchi E, Dufour S et al (2005) Hydrogeomorphic processes affecting riparian habitat within alluvial channel-floodplain river systems: a review for the temperate zone. River Res Appl 21:719–737
Stromberg JC (2001) Restoration of riparian vegetation in the south-western United States: importance of flow regimes and fluvial dynamism. J Arid Environ 49:17–34
Stromberg JC, Patten DT (1990) Riparian vegetation instream flow requirements: a case study from a diverted stream in the eastern Sierra Nevada, California, USA. Environ Manag 14(2):185–194
Sun G, Zhou G, Zhang Z et al (2006) Potential water yield reduction due to forestation across China. J Hydrol 328:548–558
Thoms MC (2003) Floodplain-river ecosystems: lateral connections and the implications of human interference. Geomorphology 56:335–349
Tockner K, Pennetzdorfer D, Reiner N et al (1999) Hydrological connectivity and the exchange of organic matter and nutrients in a dynamic river-floodplain system (Danube, Austria). Freshw Biol 41:521–535
Toda Y, Ikeda S, Kumagai K et al (2005) Effects of flood flow on flood plain soil and riparian vegetation in a gravel river. J Hydraulic Eng 131(11):950–960

Trabucco A, Zomer RJ, Bossio DA et al (2008) Climate change mitigation through afforestation/reforestation: a global analysis of hydrologic impacts with four case studies. Agric Ecosyst Environ 126:81–97

Trimble SW (2008) Man-induced soil erosion on the southern Piedmont, 2nd edn. Soil and Water Conservation Society, Ankeny

UNEP (2009) Global partnership on forest landscape restoration. http://www.unep-wcmc.org/forest/restoration/globalpartnership/. Accessed 30 June 2009

USDA Natural Resources Conservation Service (NRCS), Rhode Island (2011) American recovery and reinvestment act (ARRA) emergency watershed protection program- floodplain easements. http://www.ri.nrcs.usda.gov/news/successstories/ARRA_EWP_Floodplain_Easements.html. Accessed 30 Mar 2012

Walling DE, He Q (1998) The spatial variability of overbank sedimentation on river floodplains. Geomorphology 24:209–223

Weston NB, Hollibaugh JT, Joye SB (2009) Population growth away from the coastal zone: thirty years of land use change and nutrient export in the Altamaha River, GA. Sci Total Environ 407(10):3347–3356

Wolman MG, Leopold LB (1957) River flood plains: some observations on their formation. Physiographic and hydraulic studies of rivers. Geological Survey Professional Paper 282-C, US Geological Survey, Reston, VA

Zhang L, Dawes WR, Walker GR (1999) Predicting the effect of vegetation changes on catchment average water balance. Technical Report 99/12, Cooperative Research Center for Catchment Hydrology, CSIRO Land and Water, Canberra

Zheng H, Ouyang Z, Xu W et al (2008) Variation of carbon storage by different reforestation types in the hilly red soil region of southern China. For Ecol Manag 255:1113–1121

Chapter 5
Connecting Landscape Fragments Through Riparian Zones

Gary Bentrup, Mike Dosskey, Gary Wells, and Michele Schoeneberger

5.1 Forest Loss and Ecosystem Services

Many formerly forested regions have been largely cleared and are now important crop and livestock producing lands (Fig. 5.1). This is true of many parts of the world including United States' southeastern coastal plain, Brazil's rainforests, Northern Europe's lowlands, China's northeastern plains, Indonesia's lowlands, and floodplains of most of the world's large rivers. Through widespread conversion of forests to intensively-managed agricultural uses, these countries have created highly productive agricultural economies.

Environmental issues have arisen as consequences of the loss and fragmentation of forests, including soil erosion, water pollution, and fish and wildlife population declines (Green et al. 2005; Schröter et al. 2005; Matson and Vitousek 2006). The pre-existing forests provided the public with high levels of desired ecosystem services, including clean water, healthy fish and wildlife, biodiversity, climate moderation, wood and food products, and aesthetic qualities (Fig. 5.2). Subsequent decline of these services has resulted in lower levels of social well-being, causing public concern (MEA 2005). To regenerate them, restoration of large tracts of land back to forest may be a logical goal, but it may not be feasible. Doing so may put the supply of plentiful and affordable food at risk, and, convincing numerous farm workers, land-holders, communities, and industries to change their social fabric woven around agriculture to one centered on forestry may pose a daunting social challenge. A more acceptable alternative might be to restore forest in only the most critical portions of

G. Bentrup (✉) • M. Dosskey • M. Schoeneberger
US Forest Service, USDA National Agroforestry Center,
Lincoln, NE, USA
e-mail: gbentrup@fs.fed.us

G. Wells Retd
Natural Resources Conservation Service, USDA National Agroforestry Center,
Lincoln, NE, USA

J. Stanturf et al. (eds.), *Forest Landscape Restoration: Integrating Natural and Social Sciences*, World Forests 15, DOI 10.1007/978-94-007-5326-6_5,

Fig. 5.1 Forest clearing has produced highly-productive agricultural landscapes. However, ecosystem services provided by those former forest lands, such as clean water and forest wildlife have diminished. Restoration of forest ecosystem services to agricultural landscapes requires landscape planning that integrates knowledge of natural science and social science principles (Photo credit: NRCS)

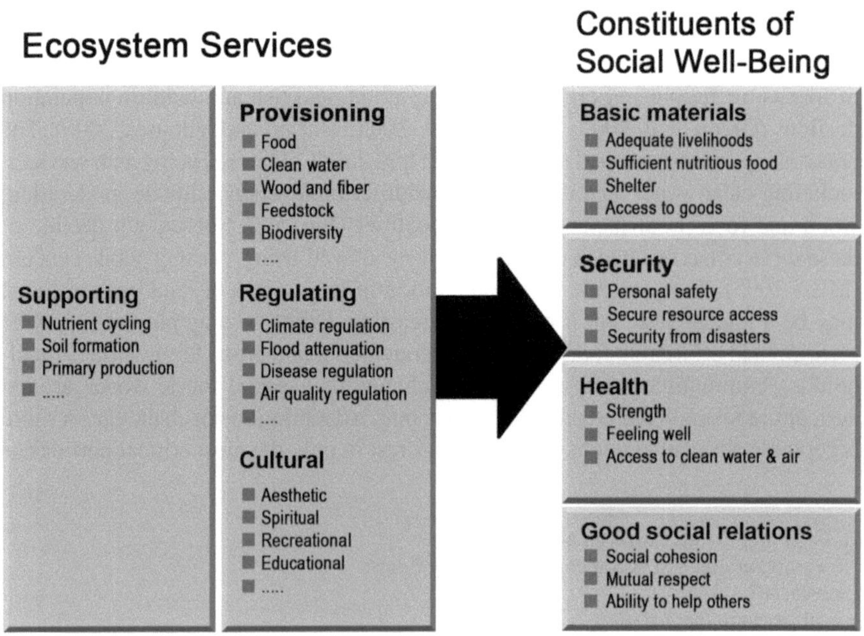

Fig. 5.2 Ecosystem services are benefits people obtain from ecosystems which in turn support components of human social well-being. Other human factors (e.g., economic, social, technological, cultural) also influence social well-being and feedback to affect ecosystems and ecosystem services (feedback not shown) (Adapted from Millennium Ecosystem Assessment 2005)

these landscapes, while maintaining most of the existing agricultural socio-ecological system. In this way, forest restoration can provide a balance between social acceptance and alleviation of environmental issues.

In this chapter, we describe how natural science and social science principles can be integrated to help resolve the trade-offs and challenges of restoring forest ecosystem services to agricultural landscapes.

5.2 Integrating Natural and Social Sciences

Restoration of forest ecosystem services in agricultural landscapes may not require restoration of large forest tracts. Small forest patches and strips, and even individual trees, restored in the right places and configurations can restore significant levels of forest functions that are associated with larger forest tracts (Garrett et al. 2000; Green et al. 2005; Nair et al. 2005; Breshears 2006; Manning et al. 2006; Benayas et al. 2008). Consequently, restoration of forest in relatively small, strategic locations may enable finding an acceptable balance among the many demands placed on agricultural landscapes.

Finding that acceptable balance, however, requires integrating natural and social science principles with a planning process whereby people set goals and make decisions that most, if not all, can agree on. Decisions must be made about where restoration should take place in the landscape, the size of the restoration zone, and specifics of vegetation design and management of these forest areas. Since successful restoration will require local landholders to be motivated to implement restoration plans, public goals for the provision of forest ecosystem services must be considered along with personal objectives of each individual landholder. Goal-setting, design development, and decision-making is facilitated by a participatory planning process involving local and public stakeholders that are informed with natural resource principles. Achieving restoration success, then, requires integrating natural and social sciences in a way that produces efficient and effective landscape management plans and encourages their implementation.

5.2.1 Natural Sciences – Riparian Zones and Continuity

Riparian areas are portions of landscapes where forest restoration can be especially effective for enhancing important ecosystem services, including cleaner water, and more fish and wildlife, among others (NRC 2002; Naiman et al. 2005). Riparian areas are lands adjacent to streams and lakes. In riparian areas, there is a high degree of interaction with the adjacent waterways. Riparian areas are flow-through zones for runoff from uplands, for channel-hyporheic interchange, and for overland flow by floodwaters that affect both water supply and water quality in adjacent waterways. Riparian vegetation contributes detritus to streams that creates structural habitat and fuels the aquatic food chain. Riparian areas have particularly high-value as habitat for terrestrial wildlife because of the close availability of water and the network pattern through landscapes that promote migration of wildlife between seasonal habitats and dispersal from population centers. For example, riparian areas constitute probably less than 5% of the total land area in the U.S., but are disproportionally effective lands for providing forest ecosystem services (NRC 2002; Naiman et al. 2005). Because of these special qualities, riparian zones are uniquely capable of producing high levels of multiple ecosystem services in otherwise nonforested landscapes.

A riparian forest buffer is a strip of forested area that separates and helps protect streams and other water bodies from negative impacts of adjacent land uses and for the provision of non-agricultural ecosystem services in agricultural landscapes (Welsch 1991). It is a restoration practice commonly designed for and managed to enhance water quality, aquatic habitat, and to increase wildlife populations (NRC 2002). Riparian forest buffers can also help to create visually pleasing landscapes and to provide erosion control among other benefits (Ryan 1998; Naiman et al. 2005). Even narrow buffers can have a large impact on water quality and wildlife in agriculture-dominated landscapes. For example, water quality and wildlife habitat can be substantially improved by forested buffers as narrow as 30 m (Welsch 1991; Sweeney 1993; Lowrance et al. 1995; Wenger 1999; Dosskey 2001; Kennedy et al. 2003). Since riparian areas occupy only a small fraction of the total landscape, forest restoration through the establishment of forested riparian buffers represents an area-efficient strategy for restoring forest ecosystem services to agricultural landscapes (NRC 1993).

5.2.2 *Connecting Fragments Using Riparian Buffers*

A key principle of enhancing ecosystem benefits using riparian buffers is the restoration of their continuity through the landscape. Continuity is critical for intercepting and filtering polluted runoff water and for providing corridors for the movement of wildlife (Welsch 1991; Crooks and Sanjayan 2006). In most agriculture-dominated landscapes, fragments of original or degraded forest remain; some in riparian areas and some in uplands. While these remnant forest patches may provide a modicum of ecosystem services, the gaps between them prevent them from achieving their full potential. By reconnecting existing forest fragments with a focus on restoring continuity through riparian zones, water-filtering and habitat-producing ecosystem services, as well as others, can be efficiently restored in a developed landscape.

A few additional ecological principles can help to identify locations for and designs of riparian buffers that will restore specific ecosystem services with even greater efficiency (Boxes 5.1 and 5.2). Individual locations vary in their capability of restoring certain ecosystem services because of topography, hydrology, or other site factors so the design of a riparian buffer can also vary from one location to another. For example, a habitat gap may represent a particularly efficient location for enhancing wildlife production (Box 5.2). However, a different location may intercept greater pollutant load and a widening of an existing buffer may be required for adequate water quality control (Box 5.1). Ecological principles for addressing other natural resource issues and ecosystem services can be added to these, if desired; the descriptions of which can be found in Dramstad et al. (1996) and Bentrup (2008). While the ecological principles outlined here indicate ***what*** can be done to efficiently restore important forest ecosystem services in developed landscapes, social science principles are necessary to determine ***how*** to encourage landholder acceptance and adoption in order to achieve implementation and sustainable results.

Box 5.1 Principles for Guiding Riparian Forest Restoration for Water Pollution Reduction

- Locate restoration areas where they will connect existing riparian forest fragments and extend the length of continuous forest along waterways and shores.
- Size restoration areas to be larger/wider at locations that intercept greater runoff load (Fig. 5.3).
- Size restoration areas to be larger/wider at locations that have steeper slopes or that have soils with lower infiltration capacity.
- Design forest plantings to promote denser herbaceous cover at locations that intercept greater overland flow.
- Select tree species that tolerate flooding for use on low floodplains and to stabilize eroding stream banks.

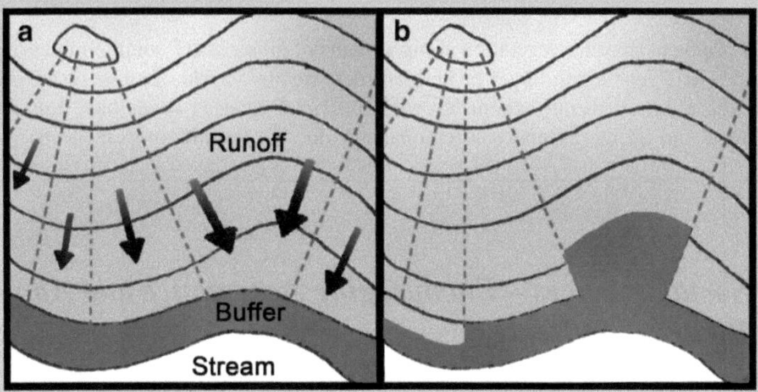

Fig. 5.3 Runoff is often non-uniform and flow is either diverging or converging due to topography, tillage practices and other factors. A fixed-width buffer will be less effective in these situations (**a**). Riparian buffer areas receiving greater runoff loads should be enlarged to intercept these greater loads (**b**)

Box 5.2 Principles for Guiding Riparian Forest Restoration for Terrestrial Wildlife Habitat Enhancement

- Locate restoration areas next to existing forest fragments to enlarge existing habitat areas and to connect fragments.
- Locate and shape restoration areas so that, when combined with existing fragments, they create block-shaped patches for promoting interior forest species, elongated patches for promoting edge species, or corridors for connecting habitat patches across the landscape.
- Select tree species, spacing, and management that create appropriate forest structure for enhancing desired species of wildlife.
- Locate forest restoration areas away from important grassland habitat areas.
- Restore gaps along larger streams first to provide the greatest overall benefit for wildlife (Fig. 5.4).

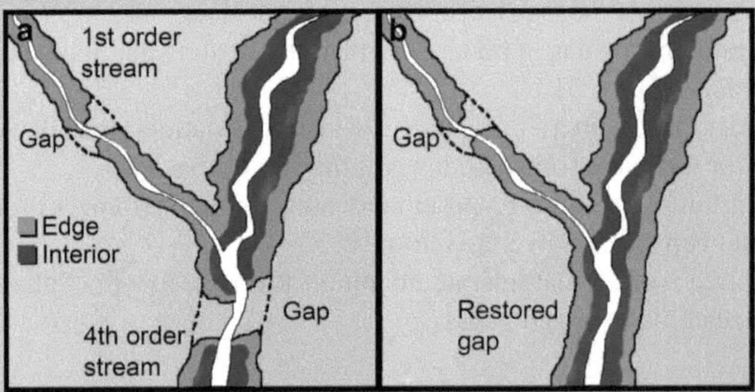

Fig. 5.4 Gaps in riparian vegetation along streams of all sizes are common in agricultural landscapes (**a**). Gaps along larger or higher-order streams should often be restored first to provide the greatest overall benefit for wildlife. These riparian zones have less negative edge effects and are more important regional corridors for wildlife movement (**b**)

5.2.3 Social Sciences – Encouraging Acceptance and Adoption of Riparian Buffers

Human values, attitudes, and perceptions play a critical role in how people create and maintain the landscapes in which they live and work. Any effort to create and maintain riparian forests on agricultural lands must appeal to this local social dimension in order to be successful (Parren and Sam 2003; Dutcher et al. 2004; Blay et al. 2008; Rosenberg and Margerum 2008; Schaich 2009). For example, one commonly held value among farmers in the United States is that good land stewardship is demonstrated by maintaining one's property in a clean and manicured manner (Nassauer 1988). Care for agricultural land is represented by visual cues such as straight crop rows, lack of weeds, mowed areas, and general landscape uniformity. Natural riparian forests with their meandering curves, downed woody debris and general lack of uniformity are often perceived by U.S. farmers as unmanaged and messy and hence do not represent the farmers' concept of good land stewardship (Ryan 1998; Ryan et al. 2003). Consequently, there is resistance from farmers to implement and maintain natural looking riparian forest buffers. To overcome this barrier, visual cues of care need to be incorporated into the design and management of a riparian forest buffer (Nassauer 1995).

Different values and concerns may exist in local agricultural communities that can lead to opposing views of forest restoration efforts (Parren and Sam 2003; Sullivan et al. 2004; Schaich 2009). For instance, riparian restoration is being used to create a network of forest corridors in West Africa to sustain isolated populations of forest elephants (*Loxodonta africana*). Some streamside villages showed strong interest in restoring riparian forest which they believe would resolve some of their water and fishing problems during the dry season while other villages in the area were opposed to any reforestation options as it means losing agricultural land (Parren and Sam 2003). In addition, some villagers have negative attitudes towards creating elephant habitat because elephants raid crop fields and can kill people (Gadd 2005). Restoration planners need to be cognizant of the full range of values, attitudes, and perceptions that stakeholders can hold towards forests and forest restoration and avoid oversimplifying their social concerns if they have hopes of creating locally supported restoration plans.

Additional social considerations may also need to be addressed in order to facilitate acceptance and adoption of riparian forest buffers (Schrader 1995; Rhodes et al. 2002; Ryan et al. 2003; Sullivan et al. 2004). Some countries have government agencies or non-profit organizations who offer financial incentives to landholders to encourage adoption. However, many landholders have concerns that riparian forest buffers will not provide any productive value after the incentives are gone and that these landscape elements will hinder farming operations. A few common social science principles related to location and design of buffers that may overcome such resistances to acceptance and adoption are listed in Box 5.3. A more exhaustive list can be found in Kaplan et al. (1998) and Bentrup (2008). By understanding these social dimensions, plans for riparian forest buffers can be modified to alleviate local social concerns while still creating a riparian forest design that is capable of providing the desired ecosystem services.

5.3 Landscape Planning to Achieve Forest Restoration Goals

Enhancing ecosystem services by restoring forest on agricultural lands often requires a larger planning area than individual farms and other agricultural landholdings. Coordinated and cumulative action on several farms is often necessary to achieve desired levels of ecosystem services. To accomplish this task, a multi-scale planning process is needed to pull together concerns and goals of individual landholders and the general public while accounting for opportunities and constraints dictated by the existing landscape. A planning process facilitates setting goals and making decisions about actions that will achieve those goals. A planning process also helps identify specific areas in the landscape to target riparian forest buffers where they will generate relatively greater ecological benefit at lower economic costs (Walter et al. 2007).

There are many ways to go about planning. In agricultural landscapes, the decision about whether to implement and maintain any restoration action often rests with many independent farmers and landholders. Even if there are public regulations concerning the placement and design of riparian buffers, an effective planning process is still necessary to reconcile and balance public goals embodied in the regulations with different goals of landholders. Some characteristics of a planning process that will do this include comprehensiveness, flexibility, scalability, and stakeholder involvement. A planning process needs to be comprehensive to address a wide range of issues and landscape conditions while being flexible enough to accommodate each decision-maker's (i.e., landholder) unique set of circumstances. For example, landholders are more willing to accept and implement a riparian restoration plan that is tailored to their needs rather than to an arbitrary and rigid set of buffer width standards (Dutcher et al. 2004). A multi-scale approach is required because each objective (e.g., farm economy, watershed water quality, landscape wildlife populations) is addressed at its own scale and each riparian buffer function operates at its own scale.

Box 5.3 Principles for Guiding Riparian Forest Restoration to Encourage Landowner Acceptance and Adoption

- Design the part of the restoration area viewable by public to be visually pleasing while the interior can be designed to achieve the desired ecological functions.
- Use selective mowing to indicate stewardship without greatly reducing the ecological functions.
- Provide visual frames to contain and provide order around the restoration area (e.g., wooden fence).
- Use interpretative signage and education programs to increase awareness and preference.
- Enhance visual interest and diversity by increasing seasonal color and by varying plant heights, textures, and forms while maintaining an overall sense of order.

(continued)

Box 5.3 (continued)

- Provide options for landowners to derive economic or personal products from the restoration area (i.e., fruit or nut products, hunting leases, decorative woody stems for floral industry).
- Allow the riparian zone to be "squared off" to facilitate farming operations (Fig. 5.5).

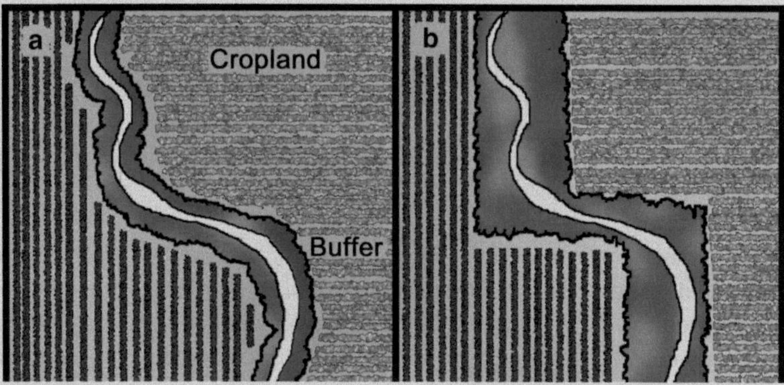

Fig. 5.5 A curving riparian buffer can hinder mechanical farming operations (**a**). The riparian buffer zone can be "squared off" to facilitate farming operations without significantly reducing ecological functions (**b**)

A key component of a planning process for forest restoration is local and public stakeholder participation throughout the planning, design, implementation, and management stages (Selin and Chavez 1995; Bentrup 2001; Blay et al. 2008). Because riparian areas by their nature cross many landholdings and influence factors well beyond their vegetative boundaries, stakeholders throughout the watershed or wildlife area need to be involved. One of the valuable aspects of a participatory-type planning process is to have face-to-face dialogue between stakeholders to learn about the commonalities and differences in their goals, expectations, and tolerances for riparian buffers (Gray 1989). This dialogue is essential because of the inherent differences between stakeholders. For example, the general public often desires wider riparian buffers while farmers desire narrower buffers (Sullivan et al. 2004). These types of differences can often be resolved through collaborative interaction and an acceptable and shared vision can be established for a sustainable network of riparian buffers (Averitt et al. 1994).

A multi-scale planning process that exhibits these characteristics has been suggested specifically for riparian buffer planning (Bentrup et al. 2003). It involves three basic components: regional reconnaissance, landscape-scale assessments, and site-scale buffer plans. A series of questions assists stakeholders through the process and provides specific but flexible guidance for analyzing resources and developing plans (Steinitz 1990; Smith and Hellmund 1993).

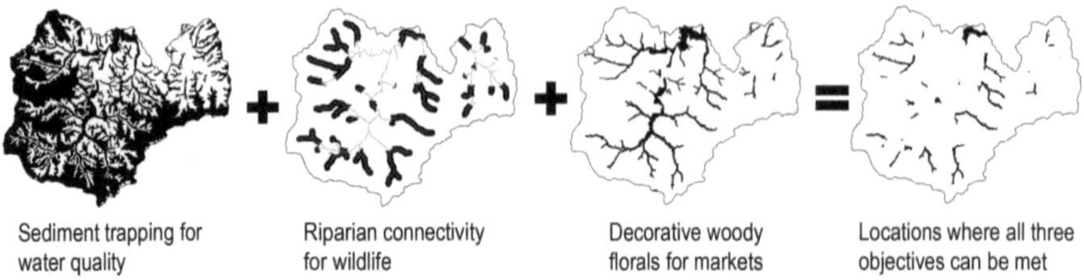

Fig. 5.6 Landscape assessments for sediment trapping, riparian connectivity, and woody florals are combined to determine where all three objectives can be achieved with a riparian forest buffer (Source: Bentrup et al. 2008)

5.3.1 *Reconnaissance and Landscape Assessments*

The regional-scale reconnaissance provides a quick overview of environmental conditions and resource issues. Often, riparian buffer planning efforts are focused on a single problem. However, by looking at the regional context, stakeholders are encouraged to consider multiple resource issues and to capitalize on capabilities of buffers to address several issues simultaneously. Some questions to answer with the reconnaissance include: What are the main resource issues in this region? What ecological and social processes are influencing these issues? What forest ecosystem services need to be restored to address these issues?

Based on the reconnaissance, more detailed landscape-scale assessments are conducted to describe existing resource conditions and trends of interest and to identify opportunities to enhance ecosystem services with strategically-placed riparian buffers. Questions that need to be answered at this stage include: Is the riparian landscape functioning well? How might the riparian and upland landscape be altered to improve functions? The natural science principles described earlier can be used during the landscape assessment process to help identify locations to target riparian buffers to achieve effectiveness and economic efficiencies. Geographic information systems (GIS) are useful for managing, processing, and analyzing spatial information in a visual manner that facilitates communication between stakeholders. With GIS, landscape assessments can be combined to identify locations where multiple objectives can be achieved with riparian buffers, allowing stakeholders to focus on potential opportunities rather than just resource problems (Fig. 5.6).

Armed with information produced through regional reconnaissance and landscape assessments, stakeholders can develop a shared vision for what they want to achieve and general options for how and where to attain their goals. These assessments provide the landscape-scale context for developing site-scale riparian buffer plans.

Fig. 5.7 Existing agricultural stream lacking a riparian forest buffer (**a**). A visual simulation of a proposed riparian forest buffer (**b**) (Photo credit: NRCS Simulation by Robert Corry)

5.3.2 Site-Scale Buffer Plans

The site-scale planning and design component blends the public goals identified in the landscape-scale assessments with individual landholder objectives and site conditions. The site-scale process is guided by the same questions used in the landscape-scale assessments but are applied to a specific landholder's site. The natural and social science principles described earlier are used to craft riparian buffer design alternatives that solve landholder resource issues and that are also acceptable to a landholder's set of attitudes, values, and perceptions. The design alternatives provide detailed recommendations on location, size, configuration, plant species and composition, and management practices.

An effective method for communicating and evaluating alternative riparian buffer designs is through photo-realistic simulations (Fig. 5.7). The communicative and non-threatening nature of simulations encourages stakeholders to actively participate in the design process and to offer feedback on the alternatives. Using simulations in participatory planning greatly increases a sense of ownership in the plan, which leads to enhanced acceptance and adoption of the proposed action (Al-kodmany 1999). If there is no regulation requiring riparian buffers, landholders maintain the right to decide if they want to implement a riparian buffer or not on their landholding. Resources and tools for planning, designing, and managing riparian buffers, including GIS-based methodologies and visual simulation software are listed in the Restoration Planner's Toolbox (Box 5.4).

Box 5.4 Restoration Planner's Toolbox

Natural and Social Science Principles

Landscape ecology principles in landscape architecture and land-use planning. Dramstad WE et al. (1996) Island Press, Wash DC

Conservation buffers: design guidelines for buffers, corridors, and greenways. Bentrup G (2008) US For Serv South Res Sta, Asheville, NC http://bufferguidelines.net

With people in mind: design and management of everyday nature. Kaplan R et al. (1998) Island Press, Wash DC

Planning, Design, and Management Resources

Riparia: ecology, conservation, and management of streamside communities. Naiman RJ et al. (2005) Elsevier Academic Press, New York

Chesapeake Bay riparian handbook: a guide for establishing and maintaining riparian forest buffers. Palone, R, Todd, A (1998) US For Serv Northeast Area, State & Private For, Nat Res Conserv Serv, Coop State Res Educ Ext Ser http://www.treesearch.fs.fed.us/pubs/10519

Conservation corridor planning at the landscape level: managing for wildlife habitat. Johnson CW et al. (2000) US Dep Agric, Nat Res Conserv Serv, Wash DC ftp://ftp-fc.sc.egov.usda.gov/WHMI/NBHpdf/nbh613.pdf

Designing greenways: sustainable landscapes for nature and people. Hellmund P, Smith D (2006) Island Press, Wash DC

The community visioning and strategic planning handbook. Natl Civic Leag Press http://www.ncl.org/pdfs/community%20visioning.pdf

Regional Reconnaissance: Online Atlas

National atlas of the United States. http://www.nationalatlas.gov/index.html

Landscape-Scale Assessments: GIS-based Methodologies

Improved indexes for targeting placement of buffers of Hortonian runoff. Dosskey M et al. (2011) J Soil Water Conserv 66:362–372

Where should buffers go? – modeling riparian habitat connectivity in northeast Kansas. Bentrup G, Kellerman T (2004) J Soil Water Conserv 59:209–213 http://www.unl.edu/nac/research/2004riparianconnectivity.pdf

Agroforestry: mapping the way with GIS. Bentrup G, Leininger T (2002) J Soil Water Conserv 57:148A–153A http://www.unl.edu/nac/research/2002agroforestrygis.pdf

(continued)

Box 5.4 (continued)

The role of GIS in selecting sites for riparian restoration based on hydrology and land use. Russell G et al. (1997) Restor Ecol 5(4S):56–68

Site-Scale Design: Resources and Tools

CanVis visual simulation kit. Software and guidebook for creating photo-realistic visual simulations http://www.unl.edu/nac/simulation/index.htm

Buffer$. An economic tool for analyzing the costs and benefits of buffers. http://www.unl.edu/nac/buffer$.htm

Riparian buffer design guidelines for water quality and wildlife habitat functions on agricultural landscapes in the Intermountain West. Johnson C, Buffler S (2008) US For Serv Rocky Mtn Res Sta, Ft Collins, CO http://www.treesearch.fs.fed.us/pubs/29201

A design aid for determining width of filter strips. Dosskey M (2008) J Soil Water Conserv 63:232–241 http://www.unl.edu/nac/research/2008bufferwidth.pdf

PLANTS. An online plant database for the U.S. and its territories. http://plants.usda.gov/

Productive conservation: growing specialty forest products in agroforestry plantings. Josiah S (2001) U of Nebraska Ext, Lincoln NE http://www.unl.edu/nac/morepublications/sfp2.pdf

5.4 Management Considerations to Achieve and Maintain Goals

Since restored riparian forests are features in an agricultural landscape that are designed to yield specific ecosystem services, some level of active management will be required to optimize and maintain these services. The type and intensity of management will depend on which services and the desired level of attainment of those services (Box 5.5). For example, obtaining a 30% reduction in sediment and nutrient transport through a riparian zone will require some harvesting and some sediment removal to achieve and maintain this level of functioning. Higher levels of sediment and nutrient reduction may require more frequent actions. Other services, like forest habitat creation, may require minimal management activity, such as occasional pruning and weed control to maintain the necessary vegetation structure. Management activities may extend into existing riparian stands to enhance their function for those services as well.

Management needs to be coordinated so that a treatment activity used to achieve one goal does not inadvertently compromise the accomplishment and sustainability of another goal. For example, harvesting biomass for fuel could negatively impact forest habitat. Temporal and spatial considerations should also be factored into the development of a management plan. Management activities may need to be restricted during certain times of the year or limited to a part of the riparian zone each year to ensure some portions remain undisturbed at all times. Management must ultimately respond to the farmer's or landholder's attitudes, values, and perceptions so that the riparian restoration compels sustained management attention over time and gains in ecosystem services will not be lost (Nassauer et al. 2001).

5.5 Conclusions

Restoring forest ecosystem services to agricultural landscapes is a daunting challenge that stems from the unfeasibility of converting large tracts of food-producing land back into forest, and, of converting farmers and farming communities to forestry. Resolving these issues requires finding a balance between public goals for food and ecosystem services as well as landholder and community goals which often include continued farming. Natural science principles suggest that an appropriate balance may be possible through the use of riparian forest buffers. Riparian areas occupy a small portion of landscapes and can produce high levels of multiple ecosystem services. Principles for guiding riparian restoration for water pollution reduction and for terrestrial wildlife enhancement are used to illustrate how natural and social science information can influence design and management. Additional ecosystem services also can be effectively restored by applying similar sets of basic scientific principles. Achieving those services, however, will require that landholders and communities accept and adopt riparian forest buffers. Coordinated and cumulative action on several farms or other landholdings is often necessary to achieve desired levels of ecosystem services. A multi-scale planning process is important for integrating both natural and social science principles in a way that produces effective restoration plans and encourages their implementation and maintenance.

5.6 Management Implications

Restoration of forest ecosystem services in agricultural regions involves many challenges and tradeoffs. Successfully navigating these difficulties and achieving success often requires careful planning that includes:

- Recognition that the ultimate goal of forest restoration is improved social well-being. Forest restoration is a means for restoring ecosystem services toward achieving that goal.

- Riparian zones can be particularly effective and efficient for restoring a wide variety of forest ecosystem services.
- A restoration plan must be based on sound natural science principles.
- A restoration plan must accommodate the needs of the farmers and landholders who will implement and maintain the restored areas.
- The optimum size, shape, and level of connectivity to which riparian zones must be restored will depend on the specific objectives, opportunities, and constraints presented by each landscape and social setting.

Box 5.5 Principles for Managing Riparian Forest Restoration for Water Quality and Terrestrial Wildlife Habitat Enhancement

- Remove any accumulated sediment that prevents runoff from flowing directly into the riparian zone (Fig. 5.8).
- Periodic harvest of green vegetation will remove nutrients captured in the riparian zone and promote vigorous new growth for sustaining nutrient uptake.
- Some overstory vegetation removal may be necessary to maintain dense herbaceous cover to sustain filtering processes.
- Avoid vehicle traffic in the riparian zone which can cause compaction and reduce infiltration capacity.
- Manage vegetation to create the vegetative structure to support the desired wildlife species.
- Avoid working in the riparian zone during peak breeding season.
- Harvesting of vegetation should occur on a rotational basis to ensure that some portion of the riparian zone remains undisturbed at all times.

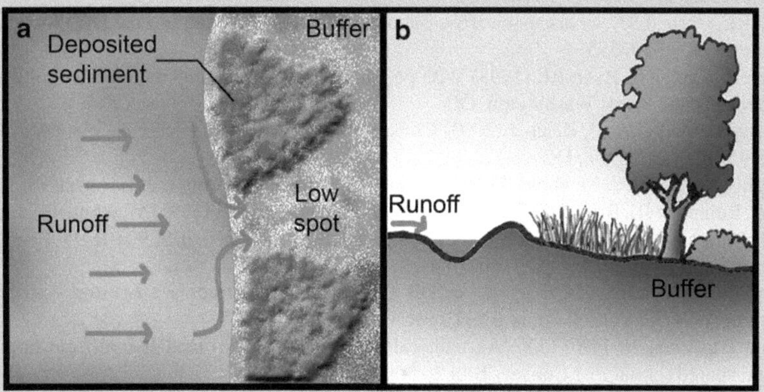

Fig. 5.8 Remove deposited sediment that concentrates runoff flows (**a**). Remove any ditch or berm that prevents runoff from flowing directly into the buffer (**b**)

References

Al-kodmany K (1999) Using visualization techniques for enhancing public participation in planning and design: process, implementation, and evaluation. Landsc Urban Plan 45:37–45

Averitt E, Steiner F, Yabes R, Patten D (1994) An assessment of the Verde River Corridor Project in Arizona. Landsc Urban Plan 28:161–178

Benayas JMR, Bullock JM, Newton AC (2008) Creating woodland islets to reconcile ecological restoration, conservation, and agricultural land use. Front Ecol Environ 6:329–336

Bentrup G (2001) Evaluation of a collaborative model: a case study analysis of watershed planning in the Intermountain West. Environ Manag 27:739–748

Bentrup G (2008) Conservation buffers: design guidelines for buffers, corridors, and greenways. GTR-SRS-109, US For Serv, South Res Sta, Asheville, NC

Bentrup G, Schoeneberger MM, Dosskey MG, Wells G (2003) The fourth P: planning for multi-purpose riparian buffers. In: Proceedings of the 8th North American agroforesty conference, Corvallis

Bentrup G, Dosskey M, Wells G (2008) Conducting landscape assessments for agroforestry. AF-39, USDA Natl Agrofor Cent, Lincoln. http://www.unl.edu/nac/agroforestrynotes/an39g11.pdf

Blay D, Appiah M, Damnyag L, Dwomoh FK, Luukkanen O, Pappinen A (2008) Involving local farmers in rehabilitation of degraded tropical forests: some lessons from Ghana. Environ Dev Sustain 10:503–518

Breshears DD (2006) The grassland-forest continuum: trends in ecosystem properties for woody plant mosaics? Front Ecol Environ 4:96–104

Crooks KR, Sanjayan M (2006) Connectivity conservation. Cambridge Univ Press, Cambridge

Dosskey MG (2001) Toward quantifying water pollution abatement in response to installing buffers on crop land. Environ Manag 28:577–598

Dramstad WE, Olson JD, Forman RTT (1996) Landscape ecology principles in landscape architecture and land-use planning. Island Press, Washington, DC

Dutcher DD, Finley JC, Luloff AE, Johnson J (2004) Landowner perceptions of protecting and establishing riparian forests: a qualitative analysis. Soc Nat Resour 17:319–332

Gadd ME (2005) Conservation outside of parks: attitudes of local people in Laikipia, Kenya. Environ Conserv 32:50–63

Garrett HE, Rietveld WJ, Fisher RF (2000) North American agroforestry: an integrated science and practice. Am Soc Agron, Madison

Gray B (1989) Collaborating: finding common ground for multiparty problems. Jossey-Bass Publ, San Francisco

Green RE, Cornell SJ, Scharlemann JPW, Balmford A (2005) Farming and the fate of wild nature. Science 307:550–555

Kaplan R, Kaplan S, Ryan RL (1998) With people in mind: design and management of everyday nature. Island Press, Washington, DC

Kennedy C, Wilkinson J, Balch J (2003) Conservation thresholds for land use planners. Environ Law Inst, Washington, DC

Lowrance R, Altier LS, Newbold JD, Schnabel RR, Groffman PM, Denver JM, Correll DL, Gilliam JW, Robinson JL, Brinsfield RB, Staver KW, Lucas W, Todd AH (1995) Water quality functions of riparian forest buffer systems in the Chesapeake Bay watershed. EPA 903-R-95-004. Technol Transf Rep, Chesap Bay Program, Annap

Manning AD, Fischer J, Lindenmayer DB (2006) Scattered trees are keystone structures – implications for conservation. Biol Conserv 132:311–321

Matson PA, Vitousek PM (2006) Agricultural intensification: will land spared from farming be land spared for nature? Conserv Biol 20:709–710

Millennium Ecosystem Assessment (MEA) (2005) Ecosystems and human well-being: current state and trends. Island Press, Washington, DC

Naiman RJ, Décamps H, McClain ME (2005) Riparia: ecology, conservation, and management of streamside communities. Elsevier Academic Press, Burlington

Nair PKR, Allen SC, Bannister ME (2005) Agroforestry today: an analysis of the 750 presentations to the 1st World Congress of Agroforestry. J For 103:417–421
Nassauer JI (1988) The aesthetics of horticulture: neatness as a form of care. Hortic Sci 23:973–977
Nassauer JI (1995) Messy ecosystems: orderly frames. Landsc J 14:161–170
Nassauer JI, Kosek SE, Corry RC (2001) Meeting public expectations with ecological innovation in riparian landscapes. J Am Water Resour Assoc 37:1439–1443
National Research Council (NRC) (1993) Soil and water quality: agenda for agriculture. Natl Acad Press, Washington, DC
National Research Council (NRC) (2002) Riparian areas: functions and strategies for management. Natl Acad Press, Washington, DC
Parren MPE, Sam MK (2003) Elephant corridor creation and local livelihood improvement in West Africa. In: Proceedings of the international conference on rural livelihoods, forests and biodiversity. Bonn
Rhodes HM, Leland LS Jr, Niven BE (2002) Farmers, streams, information, and money: does informing farmers about riparian management have any effect? Environ Manag 30:665–677
Rosenberg S, Margerum RD (2008) Landowner motivations for watershed restoration: lessons from five watersheds. J Environ Plan Manag 51:477–496
Ryan RL (1998) Local perceptions and values for a midwestern river corridor. Landsc Urban Plan 42:225–237
Ryan RL, Erickson DL, De Young R (2003) Farmers' motivations for adopting conservation practices along riparian zones in a mid-western agricultural landscape. J Environ Plan Manag 46:19–37
Schaich H (2009) Local residents' perceptions of floodplain restoration measures in Luxembourg's Syr Valley. Landsc Urban Plan 93:20–30
Schrader C (1995) Rural greenway planning: the role of streamland perception in landowner acceptance of land management strategies. Landsc Urban Plan 33:375–390
Schröter D, Cramer W, Leemans R, Prentice IC, Araújo MB, Arnell NW, Bondeau A, Bugmann H, Carter TR, Gracia CA, de la Vega-Leinert AC, Erhard M, Ewert F, Glendining M, House JI, Kankaanpää S, Klein RJT, Lavorel S, Lindner M, Metzger MJ, Meyer J, Mitchell TD, Reginster I, Rounsevell M, Sabaté S, Sitch S, Smith B, Smith J, Smith P, Sykes MT, Thonicke K, Thuiller W, Tuck G, Zaehle S, Zierl B (2005) Ecosystem service supply and vulnerability to global change in Europe. Science 310:1333–1337
Selin S, Chavez D (1995) Developing a collaborative model for environmental planning and management. Environ Manag 19:189–195
Smith DS, Hellmund PC (1993) Ecology of greenways. Univ Minn Press, Minneapolis
Steinitz C (1990) A framework for theory applicable to the education of landscape architects (and other design professionals). Landsc J 9:136–143
Sullivan WC, Anderson OM, Lovell ST (2004) Agricultural buffers at the rural-urban fringe: an examination of approval by farmers, residents, and academics in the Midwestern United States. Landsc Urban Plan 69:299–313
Sweeney B (1993) Effects of streamside vegetation on macroinvertebrate communities of White Clay Creek in eastern North America. Proc Acad Nat Sci Phila 144:291–340
Walter T, Dosskey M, Khanna M, Miller J, Tomer M, Wiens J (2007) The science of targeting within landscapes and watersheds to improve conservation effectiveness. In: Schnepf M, Cox C (eds) Managing agricultural landscapes for environmental quality: strengthening the science base. Soil and Water Conser Soc, Ankeny
Welsch DJ (1991) Riparian forest buffers: function and design for protection and enhancement of water resources. NA-PR-07-91. US For Serv, Northeast Area State and Priv For, Radnor
Wenger S (1999) A review of the scientific literature on riparian buffer width, extent, and vegetation. Univ GA Inst Ecol, Athens

Chapter 6
Understanding Landscapes Through Spatial Modeling

Michael C. Wimberly, Stephen P. Boyte, and Eric J. Gustafson

6.1 Introduction

Ecological restoration is defined as "the process of assisting the recovery of an ecosystem that has been degraded, damaged, or destroyed" (Society for Ecological Restoration International Science and Policy Working Group 2004). Restoration thus entails the identification of a reference ecosystem to serve as a benchmark and the implementation of management practices that move the forest toward a set of desired future conditions. Expanding the scale of forest restoration from individual stands to broader landscapes presents several challenges. The reference ecosystem concept must be extended to encompass reference landscapes consisting of multiple forest types and successional stages. Because of the pervasive influence of human land use, it is often difficult, if not impossible, to identify modern landscapes that can provide suitable benchmarks. Furthermore, scientists and managers often lack a clear understanding of how proposed restoration activities will impact disturbance regimes and forest succession across broad geographic areas. Because of these knowledge gaps, there is often considerable uncertainty about what the desired outcome of forest landscape restoration should be and whether management activities will actually move the landscape toward the desired state.

Forest landscapes encompass heterogeneous mosaics of physical environments, community types, disturbance histories, and land ownerships. Hierarchy theory posits that rates of change decrease with increasing spatial extent in ecological

M.C. Wimberly (✉) • S.P. Boyte
Geographic Information Science Center of Excellence,
South Dakota State University Brookings, SD 57006, USA
e-mail: michael.wimberly@sdstate.edu

E.J. Gustafson
USDA Forest Service Northern Research Station, Institute for Applied Ecosystem Studies,
Rhinelander, WI 54501, USA

J. Stanturf et al. (eds.), *Forest Landscape Restoration: Integrating Natural and Social Sciences*, World Forests 15, DOI 10.1007/978-94-007-5326-6_6,

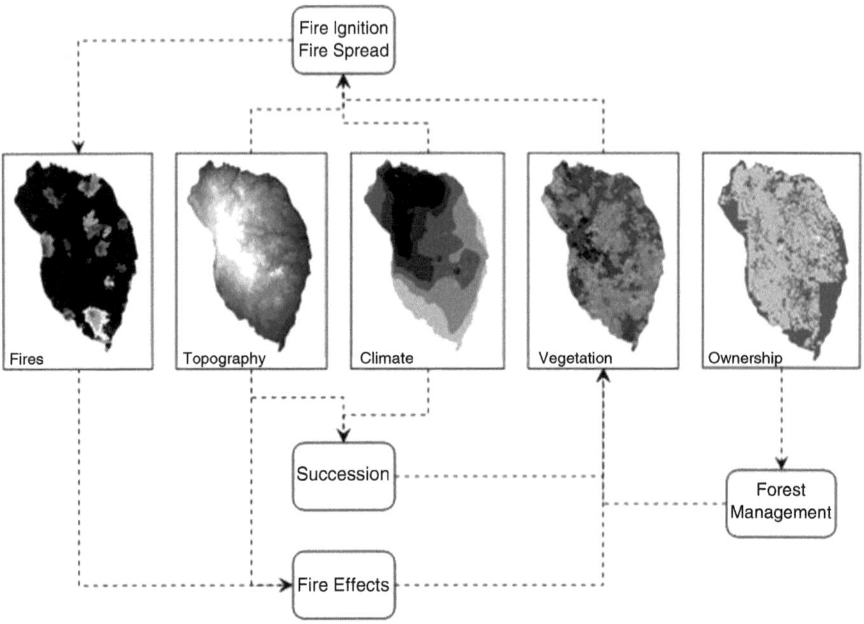

Fig. 6.1 Major landscape patterns and process simulated in forest landscape models. The physical environment and vegetation patterns influence the initiation and spread of wildfires and other natural disturbances. The physical environment also influences fire effects on vegetation and pathways of forest succession. Land ownership is a major driver of forest management practices, which in turn influence vegetation patterns

systems (Urban et al. 1987). Therefore, large landscapes must be studied over temporal extents ranging from decades to centuries. These broad spatial and temporal scales limit our ability to use traditional experimental and observational methods to study forest landscapes. In some cases, historical datasets can be used to study landscape changes (e.g., Wimberly and Ohmann 2004). However, simply extrapolating past trends into the future is problematic because future landscape changes are likely to occur in the context of climates, species assemblages, and socio-economic conditions that have no historical analogues (Hobbs et al. 2006). Despite our limited knowledge of landscape dynamics, land managers must still make decisions that will influence future forest landscapes for decades to centuries. For these reasons, landscape simulation models are increasingly being used for scientific research in the field of landscape ecology and as decision support tools to assist in the practice of forest landscape restoration.

Most forest landscape simulation models currently in use are spatially explicit, with discrete landscape units represented as spatial data structures such as raster cells or vector polygons (Fig. 6.1). The forest vegetation within each landscape unit is characterized by one or more variables such as dominant species, stand age, successional stage, or specific stand structure measurements such as tree size and density. Mathematical or rule-based algorithms are applied to model successional changes in forest vegetation. Disturbance processes such as fire, windstorms, insects, and timber harvesting change forest vegetation and are also constrained by the spatial pattern of forest vegetation. Spatial relationships are explicitly modeled. They include vertical interactions among climate, topography, soils, and vegetation within landscape units and horizontal interactions such as seed dispersal and fire spread between landscape units and adjacency constraints on timber harvests. Given an initial landscape condition, landscape simulation models generate projections of landscape dynamics that reflect the underlying data and assumptions used to specify the model and estimate parameters.

Several papers have reviewed different types of landscape models, focusing on conceptual and technical aspects of model design (Baker 1989; Keane et al. 2004; Perry and Enright 2006; Scheller and Mladenoff 2007; He 2008). In contrast, this review will examine *how* landscape models are applied in science and management. It will focus on three major applications of landscape simulation models in the field of forest landscape restoration. The first application involves using landscape models to reconstruct historical reference landscapes based on the characteristics of historical disturbance regimes. The second application uses simulation models to project forest landscape change to evaluate the potential effectiveness of forest restoration strategies. The third example applies landscape models in an exploratory framework to expand our understanding of the process of landscape change in disturbed landscapes. The three approaches will be illustrated using examples taken primarily from studies focusing primarily on the effects of timber harvesting and wildfire in temperature forests in North America. However, we also note that forest landscape models can incorporate other disturbances such as insect outbreaks (Cairns et al. 2008) and windstorms (Scheller and Mladenoff 2005; Shifley et al. 2006), and are applied globally in locations ranging from Europe (Schumacher and Bugmann 2006), to China (Bu et al. 2008; Leng et al. 2008), to Australia (Perry and Enright 2002).

6.2 Simulating Historical Reference Landscapes

Forest landscape restoration is predicated on our ability to define reference conditions to serve as benchmarks for restoration. However, because most forest landscapes are dynamic, mosaics of different forest types and successional stages, static reference conditions are often inappropriate. Therefore, the concept of a natural or historical range of variability (HRV) has emerged as a framework for land management and forest restoration (Hunter 1993; Landres et al. 1999; Allen et al. 2002). At the simplest level, the HRV concept can be implemented by defining a range or probability distribution of the relative amounts of different successional stages under the historical disturbance regime. More sophisticated assessments of HRV may also consider the spatial arrangement of successional stages across the landscape. An important implication of the HRV concept is that there is no single "correct" reference condition at the landscape scale. Instead, there may be a variety of potential restoration targets that fall within the HRV.

Although historical data can be used to reconstruct reference landscapes, "snapshots" of landscape characteristics at a single point in time are usually not sufficient for understanding the dynamics of historical landscape conditions in forest ecosystems impacted by wildfires, insect outbreaks, floods, windstorms, and other large-scale disturbances. Simulation models can be used to link data on the rates, sizes, and effects of historical disturbances with knowledge of forest succession to estimate the composition and configuration of forest landscape mosaics. Thus, the application of landscape models to simulate historical reference conditions is primarily a "predictive" approach to landscape modeling rather than an "explanatory" approach (Peck 2004; Perry and Millington 2008). However, explanation of the underlying ecological phenomena is often an important secondary goal, in which techniques such as sensitivity analysis and uncertainty analysis can be applied to examine the influences of model parameters and processes on simulated landscape dynamics (Wimberly 2004; Wimberly and Kennedy 2008).

6.2.1 HRV of Old-Growth Forests in the Oregon Coast Range

In coastal Douglas-fir forests of the Pacific Northwest, forest management controversies have focused on the logging of old-growth forests, the resulting fragmentation of the remaining old growth, and the effects that these changes have had on threatened and endangered species such as the northern spotted owl, marbled murrelet, and Pacific salmon. These concerns eventually led to the development of the Northwest Forest Plan (Forest Ecosystem Management and Assessment Team (FEMAT) 1993). A key component of this plan is a network of reserves where management objectives focus on the development and maintenance of late-successional habitat conditions. However, there has also been growing recognition that wildfires occurred for millennia prior to human settlement in coastal Douglas-fir forests, and that these historical fire regimes encompassed a wide range of fire frequencies, severities, and spatial patterns (Long et al. 1998; Long and Whitlock 2002; Weisberg and Swanson 2003). This evidence of historical wildfires has raised fundamental questions about more recent declines in old growth. Are the current low levels of old-growth forests really an unprecedented effect of logging and other human activities? Or is it possible that old growth was actually an uncommon and highly variable component of the pre-settlement landscape?.

To address these questions, the LAndscape Dynamics Simulator (LADS) was developed to estimate the range of historical variability in the amount and spatial pattern of old-growth forests (Wimberly et al. 2000; Wimberly 2002). Prior to Euro-American settlement, the Oregon Coast Range was characterized by a gradient of disturbance regimes, ranging from large, infrequent, stand-replacing wildfires in the North and along the coast to smaller, more frequent, mixed-severity wildfires to the South and in the interior. Data on historical fire return intervals, severities, and sizes was obtained from dendro-ecological (Impara 1997), paleo-ecological (Long et al. 1998; Long and Whitlock 2002), and historical (Teensma et al. 1991) studies. These data were input into the LADS model to simulate the occurrence and spread

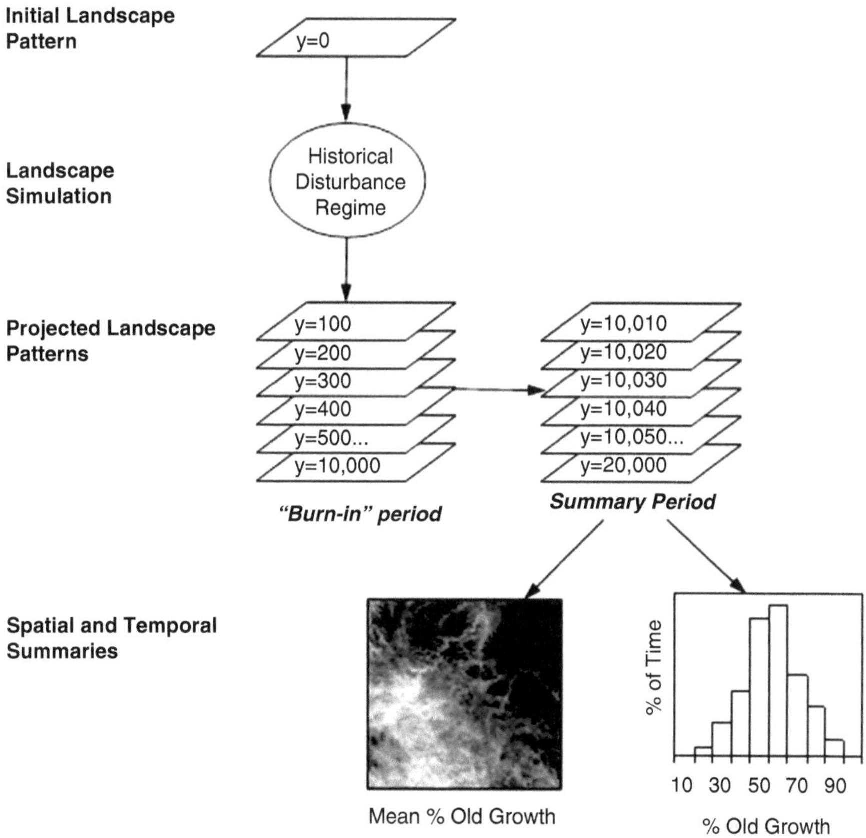

Fig. 6.2 Flowchart illustrating the process of HRV simulation. A landscape simulation model is used to simulate a time series of landscapes using parameters based on the pre-settlement disturbance regime. Following a "burn-in" period in which the simulated landscape patterns overwrite the arbitrary initial conditions, spatial and temporal variability is summarized using a variety of methods

of wildfires, the effects of these fires on forest vegetation, and the pathways of forest succession that occur after wildfires. By running a large number of model simulations, it was possible to translate available information about the historical disturbance regime into estimated probability distributions of the relative abundances old growth and other successional stages (Fig. 6.2).

The results of simulation studies using the LADS model have demonstrated that present-day forest patterns in the Oregon Coast Range are far outside the range of historical variability (Wimberly et al. 2000, 2004; Wimberly 2002). In the historical simulations, old-growth forests occupied an average of ~45% of the Coast Range, but were highly variable in both space and time (Fig. 6.3). Even after accounting for disturbance-driven temporal variability, current amounts of old growth (less than 2% of the landscape) are much lower than would be expected under the pre-settlement disturbance regime (Wimberly et al. 2004).

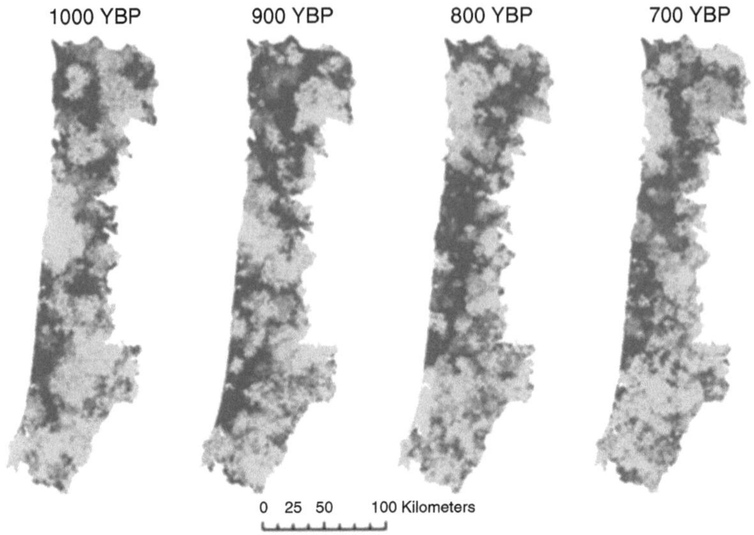

Fig. 6.3 LADS simulation of one hypothetical time series of historical landscape patterns in the Oregon Coast Range. *Dark gray* patches represent closed-canopy old-growth forests. *Light gray* patches represent other forest structure classes (including early-successional, young, and mature forests)

Two major changes to the regional pattern of forest successional stages were also evident. First, there was a shift from a historical landscape that was usually dominated by at least one large (>400,000 ha) patch of old growth to a modern landscape in which old growth mostly occurs in much smaller fragments. Second, there was also a decrease in the total number of small, old-growth patches from the historical landscape to the present. Whereas historical landscapes often had many small remnant old-growth patches embedded in areas of younger forest, there are large portions of the current landscape that are highly isolated from the nearest old-growth patch (Wimberly 2002; Wimberly et al. 2004).

6.2.2 *Modeling Landscape Departure from Historical Reference Conditions*

Landscape changes in the Oregon Coast Range and other areas of the coastal Pacific Northwest have occurred because the historical disturbance regime of relatively infrequent, large wildfires has been replaced by a forest management regime dominated by smaller and more frequent clearcuts (Wimberly et al. 2004). In contrast, many forests in the interior West had historical disturbance regimes characterized by frequent fires that maintained fuel loads at relatively low levels, leading to a fire regime dominated by patchy, low-severity fires. In the current landscape, fire suppression, selective logging, and grazing have contributed to a more homogeneous landscape of dense forests and high fuel loads, increasing the potential for uncharacteristically large and severe wildfires (Hessburg et al. 2005). The National Fire Plan, which is aimed at reducing the risk of wildfire and restoring forest health, is applying HRV concepts to support fire and fuels management (Keane et al. 2007). The prioritization of forest management activities is being based in part on the assignment of a fire regime condition class (FRCC), which assesses the degree to which the current fire regime, fuel loads, and vegetation structure differ from the conditions that occurred under the historical fire regime (Schmidt et al. 2002).

The LANDSUM model was developed as part of the larger LANDFIRE project to serve as a tool for modeling the spatial distribution of fire regimes and the resulting vegetation patterns across heterogeneous forest landscapes (Keane et al. 2006). The goals of the LANDFIRE project are to provide digital maps and datasets characterizing vegetation, fuels, and fire regimes across the United States (Rollins and Frame 2006). Two of the national products being developed by LANDFIRE are the FRCC product and the FRCC departure index product. To produce these products, the LANDSUM model is used to simulate a probability distribution of historical reference conditions under the pre-settlement fire regime, and this simulated HRV is compared with the current landscape conditions (Keane et al. 2007). These comparisons are made at the scale of relatively small (e.g., 81 ha) landscape reporting units, which allows the resulting departure from historical reference conditions to be mapped across the landscape to identify specific areas of high departure from the HRV (Karau and Keane 2007).

Although LADS and LANDSUM were both developed to carry out HRV simulation modeling and are conceptually similar in many respects, each was developed with a different application in mind. LADS was originally developed for use in regional ecosystem assessments. Therefore, many fine-scale details were sacrificed to produce a model that could efficiently simulate large areas (millions of ha) over long time frames (thousands of years) on a single-processor desktop computer. Results from LADS have primarily been used as baselines for broad-scale comparison with current and projected future landscape conditions when conducting regional assessments of forest policy (Nonaka and Spies 2005; Thompson et al. 2006; Nonaka et al. 2007; Spies et al. 2007b).

In contrast, LANDSUM is more focused on making landscape-level assessments (simulation areas encompassing tens of thousands of hectares), with more emphasis on capturing relevant local variability in the environment and the resulting spatial patterns of fire and vegetation (Keane et al. 2002; Karau and Keane 2007). This greater emphasis on local detail is necessary to support the goal of using LANDSUM to map the spatial patterns of deviation from historical reference conditions, and ultimately to apply the resulting information to help prioritize fuels management activities (Keane et al. 2007). To this end, an accompanying set of analytical tools and methods has been developed for quantifying the departure of current conditions from the modeled HRV (Steele et al. 2006). However, the cost of this additional complexity is high computational demand, necessitating the use of parallel processing to carry out simulations at regional to national levels.

6.3 Projecting Future Landscape Changes

Scenario-based landscape modeling has proven to be a valuable tool for forest planning and environmental assessment. This approach involves developing a limited set of alternative future scenarios (usually 2–6) that encompass projections of future landscape conditions based on a set of assumptions about land management policies and the resulting environmental changes (Peterson et al. 2003; Nassauer and Corry 2004). These hypothetical but plausible futures are intended to serve as structured narratives that outline the range of uncertainty about what the future may bring. Spatial simulation models are often applied as tools to project the changes in forest landscape pattern that will occur under alternative forest management scenarios. A frequent goal of these assessments is to contrast the future landscape conditions resulting from a continuation of current management practices with various alternatives strategies that aim to restore the forest landscape to a set of desired future conditions (Fig. 6.4).

These applications of landscape simulation models can be viewed as a hybrid of the predictive and explanatory modeling approaches (Peck 2004; Perry and Millington 2008). They are predictive in that the scenarios are developed for real landscapes using a realistic set of alternative management strategies with an underlying objective of projecting future landscape conditions under the alternative scenarios. However, there is typically not an expectation that the models will predict the details of future landscapes with high accuracy or precision. Instead, the emphasis is typically on comparing and contrasting the results of the alternative scenarios to gain an understanding of the relative effects of alternative forest restoration strategies. In this context, the model application can be also viewed as a heuristic exercise in which a major goal is to gain insights into how the interactions of forest restoration activities with ecological processes lead to different trajectories of future landscape change.

6.3.1 Alternative Forest Policies in the Oregon Coast Range

The Coastal Landscape Modeling and Assessment (CLAMS) project used a model called the Landscape Management Policy Simulator (LAMPS) to project regional changes under alternative policy scenarios (Johnson et al. 2007). Policy simulations accounted for different management practices in major land ownership classes, including federal forests managed by the USDA Forest Service and the Bureau of Land Management, state forests managed by the Oregon Department of Forestry, and private forests owned by the forest industry and nonindustrial private landowners. A unique characteristic of the LAMPS model, compared to most other forest landscape simulators, is that it can simulate forest dynamics over extremely large areas (millions of hectares) while at the same time providing extremely detailed information about forest conditions (individual tree lists for each forest stand) (Bettinger et al. 2005). Under each policy scenario, the LAMPS model accounts

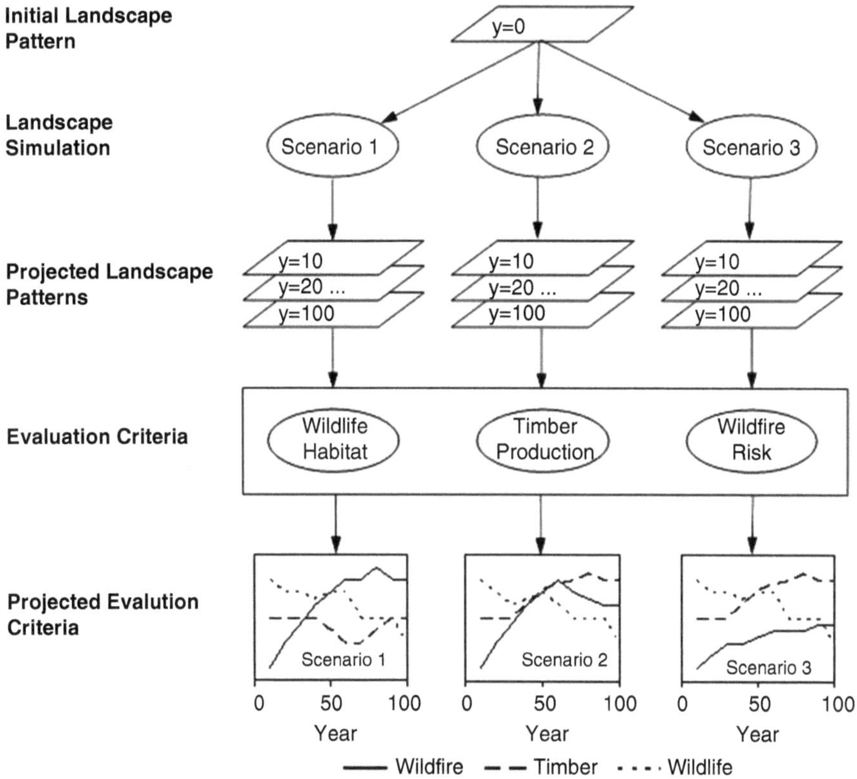

Fig. 6.4 Flowchart illustrating the process of modeling alternative future scenarios. Starting with the current landscape configuration, change is simulated for multiple scenarios based on different forest restoration strategies. The resulting time series of projected landscape configurations are evaluated using a variety of criteria, such as timber production, wildlife habitat suitability, and susceptibility to wildfire

for multiple processes including the death of trees from competition and natural disturbances, the removal of trees through management activities, growth of trees, decay of snags and logs, and the establishment of new trees through either planting or natural regeneration. Because of these characteristics, the output of LAMPS can be linked with detailed habitat suitability models that are based on size distributions of live and dead trees and the spatial arrangement of various stand types across the landscape (McComb et al. 2007; Spies et al. 2007b).

The outcomes of these types of alternative future assessments are dependent on the variety of scenarios that are examined. Two scenarios of particular interest were increasing the number of residual trees left following timber harvests on private lands, and eliminating the practice of thinning young forest plantations on federal lands (Johnson et al. 2007; Spies et al. 2007b). Both scenarios were considered to be realistic policy alternatives that could potentially be implemented through changes to the Oregon Forest Practices Act (affecting private and state lands) or the Northwest

Forest Plan (affecting federal lands). These two alternatives were compared to a baseline scenario that modeled a continuation of current forest management policies across all ownerships. Distributions of forest classes based on tree sizes and hardwood/conifer composition were similar under the baseline and the two alternative scenarios (Spies et al. 2007b). The projected area of old-growth forests in the Coast Range was not sensitive to either of the alternative policies, nor was the projected area of suitable habitat for late-successional species such as the northern spotted owl, the marbled murrelet, and the Pacific fisher (McComb et al. 2007; Spies et al. 2007b). However, habitat for several other species such as the western bluebird, the red tree vole, the olive-sided flycatcher, and the pileated woodpecker was higher under the scenario with increased live-tree retention following timber harvest on private lands. Overall, the results of the analyses indicated that modification of forest management practices on private lands has a greater potential to increase habitat for sensitive species than applying additional restrictions to timber harvesting on the federal lands.

Although there are important differences between the processes of wildfire disturbance and timber harvesting, forest management activities can emulate certain effects of fire and other natural disturbances (Perera et al. 2004). Another study of alternative future scenarios in the Oregon Coast Range used a broader range of forest policy scenarios aimed at restoring various aspects of the historical fire regime (Thompson et al. 2006). The alternatives included increased retention of live trees following timber harvest to emulate the variable severity of historical wildfires, increased rotation lengths to emulate the frequency of historical wildfires, and increased aggregation of harvest units to emulate the size distribution of historical wildfires. When comparing this set of scenarios, increasing live tree retention had a relatively small effect on the distribution of major forest structure classes. In contrast, lengthening the harvest rotation resulted in a significant reduction in the amount of early-successional forests, coupled with an increase in the amount of mature forests. Although it is unlikely that any of these "extreme" forest policy scenarios would be implemented exactly as modeled, they are still valuable for exploring the bounds of what could possibly be achieved by forest restoration efforts over the next century.

6.3.2 *Strategies for Managing Forest Landscape Disturbances*

Maintaining fire-dependent forest types while also reducing the landscape-wide risk of wildfire by managing the landscape mosaic of forest conditions is an important goal for federal land managers, but is difficult to achieve. Furthermore, public forests are inevitably surrounded by other lands over which agency managers have no control. Fire risk abatement on multi-owner landscapes containing flammable but fire-dependent ecosystems epitomizes the complexities of managing public lands. The LANDIS (LANdscape DIsturbance and Succession) model was used to evaluate the relative effectiveness of four alternative fire mitigation strategies on the Chequamegon-Nicolet National Forest (Wisconsin, USA), where fire-dependent pine and oak systems overlap with a rapidly developing wildland urban interface (WUI) (Sturtevant et al. 2009). The potential fire-risk mitigation strategies included: (1) ban debris burning (i.e., reduce fire ignition rate by 25%); (2) reduce fire ignition rates by removing understory conifers next to roads on federal lands; (3) placement of permanent firebreaks within fire-prone land types; and (4) redistribution of "risky" management treatments (i.e., those establishing pine or oak) to areas of the National Forest >1 km from housing developments (WUI). Of the risk mitigation strategies evaluated, reduction of ignitions caused by debris-burning had the strongest influence on fire risk, followed by the strategic redistribution of risky forest types away from the high ignition rates of the WUI. Other treatments (fire breaks and reducing roadside ignitions) were less effective. Simulations also showed that some form of active management is required for long-term maintenance of fire-dependent communities (i.e., pine and oak), which also represent the greatest fire risk to homes in the WUI.

Multiple global changes such as timber harvesting in areas previously undisturbed by cutting and climate change will undoubtedly affect the composition and spatial distribution of boreal forests, which will in turn affect the ability of these forests to retain carbon and maintain biodiversity. To reliably predict future states of the boreal forest it is necessary to understand the complex interactions among forest regenerative processes (succession), natural disturbances (e.g., fire, wind and insects) and anthropogenic disturbances (e.g., timber harvest). LANDIS was used to simulate various scenarios of global change on forest composition, biomass (carbon) and landscape pattern in south-central Siberia (Gustafson et al. 2010). Scenarios simulated included: (1) current climate and disturbance (HRV); (2) current climate plus timber harvest; (3) future climate (as predicted by the Hadley Global Circulation Model); (4) future climate plus outbreaks of the Siberian silk moth (currently climate limited); and (5) future climate plus timber harvest and silk moth outbreaks.

Most response variables were more strongly influenced by timber harvest and insect outbreaks than the direct effects of climate change. Direct climate effects generally increased tree productivity and modified the probability of species establishment, but indirect effects on the fire regime generally counteracted the direct effects of climate on forest composition. Harvest and insects significantly produced changes in forest composition, reduced living biomass and increased forest fragmentation. The study concluded that global change is likely to significantly change forest composition of central Siberian landscapes, with some changes taking ecosystems outside the historical range of variability. However, the direct effects of climate change in the study area are not as significant as the exploitation of virgin forest by timber harvest and the potential increased outbreaks of the Siberian silk moth. Novel disturbance by timber harvest and insect outbreaks may greatly reduce the aboveground live biomass of Siberian forests, and may significantly alter ecosystem dynamics and wildlife populations by increasing forest fragmentation.

6.4 Understanding How Landscapes Change

The previous two sections outlined model applications where a major goal is to make realistic predictions of historical or future landscape conditions within a particular landscape. Landscape models can also be applied in a more generalized experimental framework, with an objective of exploring hypotheses about landscape pattern-process relationships (Fahrig 1991). Forest ecosystems often exhibit strong feedbacks in which disturbances influence the spatial pattern of vegetation, and vegetation pattern in turn constrains fire spread and fire effects. Computer simulation models are particularly valuable for understanding these systems because they can be used to project the outcomes of complex interactions, and allow the modeler to observe outcomes that they may not otherwise have been foreseeable (Rykiel 1996). Creating a landscape model requires development of a conceptual framework for representing the forest landscape, specification of mathematical equations and rule sets for modeling interactions between system components, and estimation of the parameters that control these interactions. Based on user-supplied inputs, the computer performs bookkeeping and computational tasks to track the multitude of state variables over space and time. Therefore, landscape simulation models can be used as "assumption analyzers" that allow scientists to see how their understanding of environmental gradients, fire regimes, and forest succession plays out over large areas and long time frames (Bart 1995).

This approach to landscape simulation modeling is primarily an explanatory or heuristic exercise, where the overarching objective is to enhance understanding of complex pattern-process interactions (Peck 2004; Perry and Millington 2008). In comparison to the more predictive, scenario-based modeling approaches described in the previous section, heuristic modeling applications tend to be less realistic but have greater generality. For example, landscape models may be applied using hypothetical scenarios that would not be considered plausible alternative futures (e.g., the elimination of human influences and the restoration of historical disturbance regime). Model implementation is often carried out using artificial landscapes and may involve complex experimental designs rather than comparisons of a limited number of scenarios. These generalizations are analogous to the simplifications that are necessary when carrying out a laboratory or field experiment (Caswell 1988). Although the results of heuristic exercises are usually not directly applicable to specific landscape restoration projects, the more general knowledge gained may become an important part of the underlying science that is applied in developing landscape restoration approaches.

6.4.1 Disturbance Regimes and Landscape Patterns

DISPATCH is a GIS-based model that was developed to simulate the spread of disturbances across forested landscapes and the resulting changes in landscape patterns (Baker et al. 1991). The model was applied to study the effects of changing fire regimes in the Boundary Waters Canoe Area (BWCA) in Minnesota using a simplified framework for modeling disturbances and landscape dynamics (Baker 1992, 1993, 1994). Fires were all assumed to be stand-replacing and landscape patterns were modeled as age classes reflecting time since the most recent fire. Rather than replicating the exact patterns of environmental variability in the BWCA, DISPATCH was run on a homogeneous, rectangular landscape that had an area equal to that of the BWCA. Because of these simplifying assumptions it was not possible to link model results back to actual locations within the BWCA. However, the simplified modeling framework made it possible to generalize the results of landscape-level simulation experiments to other landscapes with similar disturbance regimes and successional pathways.

DISPATCH was used in several studies that compared alternative scenarios that considered different temporal patterns of changes in disturbance regimes (Baker 1992, 1993, 1994). The *historical* scenario combined fire regimes from three time periods when fire return intervals in the BWCA remained relatively constant: The pre-settlement period (AD 1368–1867), the settlement period (AD 1868–1910), and the suppression period (AD 1911-present). The *pre-settlement* scenario utilized the fire regime from the pre-settlement period for the entire 1,000-year simulation. The *restoration* scenario was the same as the historical scenario through 1993, at which time the fire return intervals were reset to pre-settlement levels to simulate restoration of the historical fire regime. Another study used DISPATCH to examine climate change scenarios that affected fire return intervals and fire sizes, fragmentation and restoration scenarios that considered changes in disturbance regimes resulting from forest management, and the effects of alternative landscape configurations at the beginning of the simulation (Baker 1995). A variety of landscape metrics, including mean pixel age, mean patch size, mean shape, mean fractal dimension, Shannon diversity index, mean richness, fraction of old growth, and mean angular second moment, were used to evaluate the effects of these scenarios on landscape structure.

These simulation experiments led to a number of general hypotheses about how forest landscapes respond to changes in disturbance regimes. Landscape structure does not respond immediately to an altered disturbance regime, but requires a period of time for the new regime to overwrite existing patterns and generate a new quasi-equilibrium. In general, landscape composition (e.g., Shannon's diversity index) responds more rapidly to an altered disturbance regime than landscape configuration (e.g., mean patch size) (Baker 1992, 1994). Landscape responses are spatially heterogeneous and scale dependent, with greater variability in response time as landscape extent decreases (Baker 1993). The time lag in landscape response is also contingent upon the disturbance regime and landscape patterns prior to the change, and whether the change results in increased or decreased rates and sizes of disturbances. For example, simulation experiments demonstrated that landscapes with lower patch densities responded more quickly than landscapes with higher patch densities (Baker 1995). Landscape structure also responded more quickly to climate warming scenarios in which fire frequency and size increased than to cooling scenarios in which fire frequency and size decreased. Landscapes typically adapted to new disturbance regimes within 0.5–2 fire rotations, suggesting that human activities or natural processes that change fire frequency cause landscape structure to exist in a constant state of disequilibrium with the disturbance regime because of these long response times.

6.4.2 Landscape Dynamics in the Oregon Coast Range

A more recent study used landscape simulation models to examine whether restoration of historical disturbance processes would be an effective strategy for restoring pre-settlement landscape patterns in the Oregon Coast Range (Nonaka and Spies 2005). Starting with the landscape configuration in 1996, two scenarios were simulated. The first scenario assumed a continuation of current forest management practices that were simulated using the LAMPS model. The second scenario assumed that no forest management would take place and pre-settlement fire regimes would be restored, with wildfire patterns simulated using the LADS model. Continuation of current land management practices for 100 years moved landscape patterns toward the HRV, but did not completely restore all aspects of the pre-settlement landscape patterns. Restoration of the historical disturbance regime initially increased the departure of landscape patterns from pre-settlement conditions, and still required several centuries to create patterns falling within the HRV. These results supported the main conclusions of earlier studies applying DISPATCH in the BWCA; landscape patterns may take centuries to respond to changes in the disturbance regime, with different metrics of landscape patterns responding at different rates.

Current forest management policies in the Oregon Coast Range are based on static reserve-based strategies. Late-successional reserves on public lands are projected to be eventually dominated by old forests, whereas private landscapes are expected to remain dominated by younger managed forests (Spies et al. 2007a). In contrast, pre-settlement fire regimes created a continuously shifting mosaic of forest age classes (Wimberly et al. 2000, 2004; Wimberly 2002). To explore the ecological implications of changes in the rates and patterns of landscape dynamics, the LADS model was modified to incorporate a simple species occupancy model. Experimental model runs were conducted for several hypothetical species with a range of dispersal distances, colonization rates, and extinction rates (Wimberly 2006). Experiments were designed to compare dynamic and static landscapes with similar landscape patterns and habitat amounts. Species exhibited a more rapid decline to extinction with habitat loss in dynamic landscapes than in static landscapes. However, in some cases, species occupancy was actually higher in dynamic landscape mosaics than in static landscapes with similar habitat amount and pattern. In these situations, habitat dynamics actually increased habitat connectivity over space and time, even though the habitat pattern at any single point in time was highly fragmented.

6.5 Summary and Conclusions

This review has outlined three examples of how landscape simulation models can be used to support forest landscape restoration. In the first type of application, landscape models of disturbance and forest succession are used to estimate historical variability in landscape composition and configuration based on information about the return intervals, sizes, and severities of historical disturbances. Assessments of the departure of current landscapes from the HRV can be carried out at a range of different scales, from coarse regional assessments to more detailed predictions of the spatial pattern of departure from HRV within individual landscapes. Key challenges in carrying out this type of assessment include selecting ecologically relevant landscape metrics to use in computing HRV and developing appropriate quantitative methods for evaluating the degree of departure from the HRV. Major limitations to this approach include a lack of reliable data on historical fire regimes in some landscapes; a scarcity of detailed information about historical landscapes patterns that could be used to validate model-based HRV estimates; and the fact that climate change, species invasions, and other human impacts have the potential to create novel ecosystems that have no historical analogue (Hobbs et al. 2006). Despite these limitations, evaluating departure of the current landscape from the HRV is an important starting point for assessing forest landscape restoration alternatives.

Another common application of landscape simulation models is to project future landscapes under alternative landscape restoration scenarios. One of the most crucial elements of this type of assessment is the number and characteristics of the scenarios that are examined. Often a relatively small number of scenarios are considered, and the scenarios are selected to be realistic representations of plausible forest restoration strategies. However, there may be other, more effective strategies that are outside the solution space of the chosen scenarios. Conclusions about the potential effects of forest landscape restoration activities may also depend upon the amount of variability among the alternatives considered. Scenarios that consider only minor modifications to current silvicultural practices will likely have only a minor effect on landscape structure and wildlife habitat when compared to a "business as usual" scenario. In contrast, examining a wider range of alternatives, including possible climate change effects, can help to outline the possible range of future landscape conditions. Validating the projections of landscape simulation models over time scales of decades to centuries remains a major challenge (Rykiel 1996; He 2008). However, the absolute accuracy of model predictions may be less important than the ability to realistically portray the relative effects of different scenarios.

Simulation experiments with landscape models focus less on making predictions of historical or future landscape conditions, and place more emphasis on exploring general hypotheses about pattern-process relationships. For example, simulation experiments may examine purely hypothetical scenarios such as the restoration of historical fire regimes across large landscapes (Baker 1992, 1993, 1994; Nonaka and Spies 2005), or consider a wide range of disturbance scenarios, initial conditions, and other parameter settings (Baker 1995; Wimberly 2006). Important insights gained from these studies include the recognition that changes in landscape composition and configuration lag behind shifts in disturbance regimes, and that temporal as well as spatial landscape heterogeneity is important to consider when assessing ecological responses to changing disturbance regimes. The general knowledge gained from these experiments can contribute to the conceptual foundation for developing forest restoration strategies, or can serve as a basis for developing more detailed and realistic alternative scenario assessments.

Literature Cited

Allen CD, Savage M, Falk DA, Suckling KF, Swetnam TW, Schulke T, Stacey PB, Morgan P, Hoffman M, Klingel JT (2002) Ecological restoration of southwestern ponderosa pine ecosystems: a broad perspective. Ecol Appl 12:1418–1433

Baker WL (1989) A review of models of landscape change. Landsc Ecol 2:111–133

Baker WL (1992) Effects of settlement and fire suppression on landscape structure. Ecology 73:1879–1887

Baker WL (1993) Spatially heterogeneous multiscale response of landscapes to fire suppression. Oikos 66:66–71

Baker WI (1994) Restoration of landscape structure altered by fire suppression. Conserv Biol 8:763–769

Baker WL (1995) Longterm response of disturbance landscape to human intervention and global change. Landsc Ecol 10:143–159

Baker WL, Egbert SL, Frazier GF (1991) A spatial model for studying the effects of climatic change on the structure of landscapes subject to large disturbances. Ecol Model 56:109–125

Bart J (1995) Acceptance criteria for using individual-based models to make management decisions. Ecol Appl 5:411–420

Bettinger P, Lennette M, Johnson KN, Spies TA (2005) A hierarchical spatial framework for forest landscape planning. Ecol Model 182:25–48

Bu R, He HS, Hu YM, Chang Y, Larsen DR (2008) Using the LANDIS model to evaluate forest harvesting and planting strategies under possible warming climates in Northeastern China. For Ecol Manag 254:407–419

Cairns DM, Lafon CW, Waldron JD, Tchakerian M, Coulson RN, Klepzig KD, Birt AG, Xi W (2008) Simulating the reciprocal interaction of forest landscape structure and southern pine beetle herbivory using LANDIS. Landsc Ecol 23:403–415

Caswell H (1988) Theory and models in ecology: a different perspective. Ecol Model 43:33–44

Fahrig L (1991) Simulation methods for developing general landscape-level hypotheses of single-species dynamics. In: Turner MG, Gardner RH (eds) Quantitative Methods in Landscape Ecology. Springer, New York, pp 417–442

Forest Ecosystem Management and Assessment Team (FEMAT) (1993) Forest ecosystem management: an ecological, economic, and social assessment. U.S. Departments of Agriculture, Commerce, and the Interior, and the Environmental Protection Agency, Washington, DC

Gustafson EJ, Shvidenko AZ, Sturtevant BR, Scheller RM (2010) Predicting climate change effects on forest biomass and composition in south-central Siberia. Ecol Appl 20:700–715

He HS (2008) Forest landscape models: definitions, characterization, and classification. For Ecol Manag 254:484–498

Hessburg PF, Agee JK, Franklin JF (2005) Dry forests and wildland fires of the inland Northwest USA: Contrasting the landscape ecology of the pre-settlement and modem eras. For Ecol Manag 211:117–139

Hobbs RJ, Arico S, Aronson J, Baron JS, Bridgewater P, Cramer VA, Epstein PR, Ewel JJ, Klink CA, Lugo AE, Norton D, Ojima D, Richardson DM, Sanderson EW, Valladares F, Vila M, Zamora R, Zobel M (2006) Novel ecosystems: theoretical and management aspects of the new ecological world order. Glob Ecol Biogeogr 15:1–7

Hunter ML (1993) Natural fire regimes as spatial models for managing boreal forests. Biol Conserv 65:115–120
Impara PC (1997) Spatial and temporal patterns of fire in the forests of the Central Oregon Coast Range. Oregon State University, Corvallis
Johnson KN, Bettinger P, Kline JD, Spies TA, Lennette M, Lettman G, Garber-Yonts B, Larsen T (2007) Simulating forest structure, timber production, and socioeconomic effects in a multi-owner province. Ecol Appl 17:34–47
Karau EC, Keane RE (2007) Determining landscape extent for succession and disturbance simulation modeling. Landsc Ecol 22:993–1006
Keane RE, Parsons RA, Hessburg PF (2002) Estimating historical range and variation of landscape patch dynamics: limitations of the simulation approach. Ecol Model 151:29–49
Keane RE, Cary GJ, Davies ID, Flannigan MD, Gardner RH, Lavorel S, Lenihan JM, Li C, Rupp TS (2004) A classification of landscape fire succession models: spatial simulations of fire and vegetation dynamics. Ecol Model 179:3–27
Keane RE, Holsinger LM, Pratt SD (2006) Simulating historical landscape dynamics using the landscape fire succession model LANDSUM version 4.0. General Technical Report RMRS-GTR-171CD, USDA Forest Service Rocky Mountain Research Station, Fort Collins, CO
Keane RE, Rollins M, Zhu ZL (2007) Using simulated historical time series to prioritize fuel treatments on landscapes across the United States: the LANDFIRE prototype project. Ecol Model 204:485–502
Landres PB, Morgan P, Swanson FJ (1999) Overview of the use of natural variability concepts in managing ecological systems. Ecol Appl 9:1179–1188
Leng WF, He HS, Bu RC, Dai LM, Hu YM, Wang XG (2008) Predicting the distributions of suitable habitat for three larch species under climate warming in Northeastern China. For Ecol Manag 254:420–428
Long CJ, Whitlock C (2002) Fire and vegetation history from the coastal rain forest of the western Oregon coast range. Quat Res 58:215–225
Long CJ, Whitlock C, Bartlien PJ, Millspaugh SH (1998) A 9,000-year fire history from the Oregon Coast Range, based on a high-resolution charcoal study. Can J For Res 28:774–787
McComb BC, Spies TA, Olsen KA (2007) Sustaining biodiversity in the Oregon Coast Range: potential effects of forest policies in a multi-ownership province. Ecol Soc 12(2):29
Nassauer JI, Corry RC (2004) Using normative scenarios in landscape ecology. Landsc Ecol 19:343–356
Nonaka E, Spies TA (2005) Historical range of variability in landscape structure: a simulation study in Oregon, USA. Ecol Appl 15:1727–1746
Nonaka E, Spies TA, Wimberly MC, Ohmann JL (2007) History range of variability (HRV) in live and dead wood biomass: a regional-scale simulation study. Can J For Res 37:2349–2364
Peck SL (2004) Simulation as experiment: a philosophical reassessment for biological modeling. Trends Ecol Evol 19:530–534
Perera AH, Buse LJ, Weber MG, Crow TR (2004) Emulating natural forest landscape disturbances: a synthesis. In: Perera AH, Buse LJ, Weber MG (eds) Emulating natural forest landscape disturbances. Columbia University Press, New York, pp 265–274
Perry GLW, Enright NJ (2002) Humans, fire and landscape pattern: understanding a maquis-forest complex, Mont Do, New Caledonia, using a spatial 'state-and-transition' model. J Biogeogr 29:1143–1158
Perry GLW, Enright NJ (2006) Spatial modelling of vegetation change in dynamic landscapes: a review of methods and applications. Prog Phys Geogr 30:47–72
Perry GLW, Millington JDA (2008) Spatial modelling of succession-disturbance dynamics in forest ecosystems: concepts and examples. Perspect Plant Ecol Evol Syst 9:191–210
Peterson GD, Cumming GS, Carpenter SR (2003) Scenario planning: a tool for conservation in an uncertain world. Conserv Biol 17:358–366
Rollins MGC, Frame K (2006) The LANDFIRE Prototype Project: nationally consistent and locally relevant geospatial data for wildland fire management. General Technical Report RMRS-GTR-175, USDA Forest Service Rocky Mountain Research Station, Fort Collins, CO
Rykiel EJ (1996) Testing ecological models: the meaning of validation. Ecol Model 90:229–244

Scheller RM, Mladenoff DJ (2005) A spatially interactive simulation of climate change, harvesting, wind, and tree species migration and projected changes to forest composition and biomass in northern Wisconsin, USA. Glob Chang Biol 11:307–321
Scheller RM, Mladenoff DJ (2007) An ecological classification of forest landscape simulation models: tools and strategies for understanding broad-scale forested ecosystems. Landsc Ecol 22:491–505
Schmidt KM, Menakis JP, Hardy CC, Hann WJ, Bunnell DJ (2002) Development of coarse-scale spatial data for wildland fire and fuel management. General Technical Report RMRS-87, USDA Forest Service Rocky Mountain Research Station, Fort Collins, CO
Schumacher S, Bugmann H (2006) The relative importance of climatic effects, wildfires and management for future forest landscape dynamics in the Swiss Alps. Glob Chang Biol 12:1435–1450
Shifley SR, Thompson FR, Dijak WD, Larson MA, Millspaugh JJ (2006) Simulated effects of forest management alternatives on landscape structure and habitat suitability in the Midwestern United States. For Ecol Manag 229:361–377
Society for Ecological Restoration International Science & Policy Working Group (2004) The SER International Primer on Ecological Restroation. Society for Ecological Restoration International, Tucson
Spies TA, Johnson KN, Burnett KM, Ohmann JL, McComb BC, Reeves GH, Bettinger P, Kline JD, Garber-Yonts B (2007a) Cumulative ecological and socioeconomic effects of forest policies in Coastal Oregon. Ecol Appl 17:5–17
Spies TA, McComb BC, Kennedy RSH, McGrath MT, Olsen K, Pabst RJ (2007b) Potential effects of forest policies on terrestrial biodiversity in a multi-ownership province. Ecol Appl 17:48–65
Steele BM, Reddy SK, Keane RE (2006) A methodology for assessing departure of current plant communities from historical conditions over large landscapes. Ecol Model 199:53–63
Sturtevant BR, Miranda BR, Yang J, He HS, Gustafson EJ, Scheller RM (2009) Studying fire mitigation strategies in multi-ownership landscapes: calancing the management of fire-dependent ecosystems and fire risk. Ecosystems 12:445–461
Teensma PD, Rienstra JT Yeiter MA (1991) Preliminary reconstruction and analysis of change in forest stand age class of the Oregon Coast Range from 1850 to 1940. T/N OR-9, U.S. Bureau of Land Management, Portland, OR
Thompson JR, Johnson KN, Lennette M, Spies TA, Bettinger P (2006) Historical disturbance regimes as a reference for forest policy in a multiowner province: a simulation experiment. Can J For Res 36:401–417
Urban DL, O'Neill RV, Shugart HH (1987) Landscape ecology: a hierarchical perspective can help scientists understand spatial patterns. Bioscience 37:119–127
Weisberg PJ, Swanson FJ (2003) Regional synchroneity in fire regimes of western Oregon and Washington, USA. For Ecol Manag 172:17–28
Wimberly MC (2002) Spatial simulation of historical landscape patterns in coastal forests of the Pacific Northwest. Can J For Res 32:1316–1328
Wimberly MC (2004) Fire and forest landscapes in the Georgia Piedmont: an assessment of spatial modeling assumptions. Ecol Model 180:41–56
Wimberly MC (2006) Species dynamics in disturbed landscapes: when does a shifting habitat mosaic enhance connectivity? Landsc Ecol 21:35–46
Wimberly MC, Kennedy RSH (2008) Spatially explicit modeling of mixed-severity fire regimes and landscape dynamics. For Ecol Manag 254:511–523
Wimberly MC, Ohmann JL (2004) A multi-scale assessment of human and environmental constraints on forest land cover change on the Oregon (USA) Coast Range. Landsc Ecol 19:631–646
Wimberly MC, Spies TA, Long CJ, Whitlock C (2000) Simulating historical variability in the amount of old forests in the Oregon Coast Range. Conserv Biol 14:167–180
Wimberly MC, Spies TA, Nonaka E (2004) Using criteria based on the natural fire regime to evaluate forest management in the Oregon Coast Range of the United States. In: Perera AH, Buse LJ, Weber MG (eds) Emulating Natural Forest Landscape Disturbances. Columbia University Press, New York, pp 146–157

Part III
Social Science Perspectives

Chapter 7
Forest Landscape Restoration Decision-Making and Conflict Management: Applying Discourse-Based Approaches

Jens Emborg, Gregg Walker, and Steven Daniels

7.1 Introduction: The Potential for Conflict in Forest Landscape Restoration

Researchers and managers define forest landscape restoration (FLR) as

> A process that aims to regain ecological integrity and enhance human well-being in deforested or degraded forest landscapes. (Maginnis et al. 2005)

FLR incorporates a broad array of highly complex issues, values, and no small amount of uncertainty. As an emerging field of research and practice there is an encouraging spirit, optimism, and enthusiasm regarding FLR. Certainly it is difficult to oppose FLR as described in the definition above. Who would not like to regain ecological integrity? Who would not like to enhance human well-being? Who would not like to turn degraded landscapes into green forests? Indeed, FLR is a noble endeavour with expansive aspirations. As laudable as the goals and values are in this definition, the real benefits stem from applications of FLR concepts and tools.

While forest managers and researchers may agree in principle when defining FLR, there are many potential sources of conflict related to FLR in practice,

J. Emborg (✉)
Forest and Landscape, University of Copenhagen,
Rolighedsvej 23, 1958 Frederiksberg, Denmark
e-mail: jee@life.ku.dk

G. Walker
Communication Studies Program, Oregon State University,
104 Shepard Hall, Corvallis, OR 97331-6199, USA
e-mail: gwalker@oregonstate.edu

S. Daniels
Department of Sociology, Social Work, and Anthropology
and Department of Environment and Society, Utah State University,
UMC 0730, Old Main Hill, Logan, UT 84322, USA
e-mail: steve.daniels@usu.edu

J. Stanturf et al. (eds.), *Forest Landscape Restoration: Integrating Natural and Social Sciences*, World Forests 15, DOI 10.1007/978-94-007-5326-6_7,

including tensions over specific restoration activities, landscape boundaries, budget priorities, and the role of private property owners. Consequently, forest managers' and scientists' FLR proposals may encounter more resistance than they might anticipate. FLR plans, while grounded in strong science, may generate controversy and scepticism. Moreover, if FLR advocates do not thoughtfully navigate the minefield of social and political controversy, the likelihood of successfully implementing their projects is reduced.[1]

The potential for conflict in relation to FLR is recognized and highlighted – and these conflicts, like other conflicts, can be founded in truly conflicting interests – but also in factors such as miscommunication, strained relationships or struggles over power. Sayer (2005) notes: "There are countless examples of attempts at restoration failing because one person's "restoration" is another person's "degradation"" (p. 101). Similarly, Brown (2005) states that in most real-life FLR situations some will stand to win while others will lose in the wake of a given project. In addition, Jones and Dudley (2005)[2] address the issue of negotiation and conflict management in relation to FLR, and provide some basic building blocks for the conflict management process, principles for successful negotiation, analytical tools for conflict management, and a selection of practical hints and advice for capacity building and effective communication. Their brief presentation of the topic provides a good point of departure for our endeavour. In this chapter we present a comprehensive discussion of natural resource conflict situations and offer conflict management principles and practices relevant to FLR.

The original charge for the chapter (and indeed the focus of the earlier version delivered at the 2007 Seoul conference, see Emborg 2007) was to look at the relevance of *conflict management techniques* to FLR. This chapter incorporates ideas from that earlier conference presentation but offers a more expansive view of conflict and forest landscape restoration. In doing so, the discussion introduces "discourse" (the various forms of communication – verbal and otherwise-between and among the parties in a decision process) as a key construct in the management of FLR conflicts. Our commentary highlights that a *discourse-based approach* to conflict and policy formation can be useful and productive. By discourse-based approaches we refer to processes in which stakeholders and actors (e.g., citizens, NGOs, policy makers, agencies) are involved in dialogue and deliberation[3] in order to address

[1] Whether conflict is likely to emerge depends on many factors, e.g.: interdependence, interests, stakes, history, quality of relationships, trust, culture, values, land tenure, property rights, perceptions, benefits, burdens, justice, jurisdiction, power, strategies. (c.f. Deutsch and Coleman 2000; Daniels and Walker 2001; Pruitt and Kim 2004).

[2] Sayer (2005), Brown (2005), and Jones and Dudley (2005) are all chapters in: "Forest Restoration in Landscapes. Beyond Planting Trees" (Mansourian et al. 2005b). This is a state-of-the-art book on FLR in which many contributors from research and practice provide insight and identify future needs for this new and exciting field.

[3] We distinguish between two kinds of communication and interaction among the parties: *Dialogue* implying that the parties exchange their respective views and interests – implying a good deal of listening and sincere effort to try to mutually understand and empathize with each other; and *deliberation* implying efforts to develop possible options for action, identify constructive ways to move forward, succeeded by careful consideration of options in consecutive steps towards making decisions. For a more developed discussion, see Daniels and Walker (2001).

conflicted situations or help public decision making (Daniels and Cheng 2004). This is a broader frame than common conflict management paradigms, and their more narrow focus on the conflict issues and episodes, which may limit how creatively and proactively we conceptualize the role for meaningful discourse in our efforts to restore forested landscapes. So by adopting a discourse-based focus, this chapter not only addresses conflict management issues, but also increases the potential for conflict prevention.

This chapter pursues four tasks. First, we highlight the nature of discourse in a natural resources management context. Second, we provide a rationale for viewing forest management as a deeply social and political process, not merely one of applied ecology and vegetation management. Doing so draws attention to human dimensions as well as ecological features, including sources of controversy. Third, we address specific features of FLR that are probable sources of controversy and their manifestation in conflict and decision situations. With the first tasks providing the foundation, we lastly consider discourse-based approaches for managing conflict and negotiating natural resource management and environmental policy decisions.

7.1.1 Discourse Defined

In daily use "discourse" means verbal communication, talk, or conversation. "A discourse," Australian environmental scientist John Dryzek explains, "is a shared way of apprehending the world. Embedded in language, it enables those who subscribe to it to interpret bits of information and put them together in coherent stories or accounts" (2005, p. 9). Jürgen Habermas highlights that an important aspect of discourse is that everybody is allowed to express their viewpoints and ideally all participants intend to reach an un-forced consensus (Habermas 1984). Discourse implies that reasons and arguments run back and forth to create shared meaning (Latin: *discursus*- running back and forth) – a process where the better arguments (from either a rational or moral position) will guide emerging consensus. In a social constructionist view, discourse refers "to a systematic, coherent set of images, metaphors and so on that construct an object in a particular way" (Burr 2003, p. 202). A major point in the social constructivist view is that many discourses can evolve around a certain event, as Burr (2003) explains:

> If we accept the view, …, that a multitude of alternative versions of events are potentially available through language, this means that, surrounding any one object, event, person etc. there may be a variety of different discourses, each with a different story to tell about the object in question, a different way of representing it to the world… Each discourse claims to say that the object really is, that is, claims to be the truth. (p.64–65)

This view certainly applies to a FLR context. At an aggregate level, decisions about forests (or health care or education or global warming) are a social negotiation between the competing views of what a good decision would be; the various social constructions compete in the political marketplace of ideas for dominance. Communication and political science scholars (e.g., Foucault 1972; Fischer 2003) have actually used the term *discourse* to refer to this kind of process in recent decades. Understanding policy processes (successful implementation of FLR among them) is therefore a process of understanding the discourse, and – by extension – improving the discourse is a means of improving the policy outcome.

7.2 The Nature of Discourse

Any public policy situation may feature multiple discourses. As social constructions built collectively, discourses typically include (1) entities (e.g., objects, processes, rules) that are promoted or emphasized, (2) assumptions about relationships, natural and created, (3) agents and their motives, and (4) key metaphors and other rhetorical forms (adapted from Dryzek 2005, pp. 17–18).

Discourses are bound up with political power, philosophy, and ideology. They compete for dominance and control; a preferred discourse influences actions and decisions (Foucault 1972). "In every society," French philosopher Michel Foucault wrote, "the production of discourse is at once controlled, selected, organized and re-distributed according to a certain number of procedures, whose role is to avert its powers and its dangers, to cope with chance events, to evade its ponderous, awesome materiality" (1972, p. 216). Discourse is the process through which reasons and arguments are exchanged to create shared meanings that will guide emergent consensus. One discourse may emphasize technical work and economic benefits while another discourse may feature stewardship practices and conservation values, and while both are arguably incomplete, neither is wrong.

7.2.1 Understanding "Good" Forestry as a Social Construction

When viewed as an application largely of technical concepts, FLR may generate consensus among scientists and managers. When viewed as incorporating social/cultural, ecological, as well as economic factors – three dimensions of sustainable forest management – FLR potentially becomes more controversial. The complexity and controversy are fuelled by the diverse perspectives, values, and interests of the stakeholders. Multiple stakeholders in a FLR project generate multiple interpretations of forest landscape restoration both in general and site-specific terms. Each stakeholder constructs a reality of the forest management situation that emphasizes some features of the situation while minimizing others. For example, a forest products business owner may look at a landscape and see the potential for selective timber harvest. A local environmentalist may draw attention to the forest's habitat conditions and species diversity. Both these constructions are legitimate and need to be communicated as part of the decision-making discourse.

Understanding good or bad forestry as a social construction casts an ecosystem as a human activity system in which the physical, biological, social, cultural, economic, and political dimensions interact. An ecosystem as a human activity system is "discourse-based;" that is, it emphasizes human interaction and learning as part of ecosystem management. This integrated socio-ecological perspective and approach (which we explain in more depth later in this chapter), has the potential to create and sustain integrated socio-ecological solutions in practice. We have seen this approach work well in many natural resource management contexts. While these applications may not be easy or unproblematic, in appropriate situations they offer a constructive and promising way to meet an extremely difficult task.

The core task of a discourse-based approach is to create a platform that stimulates constructive communication interaction and informal problem solving, where all interested and relevant stakeholders have access to express their views and arguments. Part of the strategy is to improve the discourse and create a foundation of common understanding of a given problem situation (e.g., the need for a FLR project) and hopefully improve relationships as well.

7.2.2 Multiple Views of "Good" Forestry

Forest management is not just about technical aspects; ecosystems incorporate human dimensions as well as the physical and biological. By recognizing that forest management involves managing human activity on landscapes, insights from both natural and social sciences become paramount to good practice. As the earlier example illustrates, two parties can look at the identical landscape and see different forests (and advance different discourses). There is no single or best social construction of a forest or landscape. Quite the contrary; there may be as many constructions or interpretations of a forest and its best management practices as there are invested parties or stakeholders. As Ingram and colleagues note, "social construction is a world-shaping exercise or, at least, encompasses various ways in which the 'realities' of the world are defined" (2007, p. 95).

Each party interprets reality and finds meaning through knowledge and experience in a social context at a given time and place. The constructions are influenced by social norms, what others think, and one's own values and beliefs. Through discourse parties can share, negotiate, and coordinate their constructions; what a forest means to them and how they think it should be managed.

Social construction offers insights into forest landscape restoration work. Even the term "restoration" is a highly value-laden concept, because it implies the current situation is somehow bad, damaged, or diminished. A construct of good forestry practice derived from the domain of ecologists may not be the same as one derived from the timber industry. The views and priorities of a conservation biologist working for a multi-national environmental NGO would likely be very different from those of a subsistence hunter/harvester. It is therefore necessary to move beyond a 'one right view of good forestry' perspective to one recognizing the legitimate existence of many different ways to value forests.

A case study of ecologically-based forest planning in Finland (Leskinen 2004) illustrates the role social construction plays in the success of FLR. The Finnish situation is typical for many countries: the regional forestry agency offered to prepare management plans for private forest landowners in order to achieve a broad range of ecological, financial, and timber supply objectives. Leskinen studied how various groups of people – forest owners, environmentalists, and forest professionals – could be viewed as distinct *discourse communities*; each with its own language and meanings regarding the very same objects. One common feature of the management plans was "retention trees," which were to be left in cutting areas to provide habitat and structural diversity in the next stand. The study showed that within a very few years after implementing the plans, most of these retention trees had been removed by the landowners, thereby defeating a significant ecological objective. Leskinen learned through landowner interviews that they never really adopted the value of the trees as ecological legacies and were more likely to view them as firewood, household wood, as messy or aesthetically unappealing ('bad' forestry), or even a waste of money.

Speaking more broadly, a tree can be variously viewed as an invasive species, fire hazard, part of my retirement savings, a cultural/historic artifact, a carbon sink, part of residential landscaping, streamside stabilizer, intermediate host for pests, and a religious symbol– or simply "fibre." The combination of values we assign to trees is a process of socially constructing their meaning. If stakeholders (including forest managers) are not aware of such fundamental differences in "worldviews," values, or perspectives, then miscommunication and misunderstanding seem likely, often without the parties being aware of any communication problems. Differing perceptions among stakeholders and succeeding miscommunication can easily cause or re-enforce conflict, as e.g. in this volume's case-study of Indonesian conservation forest in Jambi (Siregar et al. 2012).

> This analysis confirmed there were quite different perceptions among stakeholders about the nature of forest resources. These depended on the nature and intensity of the interaction each group had with the forest. Stakeholders' perceptions were affected by past experiences as well as their knowledge, interests and values.

Within a given social context there may well be many different perceptions of possible costs, burdens, and benefits of a particular FLR project and these differing perceptions are a seedbed for misunderstanding and potential conflict. Leskinen's study (2004) illustrates that ecological value alone is not sufficient to ensure successful restoration projects – the work must have social value and political legitimacy as well (which again are perceived differently from person to person – each of us simply has different values and thus perceptions of what is legitimate). The Finnish foresters' failure to understand the landowners' values and to help them appreciate the retention trees resulted in a far less innovative and ecologically valuable outcome than might have been hoped for. Although the conflict may have been latent rather than overt, meaningful discourse could have fostered learning, understanding, and shared action. Similar patterns and effects seem to be in play in restoring broadleaved forests in southern Sweden (Löf et al. 2012) where highly ambitious environmental goals for planting broadleaved trees turned out to be in conflict with the perceptions and attitudes among forest owners who preferred conifers for economic reasons. Such differences in the perceptions of 'good' and 'bad' forestry also explain the disappointing rate of restoration found in southern Sweden (Löf et al. 2012).

7.2.3 Changing Views of 'Good' Forestry

Some of the increasing interest in forest restoration stems from the changing social construction of the purpose of forestry. Our notions of what constitutes 'good forestry' change more frequently than does the underlying production period. There are many forests around the world that might need to be 'restored' that also represent what was at one time the most contemporary forestry thinking of the day (e.g., plantations with little species diversity that replaced more complex native forests). There are few other endeavours where the legacy of one's decisions is as long lived and as visible as in forestry.

The construct of 'sustainable forestry' provides an illustration. The broad and deep discussion of the concept of *sustainability* emerged after the Brundtland Report (WCED 1987) was published. More specific interpretations of *sustainable forestry* were coined for various regions of the world after the United Nations Conference on Environment and Development (UNCED) in 1992. As a result of regional international forest processes (e.g., the Montréal Process and the Pan-European Forest Process), a common framework was established across all regions suggesting that sustainable forest management should balance ecological, social, and economic aspects into a more holistic approach to forestry. Sets of criteria and indicators for each of the three dimensions were developed for each region through regional processes involving experts as well as policy makers. Many forest restoration projects would indeed be legitimized by the idea of sustainability and address those criteria and indicators. A social constructionist perspective of forestry weaves into the sustainable forestry movement in at least three ways:

First: In recent years stakeholders' perceptions of forestry have broadened. In Denmark for instance, the scope of forestry today includes wood extraction, economic revenue, hunting, nature protection, ecological function, recreation, amenity values, and much more. This Danish re-interpretation of forestry has implied an increased emphasis on biodiversity protection and ecological functioning of the forests. Consequently, non-intervention forest reserves have been established and nature-based forest management practices have evolved. Even though this might represent what most Danes consider as 'good forestry practice,' each individual has her/his own personal (maybe quite different) interpretation of 'good forestry practice' (e.g., influenced by personal values, knowledge, judgment). Such general and individual interpretations of 'good' and 'bad' reflect current social and political contexts. Accordingly, rates of change in these attitudes likely will vary geographically (e.g., the discourse communities of Leskinen (2004) or the stakeholder categories of Siregar et al. (2012). In Denmark the concept of sustainable forestry was constructed among scientists and policymakers years before it actually hit the ground among forest practitioners.

Second: Changes in our understanding of the role and function of forests and forest management actually form part of the background for the increasing focus on FLR. The rationale for this postulate is that this broader scope of forestry in many cases leads to the perception of 'something missing' in a particular forest (e.g., biodiversity in the Danish forests – see Christensen and Emborg 1996), or the almost complete lack of *wilderness* in the UK described by Convery and Dutson (2012). In other places, the social and/or economic aspects of forestry have come more into focus (e.g., community forestry, timber extraction, forest-based tourism). In any case, with the perception of 'something missing' a need for change arises and a need to restore functions or services provided from a given forest.

Third: Another significant consequence of the changing social construction of forestry is that forests are increasingly perceived as embedded in broader social and landscape contexts. Forests are now mostly considered integrated elements in larger multi-functional cultural and natural landscapes. A straightforward result of this trend is that more people and groups perceive they have a stake in forests – and accordingly want to influence and give input to forestry decision making. Establishing user boards, inviting the public to participate in forestry planning, and at least tolerating citizens who engage in public protests have become standard activities in the forest policy arena. Forest management has become ecosystem management in a social context. People and social situations have entered the forestry scene – and ecosystems and social systems are being linked more closely to one another. These changes are also manifested in the international forest policy arena, as exemplified by suggested criteria for public involvement in the Pan-European Criteria and Indicators for Sustainable Forest Management (MCPFE 1998, e.g. Concept area: "Public awareness" under Criterion 6 in the Resolution L2).

7.2.4 A Contemporary View of Good FLR Practice – An Integrated Socio-ecological Approach

The above discussion shows how forestry has become increasingly social in two important ways: *First,* there is recognition that a broader set of social values (including biodiversity protection, multiple-use, and timber extraction) needs to be incorporated into our management regimes. This consideration of social values clearly needs to be consistent with the ecological attributes of the site; there are few advocates for practices that are ecologically unsustainable (although there certainly are places where they are occurring[4]). *Second,* there is more social visibility of forestry (and more broadly natural resource management) and people have a greater desire to become involved in it. Forestry is no longer the exclusive domain of 'Herrn Forstmeister' conducting management on behalf of the lord of the manor. Managing forests at a landscape scale impacts our everyday environments; relevant stakeholders scrutinize management practices. FLR projects must include and relate to various social and political processes – very similar to the case of sustainable coastal development presented by Burbridge (2012).[5] This increasing social orientation of forestry

[4] In this volume, Xi et al. (2012) provides an example from 'the wood-basket' of the USA (the extensive pine plantations of SE US) where the southern pine beetle is spreading rapidly, causing massive economic loss and ecological destabilization. Löf et al. (2012) is another example that shows how poorly adapted tree species in combination with climate change and storms can cause ecologically unstable forests e.g., prone to wind-throw and other calamities (e.g., insects and root-rot).

[5] This parallel illustrates well the potential for transfer of knowledge from one field of knowledge and experience to another – e.g., from coastal management to forestry, or from conflict management and policy formation to FLR.

is clearly reflected as a need for a holistic socio-ecological approach to FLR, most clearly stated in the 'Key points to retain' of Chokkalingam et al. (2005):

"Three key lessons have emerged from a Centre for International Forestry Research (CIFOR)-led study on reforestation/rehabilitation/restoration in six countries:

1. It is necessary to strengthen local organisation and participation in restoration projects.
2. It is necessary to consider local socio-economic needs in choices of approaches and options.
3. In the long run, it is necessary to ensure that clear and appropriate institutional support and arrangements are in place." (p.405)

Similar integrated socio-ecological perspectives are echoed in the concluding chapters of Dudley (2005b) and Mansourian et al. (2005a). These conclusions in combination with this book well express the current dominating perception of best FLR practice – requiring a combined ecological and social approach to be sustainable and successful in the long run. An integrated ecological and social approach is essential to effective FLR work, but as the social construction discussion highlights, where there are diverse interpretations among stakeholders there is the seeming inevitability of correspondent conflict. Conflict – tensions and incompatibilities among interdependent parties (Folger et al. 1997) – is inherent in forest management generally and FLR projects specifically.

7.3 Controversy in FLR

Many chapters in this book reflect the social construction/conflict emergence dynamic that we argue characterizes FLR. The case study from Wild Ennerdale (Convery and Dutson 2012) clearly illustrates how different perceptions of 'good' and 'bad' land use (e.g., forestry, farming, or wilderness) are socially constructed in a typical FLR case. The Wild Ennerdale Initiative launched a new vision of this rural valley community, namely "to allow the evolution of Ennerdale as a wild valley for the benefit of people, relying more on natural processes to shape its landscape and ecology," which indeed is the core of many FLR projects.[6] How such a vision makes sense strongly depends on who you are, your values, goals, profession, lifestyle, and so on. The case documents how farmers had quite different perceptions of the

[6] Naturalness and authenticity are considered important measures of restoration success, as e.g., explained by Dudley (2005a): "Impacts on authenticity or naturalness: On an ecosystem scale, measuring impacts on overall naturalness of forests is easier than surveying biodiversity and acts as a partial surrogate: generally the greater the naturalness of a forest, the more of its original constituent species are likely to survive". Worldwide forest authenticity is declining fast. In most West European countries, less than 1% of forests are classified by the United Nations as "undisturbed (UNECE and FAO 2000)." This partially explains the great interest for natural forest dynamics, 're-wilding' in Europe and the force and enthusiasm by which re-establishment of (semi-)natural forests, non-intervention forests and strict forest reserves have taken place (e.g., Cost Action E4 and E33, Parviainen et al. 2000).

project than did the people behind the Wild Ennerdale Initiative, as reflected in two farmers' statements:

> Really we're just paid to be park keepers aren't we. Keep the place looking nice...we're just paid to have it look nice for the tourists, but the thing is they've made us take all the sheep off the fell.
>
> I think the farming activity in the valley is now considered to be fairly peripheral to the general sort of aim of Wild Ennerdale ... we all get the impression that they would quite like us to go away.

This case shows that the ability of planners and managers to capture and understand such differences in stakeholder perceptions (including their own) is crucial for crafting effective FLR process design and successful implementation. Lack of such skills may lead to miscommunication, misunderstandings and possibly conflict. This is fundamental to competent conflict management: to listen, to understand the perspectives and worldviews of others (and one's self), and to live with and handle more than one truth.

Han et al. (2012) present a case with significant potential for conflict. Clearly, the goal of recovering the populations of Amur tigers is a noble one. But as is often the case with species preservation efforts, the distribution of benefits and costs are not equal: current and future generations around the world receive the benefits, while the costs are borne by the residents currently living in the region. To the extent that farmland needs to revert to native vegetation, or that traditional forest uses must change, or that increased numbers of tigers create a safety issue for humans, or pastoral activities are more difficult (e.g., tigers prey on domesticated animals or the tigers' prey species compete with domesticated animals for forage) then the local support for tiger recovery will diminish. In short, it will be necessary to craft a new ethic of co-existence between the local residents and tigers, and community-level discourse could be a big part of the social process through which such an ethic might emerge. But if government conservation officers and biologists attempt to institute sweeping changes based upon satellite imagery of forest canopy densities and computer models of predator–prey relationships – but no sensitivity to local life patterns and the cultural significance of forest-based activities – then their prospects for successful tiger recovery are diminished.

7.4 A Discourse-Based Approach to FLR

Ecological science undoubtedly serves as a guidepost for FLR by informing both the analysis of current conditions as well as providing the basis for restoration strategies. Without a theoretical foundation, the questions of what is somehow undesirable and what would be preferable become a value-driven popularity contest lacking rigor and rationale. By the same token, addressing the social conflict dimensions of FLR requires a similar theoretical foundation. Without comparable theory, our management of the social components of FLR cannot match the rigor of our ecological thinking. Absent theory, conflict management techniques run the risk of devolving into a set of disjointed facilitator tactics, with little guidance informing the decision of when or how to use them.

At an aggregate level, decisions about forests involve social negotiation as the various social constructions compete for dominance in the political marketplace of ideas. Understanding policy processes (successful implementation of forest landscape restoration among them) relies on understanding the pertinent discourses, and – by extension – improving the dominant and integrating discourse as a means of improving the policy outcome (Fischer 2003).

7.4.1 What Are 'Discourse-Based Approaches' to Natural Resource Decision Making?

Two essays (Daniels and Cheng 2004; Walker and Daniels 2005) distinguish 'discourse-based' conflict-management and decision-making approaches from 'technical-regulatory' approaches that minimize or control discourse. While a discourse-based strategy emphasizes multi-stakeholder participation and communication interaction, a technical-regulatory method (or 'tech-reg') relies on technical solutions that are subsequently routinized and enforced through regulations. Daniels and Cheng contend specifically that tech-reg has been the dominant natural resources decision-making and management paradigm over the past 50 years – seeking technical solutions to natural resource problems and then implementing those solutions through the regulatory authority of government bodies. Consider these examples: Are tropical forests in Borneo being high-graded? No problem – Foresters will develop harvest guidelines and caps, the state will pass regulations, the people will comply, and the forests will recover. So the birds in Greenland are being over-harvested? No problem: Biologists will determine the right harvest level, the state will pass requirements, the people will comply, and the birds will recover. The countless replications of this sequence around the world often fall short of management objectives; the tech-reg approach marginalizes stakeholders, and attempts to apply a management prescription that largely ignores the extent to which legitimacy is a precondition to effective implementations and is also socially constructed. When tech-reg approaches fail, it is often some combination of: (1) seeking a technical solution to what is fundamentally a question of values (i.e., the social construction discussed above), (2) relying solely on science to predict cause-and-effect relationships in complex, controversial, and dynamic contexts, and (3) assuming that the State has sufficient power to successfully impose regulations that people will readily comply with. That said, tech-reg approaches certainly seem appropriate in some circumstances, for example, in cases where immediate restoration action is needed in the wake of natural disasters (Conner et al. 2012). The case of the Southern Pine Beetle (Xi et al. 2012) is an example of a common variation of the tech-reg approach in which education initiatives and information campaigns substitute for the regulatory part of the program. Similarly, interest in alternatives to tech-reg has grown, as reflected repeatedly in this volume c.f., Indonesia (Siregar et al. 2012), and the Netherlands (Hendriks et al. 2012). Even as tech-reg strategies have dominated the natural resources decision-making landscape, various discourse-based methods have emerged in the last two decades. Although these approaches go by many different names and have distinctly different formats and objectives, they all serve as platforms for bringing scientific knowledge and social values together in a process that promotes innovation, joint learning, and integrative problem solving. Some of these methods are best used before conflict becomes pronounced, yet others have been crafted specifically in response to highly escalated conflict.

7.4.2 Forest Landscape Restoration as a Conflict Situation

How might conflict manifest itself in FLR projects? It does so by exhibiting key attributes of conflict situations. *First,* FLR involves interdependence and interference. Interdependence implies that what one party decides to do will affect what the other party will be able to do – the parties cannot act independently. Goal interference implies the immediate goals of both parties cannot be achieved simultaneously. Further, this definition presumes peoples' perceptions (rather than some hypothetical truth) actually determine their chosen behaviour in a given situation. *Second,* parties often view conflict as negative. To most people conflict is something unpleasant, expensive, and stressful that distracts from more constructive endeavours (Carpenter and Kennedy 2001). Additionally, conflict likely features some degree of miscommunication among the parties (Bush and Folger 2005). *Third,* conflict may draw on a range of sources. Wehr (1979) characterizes the nature or source of incompatibility in conflict situations by distinguishing between: fact-based, values-based, interest-based, jurisdiction-based, person-based, history-based, and culture-based conflicts (see also Daniels and Walker 2001, p.30).

Environmental and natural resource conflicts have some special characteristics that should be added to the three identified above. Across huge differences in the social, economic and ecological contexts, environmental/natural resource and other land-use conflicts have some common features (Daniels and Walker 2001):

- Many stakeholders
- Multiple issues
- Strong and conflicting interests
- Complex ecological and social settings and dynamics – expert requirement
- Formal/legal rights and informal historically agreed-to land-use rights/practices
- Differing values and worldviews – some deeply held
- Cultural differences
- Overall and local concerns
- Expert knowledge and traditional knowledge

These common features seem to echo what can be observed in numerous FLR cases presented in Mansourian et al. (2005b; IUFRO 2007), as well as in this volume.

In light of the complex nature of conflicts generally and natural resource conflict specifically, the question is: How can a party (such as a decision authority) make decisions, induce change, implement plans and make progress in such a chaotic environment of multiple issues, people, and groups with differing values, conflicting interests, and competing expertise? Answering this question includes two steps: first, determining the nature of the specific natural resource conflict, and second, applying an appropriate discourse-based conflict management approach. These two areas are addressed in the remaining sections of this chapter.

7.4.3 *Conflict Types Relevant to FLR*

It is possible to categorize the nature of a specific conflict into different types of conflict, which can be helpful for example when designing appropriate FLR decision making processes. Examples of such different types of conflicts could be:

7.4.3.1 Interest-Based Conflict

Interest-based conflict arises when an FLR proposal will have negative impacts on individuals' personally held goals (e.g., economic prospects). Some examples of interest-based conflicts include: restoring a riverine system requiring that some farm land be taken out of cultivation; elimination of fertilizer or herbicide application to protect an endangered species; elimination of traditional forest uses (e.g., high-grading tropical forests, fuel-wood gathering) to promote organized commercial forestry. Economic interests are part of the concerns of the farmers in the UK case provided by Convery and Dutson (2012).

7.4.3.2 Values-Based Conflict

Values-based conflict emerges when there are different views of what comprises a 'good' landscape. The conflict may be between science and local knowledge, or between different types of technical disciplines. It often develops from different value-weighting (e.g., private vs. public interest, current vs. future generations). Traditional ways of life can be extremely important to people, but a scientific forest restoration policy grounded in ecology may not incorporate them. This type of conflict is certainly an ingredient in the case-example from the UK (Convery and Dutson 2012).

Examples of values-based conflict include reintroduction of wildfire as a disturbance element (scientists might think it is good, even inevitable, and local residents might think it foolishly dangerous); reintroduction of pest/predator species (prairie-dogs and wolves in the western United States); proposed conversion of existing plantations of non-native species back to native forests. Efforts at reforestation in Ireland generated considerable local conflict largely over values about landscape appearance (Carroll 2007).

7.4.3.3 Authority/Jurisdiction Conflict

Authority/jurisdiction conflict is salient when there are unresolved questions about which agency, level of government, or civic sphere (public v. private) has the appropriate authority to make the decisions. Often there are different management structures in place to produce different outputs, and their mandates and procedures may be incompatible to a greater or lesser extent. One common example is a situation in which one government agency might manage the forest, another controls wildlife, yet another regulates the waterways. A second example of jurisdictional conflict arises when there are multiple levels of government (municipalities, states, regions, nations, or international treaties) all of which have enacted standards they expect to be met. A third example is when different branches of government (judicial, legislative, and executive) are simultaneously involved. Finally, there can also be the broader questions about whether the decision rightfully belongs in the public sphere at all, or whether private entities should be in control. Oftentimes, some or all of these four different types of jurisdictional conflict may be in play at the same time; certainly the American forest policy situation in recent years has been confounded by them all.

7.4.3.4 Legitimacy Conflict

Legitimacy conflict arises when the citizenry does not agree with the governance approach of the government - due to concerns based in ethics, rules, legislation, agreements, contracts or social norms. This is often a blend of values-based and authority-based conflict. Examples include when the government is viewed as corrupt and out of touch; when government is believed to be serving narrow private interests over the broad public interest; or when policy emerging from the government does not have the support of the citizenry.

7.4.3.5 Cultural/Historical Conflict

Cultural/historical conflict can be considered as merely the latest chapter in a problematic on-going relationship. Tribal or ethnic differences would be typical elements of this conflict type. The decision about the forest could actually be a way for one group to exert dominance over others. The forest may be a proxy issue for the real long-term on-going relationship conflict. To illustrate, two villages of different ethnic origins compete over various issues and have been engaged in a long-standing power-struggle; a foreign aid FLR-project around the villages turns into a new arena for the power-struggle; newcomers and old-time (over generations) residents in a rural community disagree on the future management principles for forests surrounding the community – should they be protected as non-intervention wilderness or should they be utilized to sustain the local supply of firewood and other material goods?

These examples illustrate the range of possible FLR-conflicts. Despite their differences, they have certain characteristics in common; they typically involve many individual actors, stakeholders and shareholders, individuals and organizations, private enterprises, governmental agencies, non-governmental organizations, organized as well as non-organized individuals. All these individual persons (each with a specific personality, background, attitudes, beliefs, values and emotions) will influence the process, which must be taken into account.

Forest managers and other relevant stakeholders could use a tech-reg approach in these sample situations, but all the examples call for improving communication interaction among the parties listening to and learning from one another as they move on together towards joint decisions and actions. Such a process involves facilitated communication and mutual gains negotiation activity among the parties. Discourse-based methods emphasize constructive communication processes such as dialogue and deliberation, mutual learning, informal problem-solving, collaborative negotiation, and shared decision-making.

7.4.4 Discourse-Based Principles and Approaches: Methods for FLR

In this section, we take a closer look at various discourse-based approaches and how they work in practice. We have chosen the term discourse-based approach carefully. Doing so differentiates pluralistic public participation in forest restoration work from more technical and regulatory approaches. The former emphasizes shared decision space, meaningful communication interaction, the integration of technical and traditional knowledge, active learning, collaborative negotiation, and systems thinking. In contrast, "tech-reg" work exhibits command and control in decision-making, participation, and communication; reliance on technical experts, adversarial relationships, and linear thinking (Walker and Daniels 2005; Daniels and Cheng 2004). Discourse-based approaches seek to incorporate the best features of a technical method (such as information from technical experts) within an accessible and transparent decision process and takes measures to make sure that the dialogue and learning goes multiple ways – is mutual. It is crucial that the technical experts listen to the concerns and ideas of the citizens and stay open to learn about other views and values and are able to meaningfully revise their thinking and planning.

7.4.4.1 Discourse-Based Principles

When mitigating a FLR conflict by a discourse-based approach, dialogue takes a central role. Dialogue is a means to unravel and understand the different ways of valuing and understanding the landscape. The stakeholders' cognitive frames regarding the situation are critical in this context and the approaches for dealing with seemingly intractable conflict in Lewicki et al. (2002) and Coleman (2000) are particularly relevant. Approaches to stimulate dialogue in order to increase the mutual understanding of others' perspectives and interests could include scientist-citizen dialogues and the collaborative learning approach (Daniels and Walker 2001). The principles of interest-based (or principled) negotiation techniques introduced by Fisher and Ury (1992) can be very helpful because the underlying interests are probably not as conflicted as are the parties' rhetorical positions.

We believe that discourse-based approaches, while quite varied in form and practice, address a set of core principles that apply well to FLR efforts. A selection of those principles (or deeds) comprises a pneumonic: FAAITH – Fair, Accountable, Accessible, Inclusive, Transparent, and Honest – together forming a robust set of guiding principles for the design and facilitation of discourse-based processes (Walker et al. 2007):

Fair: The design and implementation of a discourse-based FLR process embodies fairness; offering participants an impartial and egalitarian way to get involved in FLR activities.

Accountable: A pluralistic FLR effort demonstrates to stakeholders who are the parties accountable and responsible for the project. Accountability is essential for building trust among the FLR parties.

Accessible: For a FLR project to be collaborative and participatory, it must provide access to relevant and interested stakeholders. Stakeholder voices should be attainable without difficulty.

Inclusive: FLR work should strive to include otherwise marginalized communities, including indigenous voices. Doing so should provide all parties with standing (Senecah 2004).

Transparent: Discourse-based FLR approaches exhibit procedures that stakeholders can understand, critique, support, and improve. Information and its sources are provided without qualification. Sound, valid, and reliable information provide part of a necessary foundation for FLR work. Relevant technical and traditional knowledge should be featured in any FLR project.

Honest: In a discourse-based FLR project, agencies, businesses, and other stakeholders are honest with one another and their community about their work. Honesty is the cousin of transparency; FLR projects are highly visible and conducted without hidden agendas.

Adhering to these core principles should increase chances for success, while violation of one or more principles may cause problems and/or conflict escalation. As the emphasis on pluralism in environmental policy and natural resource management has increased in recent years, so have methodologies that qualify as discourse-based approaches. In the following section, we present a selection of discourse-based methods we find particularly relevant in a FLR-context.

7.4.5 Discourse-Based Methods

Many frameworks exist for conducting discourse-based FLR agendas. We highlight a small set of well documented and tested methods that all seek to encourage dialogue, improve discourse, and assist deliberation. All of these methods are capable of dealing with the range of conflict types discussed above.

7.4.5.1 Collaborative Learning

Collaborative learning (CL) is a framework for public policy conflict management and decision making. Its specific applications to date have been in the natural resource arena including forest planning and ecosystem management. Collaborative learning (Daniels and Walker 2001) is a hybrid of soft systems methodology (SSM), alternative dispute resolution (ADR), adult and experiential learning theory, and participatory communication. It encourages systems thinking, joint learning, open communication, and focuses on appropriate change. It emphasizes continual, significant improvements in the management situation, through assessment, training, project implementation, and monitoring.

7.4.5.2 Mediated Modeling

Mediated modeling is an approach that uses models and simulations guided by facilitators. A mediated modeling process employs systems dynamics thinking and promotes "the integration of expert information and stakeholder participation in a dynamic framework to address complex problems" (van den Belt 2004, p. 15). A mediated modeling process: (1) increases the level of shared understanding among stakeholders; (2) builds consensus about the structure of a complex issue and its dynamic nature; (3) provides a strategic and systematic foundation for investigating management alternatives; and (4) serves as a mechanism for sharing and disseminating insights stakeholders have generated (van den Belt 2004, p. 17).

7.4.5.3 Constructive Confrontation (CC)

Designed to address intractable conflict situations, constructive confrontation employs a medical metaphor. This method "follows a medical model," Burgess and Burgess report, "in which destructive conflict processes are likened to diseases–pathological processes that adversely affect people, organizations, and societies as a whole" (1996, p. 307). As in medicine, Burgess and Burgess explain, CC utilizes an incremental approach. "Constructive confrontation alerts parties and intermediaries to pitfalls to be avoided, pathologies to be corrected, and opportunities to be exploited," without specifying a specific agenda or end result (1996, p. 308). CC consists of three general steps: diagnosis, treatment, and monitoring. Diagnosis starts with the development of a conflict map; treatment follows conflict diagnosis. According to Burgess and Burgess (1996), treatment involves "the identification and implementation of realistic, incremental steps for reducing as many of the overlay problems as possible" (p. 309). The last step, monitoring, evaluates the intervention and guides adjustments "as the conflict continues and changes over time" (p. 309).

7.4.5.4 Structured Decision-Making (SDM)

Employed by United States natural resource management agencies, SDM methods follow a linear decision-making path. It involves the following steps: (1) identify management objectives; (2) develop management alternatives; (3) generate models of potential outcomes; (4) populate models with appropriate scientific data and information, (5) test the models' credibility; and (6) monitor the program to assess its effectiveness (Kimball 2007). As a discourse-based method, SDM in natural resource management situations (such as forest landscape restoration) displays a number of attributes. It provides a framework to integrate diverse perspectives, it relies on scientific data while increasing internal and external stakeholder involvement, it is policy relevant and requires transparency, it addresses uncertainty as it responds to complex problems, and it improves planning and the efficient use of human resources (Kimball 2007).

7.4.5.5 Search Conferencing and Participatory Design (SC-PD)

Developed by Diemer and Alvarez (1995) as a combination of two techniques, SC-PD is presented as an adaptive social process that can respond to value conflicts in constructive ways. Neither search conferencing, or participatory design as techniques are new, but Diemer and Alvarez see their combination as a public participation innovation compatible with ecosystem management and sustainable forestry. Search conferencing is designed to generate a "planning community" in three phases. First, SC participants brainstorm significant events, both globally and locally. Second, participants examine their particular system (e.g., organization, community, issue) and generate a "communal history." They critique their "system" and determine its most desirable future. Third, parties integrate the information compiled during phases 1 and 2. They identify "desirable and achievable futures" and detailed action plans for reaching their goals (p. 13). After the search conference has produced a strategic plan, community members work together in a PD workshop to learn about organizational design principles necessary to organize for the long-term. Diemer and Alvarez emphasize that SC and PD need to occur consecutively; the search conference provides "adaptive relations between system and environment" (p.11) and the PD workshop contributes the organizational knowledge needed to sustain the adaptive strategic plan.

These discourse-based methods are illustrative of the innovative approaches being applied in natural resource management and environmental policy decision situations.

For example, Fischer (2000) discusses the "consensus conference" and "participatory resource mapping," while Weber (2003) features "grassroots ecosystem management." FLR projects can work through challenging conflict situations and implement sound policies by involving stakeholders via an appropriate discourse-based approach.

7.5 Summary, Conclusions and Policy Implications

Restoration projects are fundamentally about change. Broadly speaking, the typical challenge of FLR is to change an undesired ecological situation (and its social dimensions) for the better (by some measure). Often the reform implies a change in current practices that produced the undesired situation. Therefore most restoration projects imply changing social systems, peoples' behaviours, and current practices. Generally, people resist change – for example, due to increased uncertainty, lack of knowledge, potential loss of privileges, increased immediate burdens or many other possible good reasons. Occasionally opposition to change occurs simply because people are comfortable and happy with the current conditions. In short, the status quo always has a constituency.

Many foresters in research and practice have reflected over the recent changes in forestry and realized that nothing is certain except for change. As a forester it is not enough to know everything about trees, you have to know about people if you want your ideas manifested in the real world. There is a growing need for knowledge about how to handle the emerging social and human dimensions of forestry. This is a significant challenge for foresters as well as forestry research and practice. As a first step the forest management community can draw on theory and experience from other fields (e.g., conflict management, watershed management, negotiation, policy) to fill in some of this gap.

Given the complexity of landscapes and the controversies surrounding their management, FLR progress will benefit from a coalition of partners. It is a rare case where a single landowner has both complete control over a landscape and also the means to accomplish meaningful restoration. Large-scale restoration is more often going to require developing a cast-of-thousands approach to mobilize resources and build a broad-based constituency for a new vision for that landscape. That process cannot be based on ecological science alone. It must be as thoughtful and insightful in its treatment of social dynamics as it is in stand dynamics. It must be as cognizant of social and attitudinal diversity as it is in species diversity. With that in mind, three summary points encapsulate much of our message:

1. *FLR conflicts are often deeply rooted in existing social systems:* Conflict management is about transforming analysis and understanding of social systems into appropriate and effective actions – to achieve desired outcomes and improve relations through fair procedures. Often conflicts are deeply rooted in very old systems of social interactions that have to be addressed in order to achieve desired long-lasting change (Lederach 1997; Pruitt and Kim 2004). Masters of conflict

management and process design must be equally good at (a) reading the cultural-institutional context, (b) understanding people, and (c) finding ways to create the right environment of power distribution and incentives. In FLR work all of this occurs in concert with effective ecological analysis and practice.

2. *Many people have mixed feelings and some resistance to change – for good reasons:* In any case it is worth reflecting on the fact that the forces and causes that lead to the starting-point of a FLR process (e.g. degraded forest) most likely still will be active if nothing systemically is changed. Accordingly, a successful and long-lasting solution (or improvement) of the situation probably depends on the change of fundamental features and or dynamics of the current social and ecological systems. Change usually means opportunity for some and threat for others – excitement and hope mixed with fear for the future, uncertainty, sentimentality are all normal reactions to change, together with emotions, alertness and maybe some degree of intuitive resistance and caution. Such reactions and behaviours will not always play out in a logical, rational or predictable manner. Often the mix of real interests at stake, perceived differences and emotions can lead to perceived conflict by the parties. Conflict escalation can easily be induced by the prospects of change (e.g., related to a FLR project) and the derived reactions and behaviours.
3. *FLR requires broad and holistic working knowledge – and condensed extraction of theory and literature from many fields:* To change systems and design effective and constructive processes of FLR will require a solid understanding of the ecosystems in question as well as an equally solid understanding of the human and social systems involved. The processes designed should work ecologically as well as socially – which is a huge task that requires intelligent combination of knowledge and skills from indeed very different bodies of theory and practice. We believe that practitioners and researchers of FLR can learn and benefit from a rich base of scientific literature about negotiation, conflict management, environmental conflict, social conflict, mediation, facilitation, etc. The challenge is that the literature and theory we want to bring into play is so vast, deep, and rich that it is difficult to comprehend, condense, and communicate in a short and digestible format. This FLR conflict management chapter can only be regarded as an initial step in that process.

To conclude, as forest managers and stakeholders increasingly recognize the human dimensions of forest landscape restoration (and similarly, REDD projects), understanding relevant discourses and discourse-based conflict management and decision making methods seem paramount. Stakeholders want a voice in the FLR projects that affect them; tech-reg strategies may be sufficient to address ecological needs but discourse-based approaches engage both ecological and human needs. As the examples featured in this and other chapters in this volume reveal, understanding the conventional forest science of FLR is necessary but not sufficient; socio-cultural understanding and discourse-based conflict management applications are essential foundations for enacting the best practices of forest landscape restoration.

Literature Cited

Brown K (2005) Addressing trade-offs in forest landscape restoration. In: Mansourian S, Vallauri D, Dudley N (eds) Forest restoration in landscapes: beyond planting trees. Springer, New York, pp 59–62

Burbridge PR (2012) The role of forest landscape restoration in supporting a transition towards more sustainable coastal development. Chapter 13 in: Stanturf JA, Lamb D, Madsen P (eds) Forest landscape restoration: integrating natural and social ciences. Springer, New York

Burgess H, Burgess G (1996) Constructive confrontation: a transformative approach to intractable conflicts. Mediation Q 13:305–322

Burr V (2003) Social constructionism, 2nd edn. Routledge, London

Bush RAB, Folger JP (2005) The promise of mediation. The transformative approach to conflict. Revised edition. Jossey-Bass, San Francisco

Carpenter SL, Kennedy WJD (2001) Managing public disputes. A practical guide for government, business, citizens' groups. Jossey-Bass, San Francisco

Carroll MS (2007) Afforestation in County Kerry, Ireland: transformations of landscapes and lifeways in a rapidly shifting rural economy. Paper delivered at the 13th international symposium on Society and Resource Management, Park City, UT, 17 June 2007

Chokkalingam U, Sabogal C, Almeida E et al (2005) Local participation, livelihood needs, and institutional arrangements: three keys to sustainable rehabilitation of degraded tropical forest lands? In: Mansourian S, Vallauri D, Dudley N (eds) Forest restoration in landscapes: beyond planting trees. Springer, New York, pp 405–414

Christensen M, Emborg J (1996) Biodiversity in natural versus managed forest in Denmark. For Ecol Manag 85:47–51

Coleman PT (2000) Intractable conflict. In: Deutsch M, Coleman PT (eds) The handbook of conflict resolution. Theory and practice. Jossey-Bass, San Francisco, pp 428–450

Conner WH, Kraus KW, Shaffer GP (2012) Restoration of freshwater cypress-tupelo wetlands in the southeastern US following severe hurricanes. Chapter 16 in: Stanturf JA, Madsen P, Lamb D (eds) A goal-oriented approach to forest landscape restoration. Springer, New York

Convery I, Dutson T (2013) Wild Ennerdale: a cultural landscape. Chapter 12 in: Stanturf JA, Lamb D, Madsen P (eds) Forest landscape restoration: integrating natural and social sciences. Springer, New York

Daniels SE, Cheng AS (2004) Collaborative resource management: discourse-based approaches and evolution of TechnoReg. In: Manfredo MJ, Vaske JV, Bruyere BL et al (eds) Society and natural resources: a summary of knowledge. Modern Litho, Jefferson, pp 127–136

Daniels SE, Walker GB (2001) Working through environmental conflict. The collaborative learning approach. Preager, London

Deutsch M, Coleman PT (eds) (2000) The handbook of conflict resolution. Theory and practice. Jossey-Bass, San Francisco

Diemer JA, Alvarez RC (1995) Sustainable community–sustainable forestry: a participatory model. J For 93(11):10–14

Dryzek J (2005) The politics of the earth: environmental discourses, 2nd edn. Oxford University Press, Oxford

Dudley N (2005a) Impact of forest loss and degradation on biodiversity. In: Mansourian S, Vallauri D, Dudley N (eds) Forest restoration in landscapes: beyond planting trees. Springer, New York, pp 17–21

Dudley N (2005b) What has WWF learned about restoration at an ecoregional scale? In: Mansourian S, Vallauri D, Dudley N (eds) Forest restoration in landscapes: beyond planting trees. Springer, New York, pp 401–404

Emborg J (2007) Conflict management and forest landscape restoration. In: Proceedings from IUFRO conference on forest landscape restoration, Seoul, South Korea, May 2007, pp 26–31

Fischer F (2000) Citizens, experts, and the environment: the politics of local knowledge. Duke University Press, Durham

Fischer F (2003) Reframing public policy: discursive politics and deliberative policy. Oxford University Press, Oxford

Fisher R, Ury W (1992) Getting to yes. Negotiating an agreement without giving in. Random House, London

Folger JP, Poole MS, Stutman RK (1997) Working through conflict, 3rd edn. Longman, New York

Foucault M (1972) The archaeology of knowledge and the discourse on language. Pantheon Books, New York

Habermas J (1984) Vorstudien und Ergänzungen zur Theorie des kommunikative Handelns. Suhrkamp Verlag, Frankfurt

Han X, Oliver CD, Ge J, Guo Q, Kou X (2012) Managing forest stand structures to enhance conservation of the Amur Tiger (*Panthera tigris altaica*). Chapter 5 in: Stanturf JA, Madsen P, Lamb D (eds) A goal-oriented approach to forest landscape restoration. Springer, New York

Hendriks RJJ, Boot RGA, de Haas W, Savenije HJF (2012) Forest landscape restoration in the Netherlands. Policy aspects and knowledge management. Chapter 2 in: Stanturf JA, Madsen P, Lamb D (eds) A goal-oriented approach to forest landscape restoration. Springer, New York

Ingram H, Schneider A, deLeon P (2007) Social construction and policy design. In: Sabatier P (ed) Theories of the policy process. Westview Press, Boulder, pp 93–126

IUFRO (2007) Proceedings from IUFRO conference on forest landscape restoration, Seoul, South Korea, May 2007

Jones S, Dudley N (2005) Negotiations and conflict management. In: Mansourian S, Vallauri D, Dudley N (eds) Forest restoration in landscapes: beyond planting trees. Springer, New York

Kimball S (2007) Structured decision processes. Plenary presentation at the Department of the Interior SES/SL (Senior Executive Service and Senior Leadership) Conference, 24 April 2007, Fairfax VA

Lederach P (1997) Building peace. Sustainable reconciliation in divided societies. US Institute of Peace Press, Washington, DC

Leskinen LA (2004) Purposes and challenges of public participation in regional and local forestry in Finland. For Policy Econ 6:605–618

Lewicki R, Gray B, Elliot M (2002) Making sense of intractable conflict: cases and concepts. Island Press, Washington, DC

Löf M, Brunet J, Hickler T, Birkedal M, Jensen A (2012) Restoring broadleaved forests in southern Sweden as climate changes, chapter 14. In: Stanturf JA, Madsen P, Lamb D (eds) A goal-oriented approach to forest landscape restoration. Springer, New York

Maginnis S, Rietbergen-McCracken J, Jackson W (2005) Introduction. In: ITTO: Restoring forest landscapes. An introduction to the art and science of forest landscape restoration. ITTO Technical series no. 23. ITTO, Yokohama, Japan, pp 11–14

Mansourian S, Aldrich M, Dudley N (2005a) A way forward: working together toward a vision for restored forest landscapes. In: Mansourian S, Vallauri D, Dudley N (eds) Forest restoration in landscapes: beyond planting trees. Springer, New York, pp 415–423

Mansourian S, Vallauri D, Dudley N (eds) (2005b) Forest restoration in landscapes: beyond planting trees. Springer, New York

MCPFE (1998) Annex 1 of the resolution L2 of the Pan-European criteria and indicators for sustainable forest management, from the third ministerial conference on the protection of forests in Europe, 2–4 June 1998, Lisbon, Portugal

Parviainen J, Bücking W, Vandekerkhove K et al (2000) Strict forest reserves in Europe: efforts to enhance biodiversity and research on forests left for free development in Europe (EU-COST-Action E4). Forestry 73(2):107–118

Pruitt DG, Kim SH (2004) Social conflict: escalation, stalemate, and settlement. McGraw-Hill Higher Education, New York

Sayer J (2005) Goals and targets of forest landscape restoration. In: Mansourian S, Vallauri D, Dudley N (eds) Forest restoration in landscapes: beyond planting trees. Springer, New York, pp 101–108

Senecah S (2004) The trinity of voice: the role of practical theory in planning and evaluating the effectiveness of environmental participatory processes. In: Depoe SP, Delicath JW, Elsenbeer

MA (eds) Communication and public participation in environmental decision making. State University of New York Press, Albany, pp 13–33

Siregar UJ, Siregar S, Budi W, Hero Y, Suharjito D, Hardjanto (2012) Incorporating social and natural science in the restoration of an Indonesian conservation forest: a case study from Jambi, chapter 3. In: Stanturf JA, Madsen P, Lamb D (eds) A goal-oriented approach to forest landscape restoration. Springer, New York

UNECE and FAO (2000) Forest resources of Europe, CIS, North America, Australia, Japan and New Zealand. Regional economic commissions for Europe and the food and agriculture organization, Geneva and Rome

van den Belt M (2004) Mediated modeling: a systems dynamics approach to environmental consensus building. Island Press, Washington, DC

Walker GB, Daniels SE (2005) Assessing the promise and potential for collaboration: the progress triangle framework. In: Walker GB, Kinsella WJ (eds) Finding our way(s) in environmental communication: proceedings of the seventh biennial conference on communication and the environment. Oregon State University Department of Speech Communication, Corvallis, pp 188–201

Walker GB, Senecah SL, Daniels SE (2007) Reflections from the road: new and improved concepts, tools, and lessons for community-based collaboration. In: Volkening LS et al (eds) Proceedings of the 8th biennial conference on communication and the environment. University of Georgia Department of Speech Communication, Athens, pp 213–234

WCED (1987) Our common future (The Brundtland Report). United Nations, World Commission on Environment and Development. Oxford University Press, Oxford

Weber EP (2003) Bringing society back. In: Grassroots ecosystem management, accountability, and sustainable communities. MIT Press, Cambridge, MA

Wehr P (1979) Conflict regulation. Westview, Boulder

Xi W, Waldron JD, Cairns DM, Lafon CW, Birt AG, Tchakerian MD, Klepzig KD, Coulson RN (2012) Restoration of southern pine forests after the southern pine beetle. Chapter 12 in: Stanturf JA, Madsen P, Lamb D (eds) A goal-oriented approach to forest landscape restoration. Springer, New York

Chapter 8
Alternative Approaches to Urban Natural Areas Restoration: Integrating Social and Ecological Goals

Paul H. Gobster

8.1 Introduction

Ecological restoration incorporates land management principles and activities aimed at returning a damaged or degraded ecosystem back to a key historic trajectory in order to achieve goals of ecosystem health, integrity, and sustainability (Society for Ecological Restoration 2004). Please consult Chap. 1 (Lamb et al. this volume) for additional perspectives.

In the United States, many restorationists look to ecological conditions present before the time of European settlement as the key historic landscape they are seeking to restore, and employ an approach to restoration management that has been called "classical ecological restoration" (Callicott 2002). Nine attributes of successfully restored ecosystems identified in the Society for Ecological Restoration International's *Primer* (2004) conform closely to this classical management approach, and have been summarized by Ruiz-Jaen and Aide (2005) as falling along three major ecological dimensions: (1) diversity measured in terms such as the richness and abundance of native plants and other species; (2) structure measured in terms such as the age, distribution, and density of vegetation; and (3) processes measured in terms such as the presence of natural disturbance regimes such as fire. While these dimensions and their measures can help guide restoration efforts on a trajectory toward ecosystem health, integrity, and sustainability, the ultimate success of classical ecological restoration is judged by how well the measures fall within an historic range of variability found in closely matched reference sites (Ruiz-Jaen and Aide 2005). Thus in a broader sense, the overarching goal of the classical approach is authenticity or fidelity in how a restored site looks and functions like one before European settlers arrived, minimally influenced by contemporary human impacts and values (Higgs 2003).

P.H. Gobster (✉)
U.S. Forest Service, Northern Research Station, Chicago, IL, USA
e-mail: pgobster@fs.fed.us

J. Stanturf et al. (eds.), *Forest Landscape Restoration: Integrating Natural and Social Sciences*, World Forests 15, DOI 10.1007/978-94-007-5326-6_8,

While this classical approach to management has led to many successful restoration projects, ecologists and other environmental professionals are increasingly questioning its efficacy in dealing with severely disturbed landscapes (e.g., Martínez and López-Barerra 2008) and unpredictable trajectories (e.g., Choi et al. 2008). These concerns might be especially apparent in urban areas, where landscape fragmentation, soil and hydrologic alterations, and microclimatic patterns introduce novel and often substantially different effects than what may have occurred historically. Perhaps even more significant are concerns raised about people's uses, perceptions, and values of the landscape and its restoration, which may pose formidable challenges for managing urban natural areas in socially acceptable ways (Gobster 2010; Ingram 2008).

In this paper I examine these issues within the context of urban ecological restoration, with an emphasis on incorporating social goals alongside ecological ones in managing natural areas. While the *Primer's* nine attributes of restored ecosystems strongly imply the classical approach as a dominant model, its mention of additional goals suggests that other approaches could be considered as conditions warrant:

> For example, one of the goals of restoration might be to provide specified natural goods and services for social benefit in a sustainable manner. In this respect, the restored ecosystem serves as natural capital for the accrual of these goods and services. Another goal might be for the restored ecosystem to provide habitat for rare species or to harbor a diverse genepool for selected species. Other possible goals of restoration might include the provision of aesthetic amenities or the accommodation of activities of social consequence, such as the strengthening of a community through the participation of individuals in a restoration project (Society for Ecological Restoration 2004).

In light of these additional goals, it is important to examine how restoration managers and stakeholders negotiate the implementation of restoration activities and practices for different urban natural area restoration sites and programs. In what follows, I describe restoration programs in two major North American cities and suggest that there may be a number of alternative approaches to restoration that could be applied to achieve social and ecological goals. From this work I outline a framework for how appropriate approaches to natural areas restoration in urban contexts might be identified for a given site or a system of sites. This framework, adapted from the USDA Forest Service's (1982) Recreation Opportunity Spectrum, could provide restoration managers with a systematic way for matching ecological goals and management practices with people's broader desires and expectations for urban nature.

8.2 Case Studies: Key Issues and Constraints

In order to better understand the diverse goals that underlie the restoration management of urban natural areas, I examined restoration activity in Chicago, Illinois and San Francisco, California to identify the key issues faced by practitioners and public stakeholder groups when restoration programs are implemented within metropolitan areas (for details see Gobster 2000, 2001, 2007a, b). Both locations have significant amounts of protected open space within their metropolitan boundaries: there are more than 279,000 ha of open space in the 9-county "Chicago Wilderness" planning region (9% of land area; 12 ha/1,000 residents) and more than 400,000 ha (24.8% of land area, 65 ha/1,000 residents) in the 9-county Bay Area Open Space Council region (Gobster 2007a). But while extensive restorative management is happening throughout these two metropolitan areas, I focused my case studies on sites within each city and its host county because of the diverse range of social and ecological issues that are being dealt with. In Chicago, there are 49 restoration sites in City of Chicago parks and another 70 sites in the Forest Preserve District of Cook County. The sites range in size from a fraction of a hectare to 1,500 ha in size and include prairie, savanna, woodland, and wetland communities. In San Francisco, there are 30 restoration sites in City of San Francisco parks and another 12 sites in Golden Gate National Recreation Area within the County of San Francisco. These sites range in size from less than 1 ha to more than 160 ha in size and include coastal dune, scrub, grassland, wetland, and non-native forest communities (Gobster 2007a).

The fragmented character of these urban natural areas imposes significant restrictions on what ecological conditions *can* be restored through management programs (e.g. Vidra and Shear 2008). For example, a prairie restoration at the scale of even the largest of sites in Chicago or Cook County is unlikely to become home to an American bison (*Bison bison*). Instead, most restorations focus on recovering or reintroducing the key flora of a target community and hope to attract smaller fauna such as butterflies and birds. By the same token, a dune restoration in the city or county of San Francisco cannot be given the freedom to shift across a park road or into a neighbor's backyard. Instead, ecological communities are necessarily fixed in space and any movement of elements in the community must take place within site boundaries. And while prescribed burning may be used to manage the understory of open oak woodlands in Chicago or reduce woody shrub growth in coastal scrub area of San Francisco, setting back succession with an all-consuming fire is not in the urban restorationist's playbook. Thus temporal dynamics are also more or less fixed and give the impression that such communities are stable and climax in character.

Along with these structural constraints there is a host of social and political issues that further define what conditions *should* be restored in urban settings (e.g., Trigger and Head 2010). Demand for open space by a diverse range of user and interest groups not only limits the number and size of restoration projects within a program but also what other uses might take place, how sites are managed, and by whom. In San Francisco, designation of natural areas and concomitant restrictions on off-leash dog access have led to a major conflict between natural area restorationists and dog owners and threatened progress toward adoption of the city's Significant Natural Resources Area Management Plan (San Francisco Recreation and Park Department 2006). Removal of exotic trees from restoration sites, especially Australian blue gum eucalyptus (*Eucalyptus globulus*), has also been a point of conflict in plan adoption, and, along with tight air quality restrictions and strong attitudes against the use of prescribed burning, public sentiment has forced restoration managers to consider alternative ways for managing natural area sites. While restoration in Chicago has also been contentious at times (e.g., Gobster and Hull 2000), volunteer-based restoration has long been a hallmark of the metropolitan region's restoration movement and has been a model emulated in other cities nationally and internationally (Ross 1994). Nonetheless, many of the Chicago Park District's larger restoration efforts have been done under contract with professional firms, with volunteers entering the scene to assist with maintenance only after the restoration design has been implemented. The magnitude and complexity of the transformation is a major reason for this, but desire for professionalism, accountability, warranty on plant materials, and time frame for implementation are also important considerations (Gobster 2007b).

8.3 Alternative Approaches to Urban Natural Areas Restoration

Constraints can often spark creativity, and in the case of natural areas management, practitioners and scholars are beginning to advocate for a broader conception of restoration and document a diversity of restoration approaches that are more in tune with the social and ecological goals they seek to achieve (e.g., Choi 2007; Gross 2003; Low 2002; Rosenzweig 2003a). Based my case studies in Chicago and San Francisco, I have identified the following range of approaches as potential alternatives to the classical approach for restoring urban natural areas in consideration of the various ecological and social constraints and opportunities present at different sites. The approaches are not intended to comprise a mutually exclusive or exhaustive typology of possibilities, but rather to illustrate how social and ecological goals might be addressed at particular sites and, at larger scopes of concern, balanced across a system of sites.

8.3.1 Classical Approach, var. 'boutique'

The steep topography of San Francisco and broad floodplains of Chicago have been good deterrents to prior development of many of the now-designated natural areas in these two cities, and while most of these sites have been damaged by overgrazing or other past alterations to vegetation cover, soil, or hydrology, some places still retain significant remnant populations of indigenous flora (e.g., Chicago Region Biodiversity Council 1999; San Francisco Recreation and Park Department 2006). Restoration of these sites conforms most closely to the classical approach to ecological restoration, where native plant diversity is maintained and enhanced through invasive species control and other management practices. However, restoration activities on small sites are sometimes carried out in unconventional ways to deal with environmental and social constraints.

Fig. 8.1 At small urban restoration sites like Brooks Park in San Francisco, volunteers rely on "boutique" methods like hand weeding to maintain sites when other practices such as prescribed burning or the use of herbicides are risky or contentious (Credit: Paul Gobster)

For example, Natural Areas Program gardeners in San Francisco, aided by a substantial force of volunteers, often resort to "boutique," labor intensive methods (Hull et al. 2004) on many of their small sites that would be impractical in larger restorations. For invasive plant control, herbicide use is generally frowned upon by the public and prescribed fire is highly controversial as many sites are in close proximity to residential areas. Consequently, nearly all plant removal is done manually, pulling weeds by hand and using simple hand tools to remove larger specimens. These techniques, along with hand planting and direct seeding of native plants, constitute the bulk of restoration management for sites such as Brooks Park, a 1.4 ha rocky hilltop grassland natural area (Fig. 8.1). For larger sites, such as the 24 ha Glen Canyon Park natural area, managers have experimented with using goats to graze back unwanted vegetation, and have proposed using machinery such as "weed whackers" and power mowers (San Francisco Recreation and Parks Department 2006). Recent research suggests that while the classical approach's prescription for reinstating natural disturbance processes such as fire may be preferable, similar results might be achieved using these alternative methods and thus may be the most feasible in high risk situations (MacDougall and Turkington 2007).

The classical approach maintains that the success of a restoration lies in part in its ability to sustain itself and follow a historical trajectory without substantial human intervention. Such a criterion, however, is simply not realistic for sites like these (Hobbs 2007). Instead, some suggest that continued human intervention *is* the key ingredient to sustainability (Jordan 2003), and in dense urban areas like

San Francisco and Chicago restoration success lies in the ability of program managers to sustain a robust corps of volunteers to steward the sites in perpetuity (e.g., Ross 1994). Boutique restoration may sound overly labor intensive, yet when viewed as a leisure activity on par with gardening it meets desired social and ecological goals for many people (Jordan 2000). And in an urban setting, even non-participants understand the need for routine landscape maintenance to sustain the beauty and function of their yards and parks, and thus extending this level of management to urban restoration is perhaps not so unfamiliar.

8.3.2 Habitat and Sensitive Species Approaches

Many early land protection efforts were aimed not so much at preserving the diversity, structure, and function of native ecosystems and processes described by the classical approach of restoration as they were at setting aside and managing habitat such as wetlands and woodlands for timber and game production (e.g., Hall 2005; Jordan and Lubick 2011). Society's interests in wildlife and plant species diversity have since broadened considerably, but habitat conservation continues to be a dominant paradigm of natural land management. Reconceptualizing urban open space as habitat has also helped to underscore the key role that restoration can play in providing essential habitat patches and corridors to ensure the survival of species in an increasingly human dominated landscape (e.g., Lundholm and Richardson 2010).

In Chicago for example, local ornithologists and recreational birders have over the last decade been persuasive advocates for the need to manage parkland to provide bird habitat along the city's 42 km Lake Michigan shoreline. The lake is an important branch of the Mississippi Flyway and more than 300 resident and migratory bird species have been documented as using the city's shoreline across the seasons. Research has shown that in urban areas, migratory birds need habitat patches at regular intervals along their route where they can safely rest and refuel (e.g., Pennington et al. 2008; Seewagen et al. 2010). The City of Chicago has responded to this new awareness by establishing a number of bird sanctuaries along its shore and is managing the vegetation and beaches to provide essential food and cover (City of Chicago 2006). Because most of the lakefront parks were built on fill to extend land holdings into what was originally open water, the classical approach to ecosystem restoration is already a considerable conceptual stretch. This ambiguity has given natural areas managers greater license in what they plant, and they use a range of natives along with native and introduced cultivars that not only provide food and cover but are adapted to the often harsh site conditions (Gobster 2001).

The oceanfront beaches and bluffs in San Francisco are also important habitat areas for migratory birds, but the city's unique geographic, climatic, and geological characteristics make some of its natural area sites additionally critical to the protection of a number of endemic species. These include flora such as the Presidio manzanita (*Arctostaphylos hookeri ssp. ravenii*) and Marin dwarf flax (*Hesperolinon congestum*) and fauna such as the Mission blue butterfly (*Plebejus icarioides missionensis*) and

Fig. 8.2 While the presence of sensitive species often requires tight controls over recreational use, some species such as the San Francisco lessingia (*Lessingia germanorum*) at this Lobos Creek restoration site needs periodic disturbance for recovery. Understanding the conditions needed to meet ecological goals may enable managers to broaden allowable uses and allow users more direct interaction and exploration of nature (Credit: Paul Gobster)

San Francisco garter snake (*Thamnophis sirtalis tetrataenia*) (Gobster 2007a). Because of the natural rarity and threatened existence of these species, site restoration is sometimes less focused on providing a classical ecosystem makeover than on providing optimal habitat conditions for the propagation of a sensitive species. The weight these species are given in restorative management invokes a kind of "ecological primacy," which in some cases makes the existence of incompatible exotics such as blue gum eucalyptus and access for uses such as off-leash dog recreation relatively non-negotiable. Incompatibilities do not always happen, however, and in other cases sensitive species might be maintained under novel conditions (e.g., Hobbs et al. 2009). For example, a 5.3 ha dune restoration was created at Lobos Creek in Golden Gate National Recreation Area to increase the dwindling population of the federally endangered San Francisco lessingia (*Lessingia germanorum*), a tiny sunflower. While the current boardwalk design discourages off-trail use of the site (Fig. 8.2), the plant requires periodic disturbance to perpetuate itself (U.S. Fish and Wildlife Service 2003). To rectify this situation, designers have thought about scheduling fun activities like annual "dune dancing" (Terri Thomas and Michael Boland, personal communication, 5 May 2004).

As these examples show, habitat and sensitive species approaches could provide managers with a greater savings than that afforded under a classical restoration in terms of the cost or time devote to management or the maintenance of existing green space uses and functions. In this way, the approach parallels ideas described by Rosenzweig's (2003a, b) "reconciliation ecology," where the outcome of saving species is given priority, opening up a variety of different and sometimes novel ways of achieving that outcome.

8.3.3 Hybrid or "Third-Way" Restorations

In most of the larger parks in San Francisco and Chicago, the landscape has been so thoroughly modified that few vestiges of indigenous nature remain. Yet in their quest to create a human habitat for aesthetic pleasure and recreational use, the original designers of these parks developed naturalistic landscapes that often had considerable ecological value (Grese 1992; Young 2004). Restoration efforts in these parks thus sometimes attempt to integrate two (or more) periods of significance—one focusing on classical ecological restoration and another on restoration of the historic designed landscape. Successful projects of this type respect the goals and intent underlying both ideas of restoration yet can produce a hybrid landscape that is its own unique expression of human and ecological values.

One example of this "third way" restoration approach in Chicago is the Lily Pool (Fig. 8.3), a 2.5 ha naturalistic oasis in Lincoln Park designed in the 1930s by noted Prairie School landscape architect Alfred Caldwell, who used a primarily native plant palette to create a symbolic rendition of the Illinois landscape as it existed prior to European settlement. In the restoration effort, Chicago park historians worked with a diverse team of professional and civic interests to restore the integrity of this historic designed landscape while enhancing native plant diversity, bird habitat, and other ecological functions and accommodating access for disabled users as required under the Americans with Disabilities Act. The restoration received a 2001 historic preservation honor award from the American Society of Landscape Architects and an innovative docent program has been established with the non-profit Lincoln Park Conservancy to interpret the site's unique values and perpetuate Caldwell's vision of the Lily Pool as a "hidden garden of the people of Megalopolis" (Maloney 2001).

A larger scale example of third-way restoration is being realized at the Presidio of San Francisco, a 600 ha former military site now managed by the National Park Service under a new model that aims to protect and restore natural and cultural values while promoting sustainable economic development through adaptive re-use of the site's substantial built infrastructure. Much of the non-native forest cover planted by the US Army in the 1880s over approximately 20% of the naturally tree-less site was slated for removal as part of natural area restoration efforts until critics successfully lobbied to maintain it for the historic reasons why it was originally planted—as a windbreak and symbol of military presence. A revised vegetation management plan seeks to maintain and rehabilitate the structural characteristics of key historic forest stands in four highly visible areas and manage the remaining forest area to increase ecosystem health and biodiversity. Forest management

Fig. 8.3 Hybrid or "third-way" landscapes such as the Lily Pool in Chicago's Lincoln Park blend ecological restoration other site goals, in this case the restoration of a 1930s historic designed landscape by Prairie School landscape architect Alfred Caldwell (Credit: Paul Gobster)

strategies outside of the key historic stands aim to increase the species, spatial, height, and age diversity of trees; encourage natural regeneration; and promote a varied understory and mid-story layer of native grasses, forbs, and shrubs (Presidio Trust 2001). While critical natural and cultural areas are being restored to their original integrity, this third landscape between the two represents a new hybrid of nature and culture.

Green spaces perform many human-oriented uses in urban areas, from historic and recreational park areas to storm water retention basins, power and transportation rights-of-way, cemeteries and institutional grounds, among other functions. While it may not be the objective of the owners of these sites to manage them for ecosystem health, integrity, and sustainability, the hybrid approach has good potential in helping to demonstrate that such ecological goals can be successfully integrated with human goals and uses of the site.

8.3.4 Designer and Accidental Ecosystems

Humans have shaped the land for millennia, and studies of aboriginal subsistence hunting and agricultural economies have shown that land use practices in some cases expanded local and regional species diversity (e.g., Minnis and Elisens 2001). Contemporary land use usually has the opposite effect, though in a few cases human

Fig. 8.4 Designer and accidental ecosystems such as Alcatraz Island in San Francisco's Golden Gate National Recreation Area create entirely new assemblages of species and conditions, in this case habitat for the endangered black crowned night heron (Credit: Paul Gobster)

designs on the land have created novel conditions for valued species to flourish that would have never occurred under "natural" conditions (Britt 2004). By diverting from the template of classical ecological restoration, these "designer ecosystems" create an entirely new approach to nature restoration where habitat creation, endangered species recovery, or other ecological goals are a byproduct of dominant human goals such as recreation or flood prevention (Palmer et al. 2004).

In San Francisco, a famous designer ecosystem is the 9 ha Alcatraz Island (Fig. 8.4), a rocky island 1.5 km off the mainland that for more than a century had been used as a military fortress then high security prison before it was abandoned in the early 1960s. As the atmosphere of quiet isolation returned, seabirds such as Brandt's cormorants (*Phalacrocorax penicillatus*) and pigeon guillemots (*Cepphus columba*) came to re-occupy the site, but the changed conditions of exotic vegetation and foundations of old prison buildings also provided new habitat for rare black-crowned night herons (*Nycticorax nycticorax*) that was absent in the island's original landscape (Hart et al. 1996). While this example might more correctly be termed an accidental ecosystem, National Park Service ecologists who now manage the island as part of Golden Gate National Recreation Area have been keen to acknowledge that these created conditions serve an important ecological function as well as reminding visitors of the historic layers present.

Environmental philosophers such as Katz (2000) and Elliot (1997), who have argued that even the classical approach to ecological restoration is an exercise in human arrogance, would surely balk at the idea of designing partly or wholly artificial ecosystems. Yet in urban settings where human control and cultivation of the landscape have long been the dominant paradigm, the designer approach may make sense for both human and ecological reasons. With radical changes predicted for many areas of the world due to global climate change, the artificial nature of cities may make them ideal laboratories and testing grounds for new ecological assemblages (Fox 2007; Hobbs et al. 2009; Link 2008) and reservoirs for future adapted species through assisted migration (e.g., Minteer and Collins 2010). Human population growth and land use changes are also driving ecologists to search for alternative restoration approaches that can maintain their ecological resilience while accommodating human preferences and impacts (Hitchmough and de la Fleur 2006).

Under such imperatives, the designer ecosystem approach to restoration represents a bold yet serious alternative to the classical model for coming to grips species loss and continued degradation of historical ecosystems. In light of these impending changes, a growing group of ecologists (e.g., Choi et al. 2008) argues that restoration efforts should not be constrained by classical notions of historical authenticity but should look toward future-oriented approaches that will continue to sustain critical ecosystem functions.

8.3.5 Nature Garden Approaches

Contemporary urban garden design is increasingly sympathetic to classical restoration goals such as the use of native plants and other aspects that enhance site sustainability (e.g., Van Sweden 1997). Such goals, however, are often accomplished in highly "unnatural" ways, and while ecological goals may form a rationale for design, the dominant focus is on human enjoyment, learning, and artistic expression.

One such example in Chicago is the 1.2 ha Lurie Garden (Fig. 8.5) in the city's recently built Millennium Park, where designers used plant materials to create a highly symbolic landscape. "Dark" and "light" sections of plantings represent the Chicago region's marshy past and prairie-farmland present landscapes, and are embraced by a hedge of trees symbolic of the northern boreal forest shaped to invoke poet Carl Sandburg's image of Chicago as the "City of Big Shoulders." Native and introduced plants are used in combination to accentuate these themes and provide variety within and across the seasons, and native species such as purple coneflower (*Echinacea purpurea*) are juxtaposed with their cultivars of different colors and heights to reinforce the idea of the garden as a nexus of nature and culture (Amidon 2005).

While the Lurie Garden may be an uncommon example, designed and vernacular nature gardens can provide key ways of bringing the functional, educational, and symbolic values of restoration into small urban spaces. One important variant of the

Fig. 8.5 The Lurie Garden in Chicago's Millennium Park uses native plants and their horticultural variants to create a highly stylized nature garden, and exemplifies how alternative approaches can help integrate restoration goals into highly formal urban settings (Credit: Mark Tomaras)

nature garden approach can be seen in school and community gardens, where participatory involvement, skills development, and community empowerment are often key goals (e.g., Feldman and Westphal 1999). While many school and community gardens are focused on food production, native plants are sometimes used separately or in conjunction with vegetables and cultivated flowers to build small scale habitats for butterflies or other insects, or to grow natives for eventual transplanting into larger scale restorations. The connection between the two types of gardens may help in linking the ecological goals of restoration with broader social and economic goals, and could be a particularly effective way of introducing restoration to diverse urban audiences (Irvine et al. 1999; Palamar 2010).

8.3.6 Unmanaged Sites: "Explorable Nature" and "New Wilderness" Approaches

Gardens by definition are special use areas where the rules of engagement can be highly specific as to what is allowed, how, when, and by whom. The fragility of some smaller classical restoration sites often turns them into gardens of sorts, and fencing and boardwalks needed to control use impacts can also limit the degree of interaction that those not actively involved in restoration have with nature (Gobster 2007b). While these sites may have considerable aesthetic and educational value,

Fig. 8.6 Nearby neighbors and dog advocacy groups successfully lobbied for an alternative approach to restoration of the Pine Lake natural area in San Francisco that allowed for greater recreational use than was originally proposed. Adults and children need places where they can actively explore nature, and marginal sites and buffers of more intact sites might provide opportunities for a range of explorable nature activities (Credit: Paul Gobster)

places must also be available that provide for more unstructured, active exploration of nature (Miller 2005).

In San Francisco, Pine Lake natural area is a bowl-shaped 3.4 ha site surrounding a shallow, 0.7 ha lake. The rest of the park adjacent to the natural area contains a children's day camp and a popular off-leash dog play area. Local residents have long incorporated their visits to the park with a walk around the eucalyptus-shaded lakeshore trail, often accompanied by their dogs. This tradition was about to change when an endangered western pond turtle (*Clemmys marmorata*) was sighted during a lake survey. A species recovery plan was developed calling for removal of many of the trees, fencing off the shore to access, killing non-native bullfrogs, closing the day camp during mating season, and outlawing dog access. Protests by neighborhood and dog advocacy groups led to a revised plan that would allow greater access, minimize tree cutting, and relocate any endangered turtles that might be found to a larger lake nearby where a sustainable population could be realized (San Francisco Recreation and Park Department 2006) (Fig. 8.6).

Given the high-use recreation at the site, the example raises important questions about how urban natural areas should be restored to balance social and ecological goals in nature. Children as well as adults need places where they can explore, get muddy, catch insects or amphibians, and in other ways get in close contact with nature in the city. These places and opportunities for "explorable nature" might mean foregoing classical restoration ideals at smaller, recreationally-oriented sites or building opportunities for more active exploration into less ecologically intact buffers or transitional areas.

Some cities in North America and Europe have also seen the spontaneous revegetation of larger abandoned industrial sites, and while such areas offer significant opportunities for restoration, the "new wilderness" that has evolved has unique ecological and social values that also raise questions about how far the classical approach to restoration ought to be applied (Kowarik and Körner 2005). In our efforts to make the most of the open spaces we have in cities, these unclaimed areas are often programmed out of existence, but we have to realize that they, too, are important parts of the ecology and experience of nature in the city (e.g., Foster 2010; Louv 2005; Miller 2005).

8.4 Criteria for Selecting Alternative Approaches: A *Restoration* Opportunity Spectrum?

As the examples above illustrate, there are a variety of alternative approaches for how restoration might be conducted within urban settings to address particular ecological and social goals. Although they are diverse in many characteristics, the sites focused on in Chicago and San Francisco are also quite small in scale and when dealing with larger projects such as industrial and post-industrial sites there may be a fuller range of approaches than is indicated by this limited survey (Westphal et al. 2010). Thus while it may be premature to construct a comprehensive typology or spectrum of alternative approaches, the examples above provide a sufficient basis for outlining some key considerations that might go into building such a framework.

A framework already developed by the USDA Forest Service for managing the recreational use of wildlands provides a useful starting point to help guide this effort. The Recreation Opportunity Spectrum or ROS (USDA Forest Service 1982) uses various physical, social, and managerial criteria to identify which areas within national forests can best provide desired settings and experiences for recreation activities. Physical setting criteria identify the size, remoteness, and naturalness of areas, under the assumption that large, isolated, and undeveloped tracts of land have the best potential for providing wilderness type experiences for users while smaller, developed sites near population centers better serve intensive uses where nature is more of a backdrop for than a focus of the experience. Similarly, social criteria specify the uses and density of users and managerial criteria the degree of control and regimentation placed upon them. Together, these three sets of criteria are used by managers to delineate recreation opportunities within a spectrum of settings from primitive to developed.

While originally intended for wildland applications, others have adapted the ROS to more urban recreation situations (More et al. 2003; Bell 2008). Because the system attempts to match people's desired uses and experiences in natural settings with the inherent capability of sites to provide them or be managed to minimize conflicts and inconsistencies, the basic ideas of the system also have applicability

Table 8.1 Framework of ecological, social, and managerial criteria for selecting approaches for urban natural areas restoration

Ecological	Social	Managerial
Natural values	Use values	Mission values and implementation
Intactness	Recreational	Protection vs. use balance
Biophysical conditions	Sense of place	Education/research
Functionality	Traditional	Sustaining partnerships
Criticality	Other opportunities	Acceptable practices
Sensitivity	Substitutes	Scale/severity
Rarity/uniqueness	Complements	Duration/noticeability
Size/Remoteness	Adjacent uses	Communication/Control
Zoning	Residential	Design/information
Buffers	Industrial	Access regulation

to deal more generally with the integration of social and ecological goals (e.g., Raciti et al. 2006). Based upon restoration efforts in Chicago and San Francisco, I have attempted to adapt ROS physical, social, and managerial criteria for identifying appropriate alternatives for urban restoration sites. These criteria are listed in Table 8.1 and elaborated upon in the following sections.

8.4.1 Ecological Criteria

An urban site that contains remnant patches of native vegetation is a rarity in many cities and provides a powerful justification for site protection and restoration (e.g., McKinney 2006; Ranta and Viljanen 2011). Assessment of species diversity, vegetation structure, and persistence of ecological processes will help to establish current site intactness and potential for restoration within the context of the classical approach. While a site that has little or no intactness opens up options for alternative restoration approaches that emphasize greater human use or serve functional values such as stormwater retention, attempts to recreate a classical landscape may still be justified for other reasons such as research or education.

Sites with low intactness might also be important for natural area protection and restoration if they provide critical habitat for species; host a rare, endangered, or threatened species; or contain species or community types that are locally rare or unique (e.g., D'Antonio and Meyerson 2002; Shapiro 2002). As was seen in the description of the habitat and sensitive species approaches above, ecological goals for restoration might still allow considerable human activity, though this will vary from species to species. The value placed on local rarity and uniqueness may be justified ecologically to maintain genetic diversity, though in some cases managers might feel a responsibility to restore a community within their jurisdiction even though other and better sites occur nearby. Of course, there are many cases where disturbance to biophysical conditions (e.g., loss of seed bank, contamination of soil) will make some restoration goals formidable or even futile, thus managers must choose their approach to site management realistically (e.g., del Tredici 2010).

The ROS places significant weight on the size and remoteness of sites in determining how they are best programmed to serve recreational use, and these same criteria might also usefully apply to urban restoration sites. Larger, more remote sites are less subject to exotic species invasion and more capable of hosting sustainable populations and ecological process such as fire; they also tend to be further away from adjacent and on-site uses that could generate conflict. Smoke from prescribed burns is less likely to drift into neighborhoods or cause panic among residents, and people making the effort to visit a large, remote natural area generally have nature appreciation as a central goal and are less likely to find restoration management practices out of line with their expectations (Ryan 2000).

In Chicago and San Francisco, many of the larger, outlying forest preserves and regional parks are successfully managed under a classical approach to restoration with minimal social conflict. But the remoteness criterion as applied in ROS can also be used to prescribe a compatible suite of approaches through the concentric zoning of larger sites in more densely populated urban areas. As with ROS where the interior zones are identified for primitive backcountry experiences surrounded by increasingly more intensive and developed uses, in San Francisco, a similar type of zoning has been applied to some natural areas under the Recreation and Park District's (2006) Significant Natural Areas Management Plan. "MA-1" management areas containing high quality remnants are the focus of intensive restoration activity and have more restrictions placed on recreational use. These are often surrounded by buffers of "MA-2" and "MA-3" zones that are of decreasing ecological importance, which receive less intensive management and can host a wider range of uses. Similarly in Chicago, Park District natural areas are often surrounded by zones of unmowed vegetation that buffer them from more intensively used and managed areas of the park.

In these ways, size and remoteness criteria could be used along with information on site intactness and priorities for species protection to zone areas for management under different restoration approaches. While the appropriate suite of approaches would vary depending on the goals and constraints of a particular site, a typical strategy might be to manage innermost areas under a classical or boutique approach, with restrictions placed on recreational access as needed. This zone would then be surrounded by a buffer managed under a habitat approach where species help support wildlife functions while allowing a greater variety of uses. Finally the outer zone would be managed as explorable nature, where largely unmanaged vegetation would still provide natural value while catering primarily to active, unstructured nature-recreation opportunities.

8.4.2 Social Criteria

The kinds of recreational uses that take place in urban natural areas are broad and include a range of active and passive activities where the natural environment is of both direct and indirect interest. These uses include active participation in restoration stewardship activities such as planting and weeding as well as research and monitoring; sedentary activities such as picnicking and active activities such as walking, jogging, bicycling, and dog-walking on or off-leash where nature provides a desirable setting for exercise and outdoor enjoyment; non-consumptive nature oriented activities such as birding and nature photography that occasionally take visitors off trail; and highly interactive play, exploratory, and consumptive activities that may involve climbing, digging, and collecting.

Some activities such as restoration stewardship may help support ecological goals, other activities such as walking along trails are largely benign, and activities such as collecting and dog-walking may threaten ecological goals if done at the wrong place or time (Fernandez-Jurucic et al. 2001; Platt and Lill 2006). All of these activities, however, may be legitimate and desirable social goals that people look to in urban nature and providing opportunities for them can help promote learning and build support for restoration programs (Miller 2006; Ryan et al. 2001).

By understanding site capabilities and user desires, managers may be better equipped to choose an alternative restoration approach that best integrates ecological and social goals and best helps to cultivate a sense of place by restoring contact with the land. There may be times, however, when more lofty ecological restoration goals are proposed for sites where established uses will become incompatible. Without earnest public involvement and the provision of reasonable alternatives, restrictions on access and use of a site can become contentious (Phalen 2009).

A thorough analysis of the social setting should also look beyond the immediate use of the site to adjacent land uses and potential concerns. Some restoration sites in Chicago and San Francisco lie directly adjacent to residential neighborhoods where homeowners have heightened concerns about activities that might reduce visual attractiveness and privacy in addition to risks from erosion, fire, and herbicides (Gobster 2000, 2007a). Few such concerns may present themselves when adjacent lands are used for industrial or transportation functions, and thus natural area managers need to be cognizant of the social context before introducing restoration activities.

8.4.3 Managerial Criteria

An important consideration in selecting the appropriate natural area restoration approach is how restoration fits within the mission of the managing agency or institution. For example, the primary mission of the Chicago Park District is to provide high quality recreation opportunities that respond to diverse customer needs, among which include opportunities for nature exploration, appreciation, and education (Chicago Park District 2011). In contrast, the primary mission of the Forest Preserve District of Cook County is to acquire, protect and restore lands and their associated flora and fauna as near as may be to their natural state, for the purposes of people's education, pleasure, and recreation (Forest Preserve District of Cook County 2011). While missions like the latter example may give an agency or institution greater justification for managing lands according to a classical restoration approach, there may be instances where those with broader missions establish nature centers or possess certain habitats that also warrant classical restoration and restrictions on more intensive recreation. Like the ROS, the alternative approaches to restoration described above can help managers provide the best match of social and ecological goals across the various sites in their system as well as develop partnerships among different managing agencies and institutions. Other key partnerships in many urban restoration projects are with volunteer stewardship groups. These usually include ecological restoration groups but may also include recreation-oriented concerns such as off-road bicycle and birding groups, school and civic groups such as environmental and garden clubs, and even animal welfare groups such as feral cat stewardship programs. Thoughtful consideration of how these groups relate to an agency's mission can lead to building effective partnerships and minimizing potential management conflicts (e.g., Newman 2008; Petts 2007; Shandas and Messer 2008) .

Once the management objective for a site is decided, guidelines should be established for implementing management practices that best meet the mix of social and ecological goals. Some considerations here relate to the scale and timing of practices. For sites where aesthetic and recreational goals are important, restoration practices could be kept smaller in scale and implemented using less intrusive, "boutique" practices that minimize conflicts with user expectations. Change might be introduced gradually, for example, by incrementally thinning canopy trees to restore more open conditions over time and by allowing aesthetically valued trees that are nonconforming but ecologically benign to live out their natural lives. Removal, relocation, or chipping and distribution of brush might appear less offensive than having large brush piles in close view. While managers should not try to "fool the public" by hiding change behind vegetative screens and the like, practices should be implemented consistent with the perceived nature of the site and how it is used (Gobster 1999).

Along with the implementation of socially acceptable restoration practices, managers can work to help communicate ecological restoration goals to the public through design and information. The introduction of visual "cues to care" such as the planting of showy plants at entryways to restoration projects and mowing trail rights-of-way to provide a transitional edge can help to frame and call attention to the stewardship of a site that might otherwise be perceived as a product of management neglect (Nassauer 1995). Likewise, signage, other on- and off-site written material, self-guided nature trails, and hosted events are among a variety of ways in which information about a restoration project can help enhance understanding and appreciation of ecological goals that may not be directly perceivable. Finally, both design and information can help regulate access to sites to minimize ecological impact and direct user experiences. For example a narrow, wood chipped trail marked with a small sign can effectively limit access to more sensitive parts of a restoration area while broader, paved paths along the site's perimeter can still allow large numbers of joggers and bicyclists to view and experience the restoration at higher speeds (Kaplan et al. 2007; Ryan 2000).

8.5 Conclusions

As the approaches and criteria described above suggest, the restoration of natural lands can be a highly interpretive endeavor in urban environments. While the classical approach assumes there is an "original nature" out there to be restored as authentically as possible, the social and ecological goals inherent in urban restoration often requires the restorationist to seek alternative approaches that are realistic and can be successfully implemented (Hobbs 2007).

Given the examples identified in these case studies of Chicago and San Francisco, further investigation of alternative approaches to restoration is warranted. Indeed, evidence from other cities in the U.S. and other countries shows that approaches focusing on rehabilitation, utilization, and the provision of environmental services such as moderation of urban heat island effects, carbon sequestration, and phytoremediation are increasing in use (e.g., Westphal et al. 2010). By further examining the social and ecological goals and constraints inherent in urban restoration projects, it may be possible to develop guidelines to advise practitioners and policymakers on which approach might be most appropriately applied to a given site. Such a "*restoration* opportunity spectrum" could help to maximize sought-after values and minimize potential conflicts.

Should all of the different approaches described here be referred to as restoration? Some have argued that the term restoration should be reserved only for uses that most closely parallel what I have referred to here as classical restoration (Jordan 2003). But in their own unique ways each of these approaches contributes to the idea that in order to be successful, ecological restoration must respond to diverse and evolving social and ecological goals (e.g., Choi et al. 2008; Hobbs et al. 2004; Palmer et al. 2004). My aim here is to clarify rather than confuse, and together these examples suggest that there are many approaches to natural areas management that provide promising foundations for restoration in urban areas (Hull 2006).

Acknowledgments Many thanks to Mimi Wagner, Herb Schroeder, Palle Madsen and an anonymous reviewer for their helpful comments on earlier versions of this paper.

References

Amidon J (2005) Moving horizons: the landscape architecture of Kathryn Gustafson and Partners. Birkhauser, Basel

Bell S (2008) Design for outdoor recreation. Taylor & Francis, New York

Britt RR (2004) The new nature: cities as designer ecosystems. LiveScience http://www.livescience.com/environment/050110_designer_ecosystem.html. Accessed 20 April 2011

Callicott JB (2002) Choosing appropriate temporal and spatial scales for ecological restoration. J Biosci 27(Suppl 2):409–420

Chicago Park District (2011) Mission and history of the Chicago Park District. http://www.cpdit01.com/resources/planning-and-development.history/. Accessed 20 Apr 2011

Chicago Region Biodiversity Council (1999) Biodiversity recovery plan. Chicago Region Biodiversity Council, Chicago

Choi YD (2007) Restoration ecology to the future: a call for new paradigm. Restor Ecol 15:351–353
Choi YD, Temperton VM et al (2008) Ecological restoration for future sustainability in a changing environment. Ecoscience 15:53–64
City of Chicago (2006) Chicago nature and wildlife plan: a strategy to enhance natural habitats within the City of Chicago. City of Chicago Department of Planning and Development, Chicago
D'Antonio C, Meyerson LA (2002) Exotic plant species as problems and solutions in ecological restoration: a synthesis. Restor Ecol 10:703–713
Del Tredici P (2010) Spontaneous urban vegetation: reflections of change in a globalized world. Nat Cult 5:299–315
Elliott R (1997) Faking nature: the ethics of ecological restoration. Routledge, New York
Feldman RM, Westphal LM (1999) Participation for empowerment: the greening of a public housing development. Places 12(2):34–37
Fernandez-Jurucic E, Jimenez MD, Lucas E (2001) Alert distance as an alternative measure of bird tolerance to human disturbance: implications for park design. Environ Conserv 28:263–269
Forest Preserve District of Cook County (2011) The early history of the Forest Preserve District of Cook County, 1869–1922. http://fpdcc.com/about/history. Accessed 20 Apr 2011
Foster J (2010) Off track, in nature: constructing ecology on old rail lines in Paris and New York. Nat Cult 5:316–337
Fox D (2007) Back to the no-analog future? Science 316:823–825
Gobster PH (1999) An ecological aesthetic for forest landscape management. Landsc J 18:54–64
Gobster PH (2000) Restoring nature: human actions, interactions, and reactions. In: Gobster PH, Hull RB (eds) Restoring nature: perspectives from the social sciences and humanities. Island Press, Washington, DC
Gobster PH (2001) Visions of nature: conflict and compatibility in urban park restoration. Landsc Urban Plan 56:35–51
Gobster PH (2007a) Restoring urban natural areas: negotiating nature in San Francisco. Final report, 04-CO-11231300-004, USDA Forest Service Northern Research Station, Evanston, IL
Gobster PH (2007b) Urban park restoration and the museumification of nature. Nat Cult 2:94–115
Gobster PH (2010) Introduction: urban ecological restoration. Nat Cult (special symposium on urban ecological restoration) 5(3):227–230
Gobster PH, Hull RB (eds) (2000) Restoring nature: perspectives from the social sciences and humanities. Island Press, Washington, DC
Grese RE (1992) Jens Jensen: maker of parks and gardens. Johns Hopkins University Press, Baltimore
Gross M (2003) Inventing nature: ecological restoration by public experiments. Lexington Books, Lanham
Hall M (2005) Earth repair: a transatlantic history of environmental restoration. University of Virginia Press, Charlottesville
Hart J, Beatty RA et al (1996) Gardens of Alcatraz. Golden Gate National Parks Association, San Francisco
Higgs E (2003) Nature by design: people, natural process, and ecological restoration. MIT Press, Cambridge
Hitchmough J, de la Fleur M (2006) Establishing North American prairie vegetation in urban parks in northern England: effect of management and soil type on long-term community development. Landsc Urban Plan 78:386–397
Hobbs RJ (2007) Setting effective and realistic restoration goals: key directions for research. Restor Ecol 15:354–357
Hobbs RJ, Davis MA et al (2004) Restoration ecology: the challenge of social values and expectations. Front Ecol Environ 2:43–48
Hobbs RJ, Higgs E, Harris JA (2009) Novel ecosystems: implications for conservation and restoration. Trends Ecol Evol 24:599–605
Hull RB (2006) Infinite nature. University of Chicago Press, Chicago
Hull RB, Robertson DP, Buhyoff GJ (2004) "Boutique" forestry: new forest practices in urbanizing landscapes. J For 102:14–19

Ingram M (2008) Urban ecological restoration. ecological restoration (special theme: ecological restoration in an urban context). Ecol Restor 26:175–177
Irvine S, Johnson L et al (1999) Community gardens and sustainable land use planning: a case-study of the Alex Wilson Community Garden. Local Environ 4:33–46
Jordan WR III (2000) Restoration, community, and wilderness. In: Gobster PH, Hull RB (eds) Restoring nature: perspectives from the social sciences and humanities. Island Press, Washington, DC
Jordan WR III (2003) The sunflower forest: restoration and the new communion with nature. University of California Press, Berkeley
Jordan WR III, Lubick GM (2011) Making nature whole: a history of ecological restoration. Island Press, Washington, DC
Kaplan R, Ivancich JE, De Young R (2007) Nearby nature in the city: preserving and enhancing livability. University of Michigan, Ann Arbor. http://hdl.handle.net/2027.42/48784. Accessed 20 Apr 2011
Katz E (2000) Another look at restoration: technology and artificial nature. In: Gobster PH, Hull RB (eds) Restoring nature: perspectives from the social sciences and humanities. Island Press, Washington, DC
Kowarick I, Körner S (eds) (2005) Wild urban woodlands: new perspectives for urban forestry. Springer, Berlin
Link SO (2008) Climate change and natural areas. Nat Area J 28:5
Louv R (2005) Last child in the woods: saving our children from nature-deficit disorder. Algonquin Books, Chapel Hill
Low T (2002) The new nature: winners and losers in wild Australia. Viking, Victoria
Lundholm JT, Richardson PJ (2010) Habitat analogues for reconciliation ecology in urban and industrial environments. J Appl Ecol 47:966–975
MacDougall AS, Turkington R (2007) Does the type of disturbance matter when restoring disturbance-dependent grasslands? Restor Ecol 15:263–272
Maloney CJ (2001) Uncovering an oasis: in downtown Chicago, a classic Prairie School landscape is restored to life. Landsc Archit 91(11):76–81
Martínez LM, López-Barerra F (2008) Special issue: restoring and designing ecosystems for a crowded planet. Ecoscience 15:1–5
McKinney ML (2006) Urbanization as a major cause of biotic homogenization. Biol Conserv 127:247–260
Miller JR (2005) Biodiversity conservation and the extinction of experience. Trends Ecol Evol 20:430–434
Miller JR (2006) Restoration, reconciliation, and reconnecting with nature nearby. Biol Conserv 127:356–361
Minnis PE, Elisens WJ (2001) Biodiversity and native America. University of Oklahoma Press, Norman
Minteer BA, Collins JP (2010) Move it or lose it? the ecological ethics of relocating species under climate change. Ecol Appl 20:1801–1804
More TA, Bulmer S et al (2003) Extending the Recreation Opportunity Spectrum to nonfederal lands in the Northeast: an implementation guide. General Technical Report NE-309, USDA Forest Service Northeastern Research Station, Newtown Square, PA
Nassauer JI (1995) Messy ecosystems, orderly frames. Landsc J 14:161–170
Newman A (2008) Inclusive planning of urban nature. Ecol Restor 26:229–234
Palamar C (2010) From the ground up: why urban ecological restoration needs environmental justice. Nat Cult 5:277–298
Palmer M, Bernhardt E et al (2004) Ecology for a crowded planet. Science 304:1251–1252
Pennington DN, Hansel J, Blair RB (2008) The conservation value of urban riparian areas for landbirds during spring migration: land cover, scale, and vegetation effects. Biol Conserv 141:1235–1248
Petts J (2007) Learning about learning: lessons from public engagement and deliberation on urban river restoration. Geogr J 173:300–311

Phalen KB (2009) An invitation for public participation in ecological restoration: the Reasonable Person Model. Ecol Restor 27:178–186
Platt A, Lill A (2006) Composition and conservation value of bird assemblages of urban 'habitat islands': do pedestrian traffic and landscape variables exert an influence? Urban Ecosyst 9:83–97
Raciti S, Galvin MF et al (2006) Urban tree canopy goal setting: a guide for Chesapeake Bay communities. USDA Forest Service, Northeastern State & Private Forestry, Annapolis, MD.
Ranta P, Viljanen V (2011) Vascular plants along an urban–rural gradient in the city of Tampere, Finland. Urban Ecosys. doi:10.1007/s11252-011-0164-9
Rosenzweig ML (2003a) Win-win ecology: how the Earth's species can survive in the midst of human enterprise. Oxford University Press, New York
Rosenzweig ML (2003b) Reconciliation ecology and the future of species diversity. Oryx 37:194–205
Ross LM (1994) Illinois' volunteer corps: a model program with deep roots in the prairie. Restor Manage Notes 12:57–59
Ruiz-Jaen MC, Aide TM (2005) Restoration success: how is it being measured? Restor Ecol 13:569–577
Ryan RL (2000) A people-centered approach to designing and managing restoration projects: insights from understanding attachment to urban natural areas. In: Gobster PH, Hull RB (eds) Restoring nature: perspectives from the social sciences and humanities. Island Press, Washington, DC
Ryan RL, Kaplan R, Grese RE (2001) Predicting volunteer commitment in environmental stewardship programmes. J Environ Plann Manage 44:629–648
San Francisco Recreation and Park Department (2006) Significant Natural Resource Areas Management Plan. Recreation and Park Department, San Francisco
Seewagen CL, Slayton EJ, Guglielmo CG (2010) Passerine migrant stopover duration and spatial behaviour at an urban stopover site. Acta Oecol 36:484–492
Shandas V, Messer WB (2008) Fostering green communities through civic engagement: community-based enviromental stewardship in the Portland area. J Am Plann Assoc 74:408–418
Shapiro AM (2002) The Californian urban butterfly fauna is dependent on alien plants. Divers Distrib 8:31–40
Society for Ecological Restoration International (2004) The SER International primer on ecological restoration. Society for Ecological Restoration International, Tucson, AZ http://www.ser.org/pdf/primer3.pdf. Accessed 20 Apr 2011
Trigger DS, Head L (2010) Restored nature, familiar culture: contesting visions for preferred environments in Australian cities. Nat Cult 5:231–250
Trust P (2001) Vegetation management plan and environmental assessment. Presidio Trust, National Park Service, San Francisco
US Fish and Wildlife Service (2003) Recovery plan for coastal plants of the northern San Francisco peninsula. USFWS Portland, Oregon
USDA Forest Service (1982) ROS users guide. USDA Forest Service, Washington, DC
Van Sweden J (1997) Gardening with nature. Random House, New York
Vidra RL, Shear TH (2008) Thinking locally for urban forest restoration: a simple method links exotic species invasion to local landscape structure. Restor Ecol 16:217–220
Westphal LM, Gobster PH, Gross M (2010) Models for renaturing brownfield areas. In: Hall M (ed) Restoration and history: the search for a usable environmental past. Routledge, New York
Young T (2004) Building San Francisco's parks, 1850–1930. Johns Hopkins University Press, Baltimore

Chapter 9
Urban Forest Landscape Restoration – Applying Forest Development Types in Design and Planning

J. Bo Larsen and Anders Busse Nielsen

9.1 Introduction

The world is getting increasingly urbanized! By 2008 the proportion of urban dwellers across the globe passed 50% and is expected to reach 60% by 2030. In Europe, the percentage of the population living in urban areas is expected to rise from 73% in 2000 to 80% in 2030 (United Nations 2004). As urban structures grow, so does the area of forest under urban influences (Rowntree 1995; Kowarik 2005). In Denmark, afforestation close to urban areas has top priority (Jensen and Koch 2004). In Sweden, urban and peri-urban forests are estimated to cover approximately one million ha by 2008, which is more than four times the area of protected forests in the country (Hedblom and Söderström 2008). In the United States, trees in urban counties account for nearly 25% of the nation's total tree cover (Dwyer et al. 2000).

The importance of urban forests has, from a forestry perspective, been widely overlooked and, as a result, undervalued. However, evidence is rising about the various environmental and social benefits and potentials of forests and trees in populated areas to limit energy use, improve air quality, reduce noise, increase water storage, maintain fragmented ecosystems and positively contribute to society development and human health and well-being (Konijnendijk et al. 2005). Today urban forests are widely regarded as the best strategy for providing green spaces for recreation

J.B. Larsen (✉)
Danish Center for Forest, Landscape and Planning, University of Copenhagen, Rolighedsvej 23, DK-1958 Frederiksberg, Denmark
e-mail: jbl@life.ku.dk

A.B. Nielsen
Department of Landscape Management, Design and Construction, Faculty of Landscape Planning, Horticulture and Agricultural Science, Swedish University of Agricultural Sciences, Box 66, 23053 Alnarp, Sweden
e-mail: anders.busse.nielsen@slu.se

J. Stanturf et al. (eds.), *Forest Landscape Restoration: Integrating Natural and Social Sciences*, World Forests 15, DOI 10.1007/978-94-007-5326-6_9,

(Van Herzele 2006) and to preserve and develop urban ecosystems and biodiversity (Alvey 2006). Therefore, the identification of forest landscape restoration strategies focusing upon forests in an urban and peri-urban context is crucial (Larsen 2005): We have to restore forests and develop nature where people live.

Urban forest landscapes are here defined as forested ecosystems of natural, semi-natural or man-made origin wherein other landscape elements such as water, wetlands, semi-open and open nature types might be integrated. Urban forest landscapes are managed for a variety of purposes of which recreation and nature protection are the main functionalities while wood production is often secondary (Bell et al. 2005; Tyrväinen et al. 2005). However, most urban forests have developed from commercial forests. In these situations managers have sought to incorporate social and environmental functions in management systems oriented towards wood production thereby focussing upon the stand as the functional unit. When dealing with urban forests the higher functional level – the forest landscape – is much more in focus. This is necessary to develop the intended recreational and ecological functionalities (Gustavsson et al. 2005). Hence, long term goals and management strategies that give attention to forest landscapes as well as stands have to be developed when planning and managing urban forests. In this context, traditional forest management, with closed forest made up of blocks of homogeneous even-aged stands established by intensive plantings after clear-cutting, will not necessarily meet the need of urban societies to balance wood production with nature protection and recreation (Krott 1998). Therefore public forest owners are increasingly paying attention to alternative forest management strategies as a means of rehabilitating forest and meeting these challenges.

When managing urban forest landscapes for future generations we are, in essence, dealing with uncertainties in future requirements (population pressure and social values) and climate (climate change). Hence, the ambition for design and management should be to develop landscapes with a robust functionality despite changes in ecological as well as social-economical settings. To achieve this, design must be flexible and inextricably linked to management. In this way the forest landscape can evolve and adapt to changing pressures, whilst at the same time, maintain and incorporate important recreational and ecological values and elements of the cultural heritage contained in the existing resource. In this context, keeping as close to nature as possible and using natural processes as the base for design and management is generally claimed to be one of the most promising approaches (Bell et al. 2005).

Design with nature means achieving three objectives: conserving nature, establishing sustainable ecosystems, and achieving a natural appearance (Sepahi 2000). This chapter aims to illustrate how these objectives can be achieved by incorporating silvicultural, ecological and landscape architectural expertise around the development of urban forest landscapes through a design-led approach that keeps as close to nature as possible. The variation in site conditions and topography and potential natural hydrological flows are used as starting point for locating a variety of site-adapted habitats and determining their composition.

The approach is described in three main sections. The first section introduces close-to-nature forest management as a silvicultural strategy for *forest* restoration. In relation to this, the Forest Development Type (FDT) is presented as a flexible planning tool for locating site-adapted habitats and for describing the long-term goals. The second section focuses on the landscape level and presents the main vegetation types that characterises both ecologically and recreationally rich forest landscapes as basis for a design-led approach to *forest landscape* restoration. In the third section of the chapter, a case-study presenting a restoration plan for a part of Vestskoven near Copenhagen is used to illustrate how the approach can be used in practice. The plan has been developed by a group of students attending the international master course in Urban Woodland Design and Management at the University of Copenhagen.

9.2 The Forest Restoration Perspective – Management with Nature

9.2.1 Close-to-Nature Management

From a forest restoration viewpoint, silvicultural strategies are required in order to develop urban forest with a high potential for nature conservation, ecosystem protection, and a natural appearance. This can be achieved by incorporating structural and functional features of natural forest ecosystems into the restoration program. This approach can be summarised by the term 'nature-based silviculture' or 'close-to-nature forest management'. The aim is to reform current practices so that they are still profitable but are more environmentally benign and more sensitive to nature conservation and the demands of sustainability. This can be done by mimicking natural forest structures, processes and dynamics (Larsen 2000; Lindenmayer et al. 2006; Hahn et al. 2007).

Close-to-nature forestry in Central Europe promotes continuous-cover forests based upon the principle of supporting natural processes of forest ecosystems by facilitating site-adapted species mixtures through natural regeneration and making use of natural self-differentiation (Schütz 2006). The concept requires a profound understanding of natural disturbance regimes and successional processes.

The disturbances and processes in natural forest ecosystems, which cause structural heterogeneity at both large and small scale are linked to regional characteristics of climate, soil, and species compositions. These disturbances include frequent small-scale disturbances in Central-European forests, small and large-scale disturbances in boreal ecosystems caused by fires and infrequent but large-scale disturbances in some areas caused by storms. Hence, models describing the region-specific disturbance patterns should be used in the development of applied silvicultural methods for those particular regions (Hahn et al. 2005). In central and western Europe the forest cycle models have been successfully used to describe the temporal and spatial dynamics of specific forest types in natural forest reserves (Leibundgut 1959; Meyer and Neumann 1981; Mueller-Dombois 1987; Jenssen and Hofmann 1996; Emborg et al. 2000; Christensen et al. 2007; Hahn et al. 2007). Such models could serve as an adequate basis for close-to-nature forest management.

The use of natural disturbance regimes to guide human disturbance regimes (i.e. thinning and cutting regimes) must, however, be complemented with other measures to restore naturalness in forest management. Lindenmayer et al. (2006) emphasize the importance of maintaining aquatic ecosystem integrity for biodiversity protection in managed forests. Maintaining and restoring natural hydrology in forests previously subjected to stand management operations such as drainage, and promoting species and forest structures that reflects and emphasis the variation in hydrology is an integral part of close-to-nature management, that can contribute to the development of habitat richness and experiential variation both of which are highly desired in urban forest landscapes.

Restoration of forests along 'close-to-nature' forest management principles will have major impacts on the forest ecosystem and their visual appearance and thereby greatly influence recreational values as these are highly dependent on the visual aspects. The relationship between recreational preferences and forest characteristics is complex and different studies have identified different preferences, suggesting that these are strongly influenced by cultural, regional, contextual, and subjective expectations (e.g. Jensen 1999; Tyrväinen et al. 2003). However, close-to-nature forest structures and related management practices are generally found to be recreationally preferable to monocultural stands managed through clear-cuts. This is the case simply because of higher visual variety and because fewer trees are removed at the same time and thus the forest keeps a more natural appearance (Nielsen et al. 2007).

9.2.2 Forest Development Types

The complex nature of near-natural forest structures and dynamics requires integrative and flexible management frameworks and tools. The concept of a Forest Development Type (FDT) provides one such framework for advancing and describing ideas about long-term goals for stand structures and dynamics in stands subjected to nature-based forest management. A FDT describes the long-term goals for forest development on a given locality (with particular climatic and soil conditions) in order to accomplish specific long-term aims of functionality such as ecological-protection, economic-production, and social-/cultural functions (Larsen and Nielsen 2007).

A major object of FDT scenarios is to describe nature-based silviculture at the stand level. For each FDT vegetation structure, species composition and regeneration dynamics are described both qualitatively (verbal descriptions) and quantitatively (numeric descriptions) for their mutual supportiveness, and the goal is specified with respect to conservation, recreation and production. Furthermore, to support the intuitive understanding and the communication of FDT scenarios, vegetation structure and composition is illustrated by means of profile diagrams (Larsen and Nielsen 2007).

In Denmark, a participatory process described by Larsen and Nielsen (2007) resulted in the creation of 19 FDTs, which can be grouped into nine broadleaved dominated, six conifer dominated, and an additional four 'historic' types (Table 9.1). Whereas all 'nature-based' FDTs encompass a balance between productive, protective and recreational/social functions, the other four 'historical' types mainly serve the

Table 9.1 The 19 Danish Forest Development Types. The name encompasses the dominating and co-dominating species

Broadleaved dominated:	*Species name (Latin name)*
11 Beech	Alder (*Alnus glutinosa*)
12 Beech with ash and sycamore	Ash (*Fraxinus excelsior*)
13 Beech with Douglas-fir and larch	Beech (*Fagus silvatica*),
14 Beech with spruce	Birch (*Betula pendula* and *pubescens*)
21 Oak with ash and hornbeam	Douglas-fir (*Pseudotsuga menziesi*)
22 Oak with lime and beech	Hornbeam (*Carpinus betulus*)
23 Oak with Scots pine and larch	Larch (*Larix kaempferi* and *x eurolepis*)
31 Ash with alder	Lime (*Tilia cordata*)
41 Birch with Scots pine and spruce	Mountain pine (*Pinus mugo*)
Conifer dominated:	Norway spruce (*Picea abies*)
51 Spruce with beech and sycamore	Oak (*Quercus robur* and *petraea*)
52 Sitka spruce with pine and broadleaves	Scots pine (*Pinus silvestris*)
61 Douglas-fir, Norway spruce and beech	Sitka spruce (*Picea sitchensis*)
71 Silver fir and beech	Silver fir (*Abies alba*)
81 Scots pine with birch and Norway spruce	Spruce (*Picea abies* and *sitchensis*)
82 Mountain pine	Sycamore (*Acer pseudoplatanus*)
'Historic' forest types:	
91 Coppice forest	
92 Forest pasture	
93 Forest meadow	
94 Unmanaged forest	

The first digit in the FDT-number indicates the main species (1 = beech, 2 = oak, 3 = ash, 4 = birch, 5 = spruce, 6 = Douglas-fir, 7 = true fir, 8 = pine, and 9 indicating a "historic" FDT). The second digit is numbered at random. FDT 12 is described in Fig. 9.1 and FDT 11, 12, 21, 71, and 92 are illustrated in Fig. 9.7

protection of recreational, natural and cultural functions. Especially the historical Forest Pasture (FDT No. 92) and Forest Meadow (FDT No. 93) can be actively used to create habitat diversity and experiential richness in urban forest landscapes.

Each FDT is described as follows (see also Fig. 9.1):

- Name: The name encompasses the dominating and co-dominating species. The first digit in the FDT-number indicates the main species (1 = beech, 2 = oak, 3 = ash, 4 = birch, 5 = spruce, 6 = Douglas-fir, 7 = true fir, 8 = pine, and 9 indicating a "historic" FDT). The second digit is numbered at random.
- Structure: A description of how the forest structure could appear when fully developed. This description is supplied with a profile diagram depicting a 120 m transect of the anticipated forest structure at 'maturity' (In Fig. 9.7 the profile diagrams of four FDT´s are displayed: No. 11- Beech, No. 21- Oak with ash and hornbeam, No. 71-Silver fir and beech, and No. 92-Forest pasture).
- Species distribution: The long-term distribution of species and their relative importance.
- Dynamics: The regeneration dynamics described in relation to the expected succession and spatial patterns (species, size).

Forest Development Type 12: Beech with ash and sycamore

Structure: Species rich, well structured forest with beech as dominating element mixed with ash and cherry and in south-eastern Denmark additionally with hornbeam and lime. The in-mixed species occur mainly in groups. The horizontal structures arise between groups of varying size and age. Where the light demanding species such as ash, sycamore and cherry dominate, vertical structures occur periodically with shade trees (beech, hornbeam, elm, and others) in sub-canopy strata.

Species: Beech. 40 – 60 %, ash and sycamore: 30 – 50 %, cherry, hornbeam, oak, lime, and others up to 20 %

Dynamics: Beech regenerates mainly in groups and smaller stands. Ash and sycamore as gap specialists regenerate in openings later followed by beech. Hornbeam belongs to the sub-canopy stratum and regenerates under shade, whereas the pioneer species (cherry and oak) only regenerate after lager openings and/or in relation to forest edges.

Functions: Productive: The forest development type has a high potential for production of hardwood in larger dimensions and of good quality.

Protective: In most parts of the country the beech dominated forest represents the potential natural vegetation; consequently, many indigenous species are connected to this forest development type. It has a great potential for conserving biodiversity connected to the NATURA 2000 habitat type 9139 and 9150.

Recreational: Through ist mixture of (indigenous) species in combination with pronounced variation in size the forest development type gives a multitude of recreational experiences and intimacy.

Occurrence: The forest development type belongs on protected sties in the eastern and northern parts of Denmark on rich, well drained soils with good water supply as illustrated below.

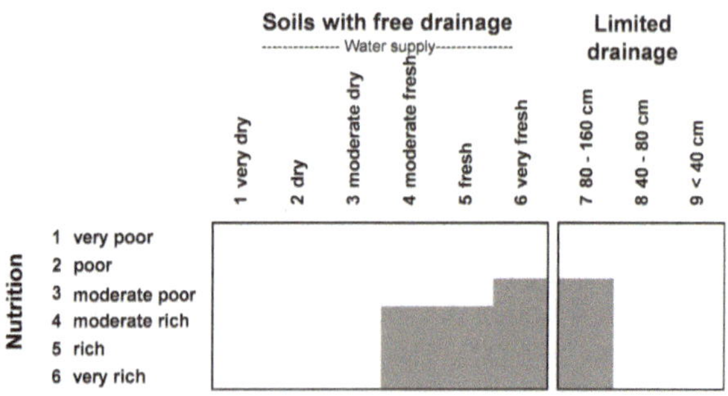

Fig. 9.1 Description and illustration of forest development Type 12: Beech with ash and sycamore

- Functionality: Indication of the forest functionality (economic-production, ecologic-protection, and social/cultural functions).
- Occurrence: Suggested application in relation to climate and soil. For this purpose the country is divided into four sub-regions with each their typical climatic characteristics. Further, the application of the specific FDT in terms of soil conditions is stated in relation to nutrient and water supply.

9.2.3 Matching Forest Development Types to Site

While different forest development types possess different site requirements it is possible to address and utilise potential heterogeneities in site conditions by matching the FDT to particular sites. This requires a thorough site survey, analyzing environmental conditions such as geology and soil types, nutrient and water supply, as well as specific site factors such as compact soil layers and as insufficient drainage. An analysis of the hydrological status of the site including existing drainage systems combined with a plan of the historical landscape with its former wetlands prior to draining could provide guidance for delineating the landscape into ecological functional units. The site classification map provides a framework for identifying appropriate FDTs for each site, thus facilitating the creation of forested landscapes where site-adapted stand and nature types reflects and emphasises variation in the landscape.

9.3 The Landscape Restoration Perspective – Planning with Design

For users of urban forests the variation between close, semi-open and open areas is important for orientating in the landscape and for the experience of unity and coherence (Sorte 1989; Kaplan 1995; Bell et al. 2005). It is also the variation in vegetation structures that is the main source of habitat diversity. According to Lindenmayer et al. (2006) maintenance of landscape heterogeneity and connectivity (habitat fragmentation) is of outmost importance for biodiversity conservation in forestry. Thus, both recreationally and ecologically rich forest landscapes are composed of five main types of vegetation structures with each their distinct experiential values and habitat qualities:

- Forest interiors *under* closed canopies.
- The semi-open areas *between* scattered trees
- Glades *inside* forested parts
- Edges *along* the boundary among forested and open parts
- Open areas *outside* forested parts

9.3.1 Forest Interiors

In a landscape perspective, the forest is unique as it is the only landscape element with an understorey environment. The intimacy one experiences when being enclosed under the canopy is a major recreational value of a forest when compared to other landscape elements. Similarly, the closed forest with its long continuity of ecological processes is of foremost importance for protecting biological values. However, emphasis is not only on the canopy trees, but rather placed on the vegetation structure as a whole. Canopy trees, undergrowth and the main characteristics of the perennial flora are all of importance for biodiversity and for the human experience of character, size and atmosphere of forest interiors (Gustavsson et al. 2005).

Most urban forests have developed from commercial production forests that mainly consisted of even-aged monoculture types with forest interiors resembling pillared halls. However, users of any urban woodland are varied and so are their preferences for stand interiors. The habitat requirements of woodland species are correspondingly diverse. Therefore, to fully develop the recreational and ecological potential of urban forests these should include a wider range of forest interiors (Gustavsson et al. 2005; Lindenmayer et al. 2006; Nielsen and Jensen 2007).

The form- and species rich forest stand types associated with close-to-nature forest management should be considered among the most valuable types to integrate timber production with high aesthetic and biological qualities (Larsen 2000; Gustavsson et al. 2005). Historical/cultural forest management regimes, such as coppice forests and unmanaged/untouched forests (Table 9.1) can also assist in diversifying forest interior rooms and habitats in urban forest landscapes (see Box 9.1) for the benefit of people, plants and wildlife (Nielsen and Møller 2008; Gustavsson et al. 2005).

Box 9.1 Main Types of Forest Interiors

Pillared hall interiors

Forest interiors resembling the pillared hall types are associated with even-aged, single-storied monocultures, often established by shelterwood regeneration or intensive plantings after clear-cuts. Its best-known European example is beech (*Fagus sylvatica*) as illustrated in Fig. 9.7 (FDT 11). The pillared hall is highly appreciated by leisure seekers. The tall trunks and the absence of understory trees and shrubs articulate the topography of the forest floor. It articulates the experience of the inner room and offers long views and free movement beneath the canopy (Nielsen and Jensen 2007). From a biodiversity viewpoint these homogeneous single-layered forests have little importance although they can encompass quite unique and visual attractive ground flora as an example the *Anemone nemorosa* 'carpet' under the beech pillared hall.

Structurally and species-rich interiors

Forest interiors with a richness of forms, sizes and species are associated with uneven-aged, species- mixed stands, managed in group or single tree selection and natural regeneration as illustrated in Fig. 9.7 (FDT 71). The variety of spatial patterns, such as areas with an under-storey of saplings and shrubs, a patchy distribution of species and stems, multilayered tree canopies and temporary canopy openings are regarded as basic requirements for the maintenance of a large proportion of the biodiversity in temperate forest ecosystems (Hahn et al. 2005). Such continuous-cover forests are also assumed to be recreationally preferable to monocultural stands managed through clear-cuts, because of their higher visual variety and the continuous tree cover where fewer trees are harvested at the same time and thus the forest keeps a more natural appearance (Ribe 1989; Hummel 1992; Lindhagen 1996; Nielsen et al. 2007).

(continued)

Box 9.1 (continued)

Low forest interiors

Low forest types are associated with various types of coppice management. The fundamental feature of coppicing is the periodic felling of small trees and shrubs, initiating re-growth from the stump or 'stool', i.e., the permanent woody base from which coppice shoots arise (Peterken 1993). This ability of many deciduous tree species to re-sprout after harvest has been widely used since ancient times, when coppice woods were important contributors of a great variety of wood and non-wood products. Coppicing has been practiced long enough to create its own ecosystem (Peterken 1993), where the periodical changes in light conditions form the basis for an exceptional flora of species which need the extra light following the coppicing to build up their resources. Yet these species also need the shaded phase of the coppice cycle to avoid competition from species of more permanent clearings (Larsen et al. 2001; Rackham 2003). After years of decline and neglect, low woodland types based on coppice management are the subject of renewed interest. In an urban context, low woodland types based on coppice are being suggested as holding potential for engaging local communities in the management of Neighbourhood Forests, where coppice can also offer inspiring environments for children's play. The 'small' scale and multi-stemmed character of coppice creates mysterious environments, where the kids can use the cut wood or break off branches themselves for their play or for craftwork activities.

Wilderness interiors

The concept of wilderness is associated with unmanaged forests. The enhancement of wilderness in urban and peri-urban forests is demanded in many places. Compared to managed forests, unmanaged forests are rich in old trees, trees with holes and cavities, dead standing trees, dead logs and branches, burned wood, and stumps with uneven-aged surfaces. These structural components and especially the continuity of ecosystem processes including dead wood are important requirements for maintenance of large parts of the endangered forest species (Hahn et al. 2005). In cities where the wilderness is an absent quality, leaving unmanaged areas in the core of larger forests or in the parts with limited use, will slowly allow the wilderness to develop.

9.3.2 The Semi-open Forest

The semi-open forest types are closely associated with the traditional cultural landscape in many parts of the world. Many species that are now endangered have adapted to semi-open, cultural landscapes with scattered trees, such as grazing forests, forest pastures and wooded meadows (Fry and Sarlöv-Herlin 1997). However, due to regeneration problems and growing demands on firewood and timber, grazing in forests was more or less abandoned in many European countries when industrialisation

Fig. 9.2 Picture from a semi-open grazing forest with oak and European bison, Eriksberg Sweden (Photo: J. Bo Larsen)

commenced about 200 years ago. Today's forests therefore often lack semi-open areas causing a lack of spatial and habitat diversity and the predominance of abrupt interfaces between open and closed parts (Gustavsson et al. 2005).

Many preference studies have shown that the semi-open forest landscape is among those most appreciated by leisure seekers. The scattered trees and open canopy offer exploring potential in the landscape and feelings of safety while walking *between* trees rather than under a closed canopy or in an open area. The prospect-refuge theory suggested in 1975 by the English geographer Jay Appleton (Appleton 1996) is widely used to explain human preference for semi-open landscapes.

The theory suggests that human responses to the landscape are linked to the evolutionary benefits of certain landscape views, especially those associated with early human evolution in the savannah landscape of sub-Sahara, a semi-open landscape with scattered trees. This landscape offers both prospect (an open view from which predators or prey/forage can be seen) and refuge (shelter and protection that shields the viewer from being seen). Preferences for semi-open landscapes are indirectly manifested in the design of many modern (i.e. post-Renaissance) park and garden landscapes, which have as their inspiration the half-open, grazed savanna-like landscape (Kaplan 1992; Olwig 2002). Depending on the grazing pressure and soil conditions, browsing can be regulated to create a wide gradient of tree and shrub densities, with their associated differences in habitats (Nielsen et al. 2005). Therefore, in urban forest landscapes restoration efforts, the re-discovery of semi-open pastoral landscapes can potentially enhance habitat diversity to be enjoyed by people and inhabited by a wide range of potential threaten biodiversity (Fig. 9.2).

Fig. 9.3 A glade in Linnebjär forest surrounded by oak trees, southern Sweden (Photo: Anders Busse Nielsen)

9.3.3 Glades

The glade is the intimate open room *inside* the forest where the solitude of the forest can be enjoyed (Fig. 9.3). In older times, the glade was the secret meeting place for young lovers, and the place where hunters waited in the sunset for the deer. In a modern urban context, the glade is beloved by children and many people going for a picnic, or it is simply the place where people take a rest before continuing their walk (Hummel 1992). Yet, the glade has hardly been given any attention in recent urban forestry research and practice.

The canopy openings created by glades allow a unique flora to develop and include plants that are shade intolerant and so cannot live under the closed canopy. However, these are also woodland species and so will not be found in the open (Gilbert and Anderson 1998; Smidt et al. 2007). In addition, glades provide habitat for many birds, insect and butterflies which rely upon their continued existence for their survival. In an afforestation perspective, the habitats of early successional stages, such as glades and open areas, can be created early in the developing woodland. Yet, if this is to be realised, the size of glades should be determined in a dynamic perspective and not as a constant which has been the traditional design approach. In young forest landscapes the intimacy and the unique glade microclimate requires rather small glades which can then be increased gradually in line with height increment of the surrounding stands. For example, a glade with a width of

60 m creates an intimate room when surrounded by mature stands with a height of 30 m (Fig. 9.3). However, if the glade is surrounded by young stands with a tree height of only 5–10 m, this sense of intimacy and the special micro climatic conditions will be absent. In a design perspective, the location of a glade has also traditionally been regarded as continuous in time. Yet, introduction of dynamic glades (ca. 0.5 ha) can enhance habitat diversity in relatively uniform areas and increase the amenity value of forests (Gilbert 1989). In forests managed along nature-based principles the small areas of group harvest, cleaned from slash could temporarily act as glades.

9.3.4 Edges

Few elements in the landscape have as high visual and ecological values per unit area as do forest edges. Forest edges and other transition zones between two adjacent ecological communities (ecotones) are generally claimed to be more species-rich than adjacent habitats, with species from the 'parent communities' (ecosystems) as well as species particular to the ecotone itself (Fry and Sarlöv-Herlin 1997). Forest edges are habitats for many light demanding tree and shrub species many of which have a high biological and aesthetic value (e.g. insect pollination, flowers). Over time, forest edges have come to play an increasingly important role as refuges for many of the species appreciated by visitors as well as plant and animal species that have become endangered because of the loss of complexity at the landscapes level including the loss of cultural landscape types such as forest pastures, forest meadows, and hedgerows (Rizell and Gustavsson 1998).

Most people visiting a forest for a picnic settle by the edge from where they have the forest at their back and have a free view over an open area. Forest edges are also a preferred zone for children and their play. Here they can find shelter and good climbing trees from where the more open surroundings can be viewed. Edges contribute positively to the aesthetic qualities of forest landscapes, and thereby also to the recreational potential. On the forest landscape level, the experiential values of forest edges are related to alternation between closed and open edges types (Tregay 1986). Some of these different types of edge are described in Box 9.2. At particular sites the recreational values are also related to the richness of blossoming species with berries and intense colours in the autumn, which attracts birds, mammals and invertebrates to be experienced by people (Gustavsson and Ingelög 1994). Additionally, the edges of many older forests often contain old solitary trees and cultural remnants such as earth and stone banks, marking old ownership boundaries. Such elements add strongly to the experience of the cultural history of the landscape (Høyring 1995).

Urban forests are often cut through by infrastructure or other urban land-uses which increases the amount of edges in comparison to most rural forests. Examples of this can be found in many urban forest landscapes such as Vestskoven at the edge of Copenhagen, Amsterdamse Bos near Amsterdam, and the Zoniënwoud near Brussels. When compared to rural forests, most urban forest landscapes are also rich on edges along internal glades and open spaces of varying size and character.

Despite the abundance of external and internal edges, little attention has been paid to edge zones in design and management of urban forests (Gustavsson et al. 2005). Edges have often been left without management which has homogenised their structure. When left unmanaged especially the outdrawn mosaic edges and the visually open one-step edges gradually disappear (these edge types are described in Box 9.2). However, through a changed attitude the abundance of edges in urban woodlands can be turned into an opportunity to work actively with the creation of varying edge structures incorporating more articulated and diverse visual and spatial qualities and richer flora and fauna diversities at both site and forest landscape level. Consequently, edge restoration and differentiation can contribute significantly to recreational as well as ecological values in urban forest landscapes.

Box 9.2 Main Forest Edges Types

The outdrawn mosaic edge

In uncultivated landscapes where the forest is allowed to expand into open parts of the landscape and in forest pastures with low grazing pressure, successional processes often mean the transitions between closed and open parts grades through intermediate stages of clumps and groups of trees, patches of shrubs or younger trees, herbaceous plants and grass in mosaic patterns. However, since the introduction of systematic forestry by the end of the eighteenth century the forest has predominately been strictly separated from the surrounding countryside by abrupt, straight edges of various profiles, as described below.

One-step edge

One-step edges are those with no shrub layer under the edge trees. They are often dominated by shade-tolerant forest tree species like beech (see Fig. 9.7, FDT 11). It has the best potential to develop in north and east facing edges (in the northern hemisphere) where the forest trees shade the edge zone. One-step edges are also associated with intimate glades, where they allow for visual contact with the interior under the surrounding canopies, and with forest pastures with high grazing pressure, where the animals elevate the canopy by browsing.

Two-step edge

Two-step edges are those with shrubs under the canopy of the edge-trees. By creating a vertical green 'wall' two-step edges can be effective in guiding the direction of the human eye. In Versailles and many other Baroque park landscapes, alleys were often designed as two-step edges to direct the eye towards the point-de-vue.

Three-step edge

In three-step edges a diversity of mostly light-demanding shrubs and tree species create dense, narrow edges with well-developed vertical structures (see Fig. 9.7, FDT 21). South and west-facing edges (in the northern hemisphere)

(continued)

Box 9.2 (continued)

have a higher potential to develop species richness than north or east-facing ones. Visually, the species richness of three step edges is especially noticeable in spring and autumn, as many of the typical edge species have attractive flowers and autumn colours. If local species are used, three-step edges can have high ecological and cultural-historical benefits. Yet, three-step edges seals off the visual contact and much flow of organisms and energy between forest interior and open areas.

Rising edge
In rising edges, trees and shrubs of decreasing size creates wide sloping edges. Rising edges are effective wind breaks for young plantations and help to maintain special microclimatic conditions in forest interiors. Because of this they have become integral parts of afforestation projects in many countries during the last part of the twentieth century. Yet, rising edges can blur the forest experience. An example of this is in 'Vestskoven' which was established to create a 'green lung' for the citizens in the urban sprawl spreading rapidly during the 1960s and 1970s in the western parts of Copenhagen, the capital city of Denmark. In Vestskoven, roads are separated from the forest by edge plantations of up to 30 m width and having the same visual appearance as the miles and miles of scrub plantations along road-sides in the open Danish landscape. Therefore, one does not have the experience of driving through a large urban forest landscape.

9.3.5 Open Areas

Integration of forested and open areas of varying sizes is a common characteristic for many of the most appreciated and well-known recreational forests landscapes in Europe (Nielsen et al. 2005). Open areas are attractive for recreation and provide habitat diversity. This is especially the case if the structure of open spaces is varied in size and management regime, such as grassland, meadows, heath, wetland and open water (Bell et al. 2005; Gustavsson et al. 2005).

General guidelines for extending open areas open areas in recreational forest landscapes exist. In the United Kingom, the extent of open spaces is generally recommended to not exceed around 30% of the area, as this is likely to diminish the feeling of being enclosed by the forest (Forestry Commission 1991). Yet, experiences from urban forest landscapes show that the structure of open areas and their interaction with the local topography, and especially the developmental stage of the surrounding forest, are more important for the feeling of enclosure and intimacy than the total amount of open areas. The size of the forest is also influential on the experience of enclosure. As woodlands increase in size, their diversity increases, usually with an increasing likelihood of incorporating some open areas.

Fig. 9.4 Aerial photo of Dyrehaven north of Copenhagen, showing the variety of open areas of different forms and sizes integrated in the forest (Photo: Peter Lassen)

Jägersborg Deergarden (Dyrehaven) covers an area approximately 1,000 ha to the north of Copenhagen and is an example of an old, well-known recreational forest landscape which is appreciated for its rich variety of forest and open areas (Fig. 9.4). Here a hierarchy of open areas of varying sizes from the Eremitage plain (150 ha) over smaller open areas (2–4 ha) to glades (0.2–1 ha) and canopy gaps (less that 200 m^2) is carefully integrated with a corresponding diversity in sizes of forested areas, tree clumps and solitary trees (Nielsen et al. 2005). The large plains are situated where the local topography is more uniform. The smaller openings and glades are found in parts with pronounced small-scale topographical variations and in areas with rather flat terrain where they add spatial diversity. Jointly, the open areas in Jägersborg Deergarden amount to 50% of the area, but because the wooded parts are dominated by 30 m tall trees, and because of the rich variation in the structure of openings and their articulation of differences in the local topography, the sense of enclosure is always present while strolling around the landscape.

9.3.6 *Designing with Nature*

Today's urban forest landscapes often lack structural complexity and are deprived of visual variation and habitat diversity (Gustavsson 1981; Kaplan 1995). This is partly because traditional silviculture has, for centuries, endeavoured to improve and homogenise site conditions for tree growth by ameliorating soils and draining or ditching (Rune 1997). On the landscape level, this has caused a pronounced loss of habitats and complexity and the disappearance of natural landscape characters and variations. In efforts to restore urban forest landscapes 'inherited' from production traditions, much of the lost complexity and landscape character can therefore be restored if the potential natural hydrology and local variations in site condition and topography are used as starting points for locating a variety of site-adapted habitats with distinct visual appearances and habitat qualities.

A thorough analysis of the landscape in its topographical variation, soil conditions, natural and cultural values, and hydrological features including potential to re-establish wetlands give an excellent basis for a 'nature-based' re-design of the landscape. In this case different Forest Development Types combined with varied open and semi-open nature types and edge types can be used as planning tools.

9.4 Case Study: Vestskoven

In 1967 the gradual conversion of approximately 1,500 ha highly productive horticultural and agricultural land at the urban fringe of greater Copenhagen commenced. The overall aim was to create a large recreational forest that could separate and structure the intense and rapid urban sprawl of the 1960s and 1970s and provide important recreational possibilities for the 300.000 new citizens in the western part of greater Copenhagen.

In the late 1990s the last cropland was acquired, and the landscape was completely converted to various woodland stands and open areas, with approximately half of the area forested and half of it being open grassland located primarily towards the centre of the area (Olsen and Wiegersma 2005). The topography in the area is only slightly undulating and to create variation, two artificial hills were constructed in the open grasslands, using the excavated soil and building materials from the surrounding urban development.

The concept for Vestskoven was to create a romantic landscape with organically formed forests contrasting with open meadows (Olsen and Wiegersma 2005) – highly inspired by the Dyrehaven area north of Copenhagen (Fig. 9.4). Yet, the concept did not take into account the process of acquiring the land. The way the farmlands were acquired became the shaping factor for the structure of the forest. Fields were planted successively as they were purchased, with little further consideration for the composition and relationship between forested and open areas. Further, once acquired, the fields were planted using monocultural stands or simple species mixtures, involving species that were then available at the nurseries. This was done without prior studies of the soil which had been intensively drained to improve the arable and horticultural production (Olsen and Wiegersma 2005). The forest is therefore composed of small stands with abrupt species transitions and edges so that there is no inter-relationship between them and between the forest and the open areas. Today the forest appears as a traditional Danish timber production forest with

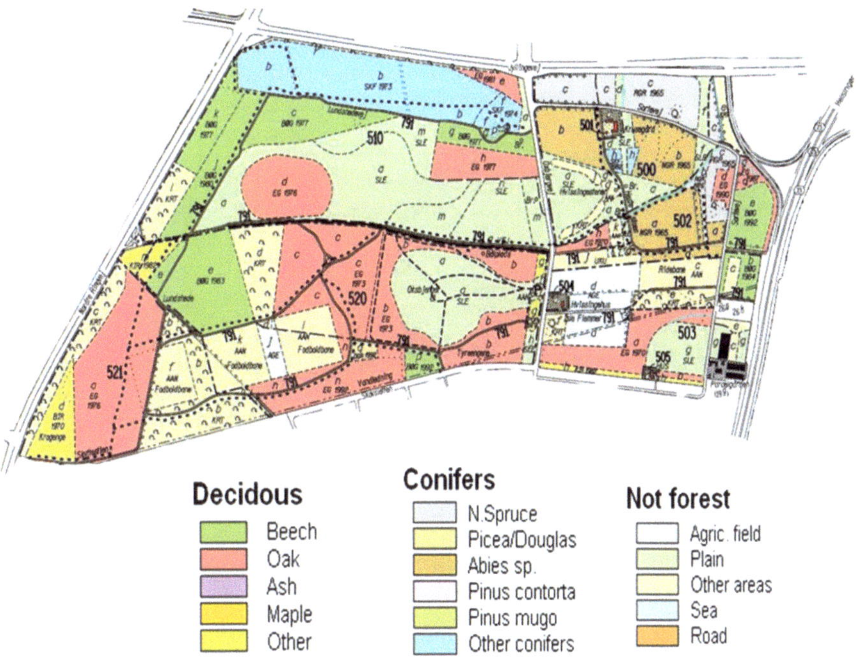

Fig. 9.5 Land-use map for the eastern part of Vestskoven, showing the fragmented composition of uniform blocks of geometrically shaped stands and open spaces

some large open areas for recreation inserted into it (Fig. 9.5). The gradual collapse of the artificial drainage system is slowly restoring the natural hydrology creating potentials for integration of ponds and wetlands.

9.4.1 Educational Project: Re-designing the Vestskoven Forest Landscape

The above description demonstrates that the Vestskoven area incorporates most of the potentials as well as the problems of urban woodlands inherited from the commercial forest management tradition. This makes it useful as a case study for urban forest landscape restoration. Hence, the eastern most part of it (approximately 150 ha) was used as the project area in the international master course Urban Woodland Design and Management at the University of Copenhagen, administered by the two authors. In what follows we give a short description of the project area with its problems and potentials and present how one of the student groups used the concept of Forest Development Types to re-design the area to create an urban forest landscape.

Fig. 9.6 Arial photo of Vestskoven seen from East, showing the abrupt transitions between forested and open areas, where the forest surrounds the foot of the artificial hill. As the trees mature the topographical effect created by the hill will be blurred

Major roads separate the eastern part of the forest from the remaining forest and from the surrounding city (Fig. 9.5). Around 70% of the area is forested with blocks of even-aged stands, whereas the remaining 30% consists of large open areas with grassland and play grounds (foot ball, riding school, enclosure for dogs etc.). The landscape is only slightly undulating, and in the middle an artificial hill rises 30 m above the surrounding plain (Fig. 9.6).

9.4.1.1 The Current Problems

- Uniform stand structures, lack of habitats within the forest
- Simple edge structures and no variation in edge types
- Fragmented blocks of geometrically formed stands and open areas
- Limited interconnection between forested and open areas and lack of connectivity between open areas
- Tree height does not match the size of the open areas causing the loss of forest 'atmosphere'
- Lack of smaller openings and glades
- The forest surrounds the foot of the artificial hill, thus blurring its topographical effect.
- Ponds and wetlands are not integrated within the forest landscape

9.4.1.2 Potential Advantages of the Site

- Stands containing many different species adapted to the site are present including native broadleaved species
- There are young stands potentially able to develop into forests with a variety of structural types
- The collapsing drainage system is slowly restoring the natural hydrology allowing for the development of ponds and wetlands

9.4.1.3 The Restoration Plan

The students' restoration plan (Fig. 9.7) includes four FDTs and is based on the existing values in the young plantations and adjacent plains. The four selected FDTs (FDT 11, Beech pillar hall; FDT 71, Silver fir with beech and spruce; FDT 21, Oak with ash and hornbeam; FDT 92, Grazing forest), have distinct experiential and ecological characteristics and unify the many small stands in larger units. The variety of sizes of open areas is increased by adding small glades in the forested parts. Some of the open areas incorporated within the plan add even further spatial variety and increase the overall coherence of the forested landscape. The larger central plain has been subdivided by additional planting of groups of trees and by breaking up some existing stands to form small groves. Visually, the tree groups are arranged to create fore-ground and middle-ground areas in the views across the plain where the surrounding forest acts as a framing back-ground. Parts of the forested and open areas have been converted to grazing forest through heavy thinning and by planting additional trees. The borders between forested parts and open areas have been re-shaped by selective cutting in some of the existing stands and planting of edge species to create a better relationship between the forested and open areas by forming more diverse and complex edge structures. The artificial hill is visually integrated in the landscape by extending the forest up on the north side of the hill. This underlines the rising topography and creates variation in the ecological conditions from the sunny and dry southern aspect to the moist and shady northern slope. Ponds are restored in areas with emerging wetlands and integrated as attractions for people and biodiversity in relation to small glades, at forest edges, and in larger plains.

9.5 Conclusions and Outlook

The chapter has described a design-led approach to close-to-nature forest management as a strategy for the restoration of forest landscapes in urbanized societies like Denmark, where the importance of recreational and ecological values surpasses the classical timber production goals. In relation to this, the concept of Forest Development Types, as developed in Denmark, has been used as a planning and

Fig. 9.7 The restoration plan for the eastern part of Vestskoven as proposed by a group of students attending the international master course in Urban Woodland Design and Management. The plan in combination with the profile diagrams of the four FDTs including examples of different edge-types gives an instant impression of the anticipated goals for the urban forest landscape as well as for the development at the different forested parts, which can be used as a off-set for participatory planning approaches

design tool for forest landscape restoration. The approach has value because it is able to describe both the development of specific forest areas and also illustrate how the anticipated landscape will integrate different ecosystems including different forest types, semi-open forests, glades, edges, open areas and water bodies. The approach recognizes existing and potential variation in topography, geology and hydrology and allows the development of robust and functional forest landscape for urban societies with high recreational, aesthetic, biological and productive values.

However, the debate about changing silvicultural practices and developing functional forest landscapes covers more than ecology, silviculture and landscape architecture. It is also linked to an on-going discussion within urbanized societies, implying that at the socio-political level urban woodland landscapes must be developed and restored in a transparent and participatory process. In this context, studies have indicated that communication of long-term goals for stand and landscape development through FDT scenarios and their illustration by means of profile diagrams are easy to comprehend by both professionals and lay persons (Larsen and Nielsen 2007; Nielsen et al. 2007). Correspondingly, they might serve to enable the active participation of local people in defining and agreeing upon long-term goals for urban forest landscape development.

This approach has recently been conducted by the municipality of Aarhus, Denmark's second largest city. The tools and considerations described in this chapter have formed the platform for the introduction of nature-based forest management in approximately 2,000 ha of urban forest. In this process the FDTs allow the development of close links between planning, design and management and supported the communication between public, politicians and professionals in the process of defining and agreeing upon long term goals for the urban forest landscape development.

References

Alvey AA (2006) Promoting and preserving biodiversity in the urban forest. Urban For Urban Green 5:195–201

Appleton J (1996) The experience of landscape, Rev edn. Wiley, New York

Bell S, Blom D, Rautamäki M, Castel-Branco C, Simson A, Olsen IA (2005) Design of urban forests. In: Konijnendijk CC, Nilsson K, Randrup TB, Schipperijn J (eds) Urban forests and trees. Springer, Dordrecht, pp 149–186

Christensen M, Emborg J, Nielsen AB (2007) The forest cycle of Suserup Skov – revisited and revised. Ecol Bull 52:33–42

Dwyer JF, Nowak DJ, Noble MH, Sisinni SM (2000) Connecting people with ecosystems in the 21st century, an assessment of our nation's urban forests. General technical report, PNW-GTR-490. USDA Forest Service, Pacific Northwest Research Station, Portland Oregon, p 483

Emborg J, Christensen M, Heilmann-Clausen J (2000) The structural dynamics of Suserup Skov, a near-natural temperate deciduous forest in Denmark. For Ecol Manage 126:173–189

Forestry Commission (1991) Community woodland design guidelines. HMSO, London

Fry G, Sarlöv-Herlin I (1997) The ecological and amenity functions of woodland edges in the agricultural landscape; a basis for design and management. Landsc Urban Plan 37:45–55

Gilbert OL (1989) The ecology of urban habitats. Chapman and Hall, London

Gilbert O, Anderson P (1998) Habitat creation and repair. Oxford University Press, Oxford

Gustavsson R (1981) Naturlika grönytor i parker och bostadsområden. Landskap 58. Swedish University of Agricultural Sciences, Alnarp (in Swedish)

Gustavsson R, Ingelög T (1994) Det Nya landskapet. Skogsstyrelsen, Jönköping, Sweden (in Swedish)

Gustavsson R, Hermy M, Konijnendijk CC, Steidle-Schwahn A (2005) Management of urban woodland and parks – searching for creative and sustainable concepts. In: Konijnendijk CC,

Nilsson K, Randrup TB, Schipperijn J (eds) Urban woodlands and trees. Springer, Berlin, pp 369–397
Hahn K, Emborg J, Larsen JB, Madsen P (2005) Forest rehabilitation in Denmark using nature-based forestry. In: Stanturf JA, Madsen P (eds) Restoration of boreal and temperate forests. CRC Press, Boca Raton, pp 299–317
Hahn K, Emborg J, Vesterdal L, Christensen S, Bradshaw RHW, Raulund-Rasmussen K, Larsen JB (2007) Natural forest stand dynamics in time and space – synthesis of research in Suserup Skov, Denmark and perspectives for forest mangement. Ecol Bull 52:183–194
Hedblom M, Söderström B (2008) Woodlands across Swedish urban gradients; status, structure and management implications. Landsc Urban Plan 84:62–73
Høyring K (1995) Cultural heritage in multiple-use forestry. In: Hytönen M (ed) Multiple-use forestry in the Nordic countries. METLA, The Finnish Forest Research Institute, Vantaa, pp 321–341
Hummel FC (1992) Aspects of forest recreation in Western Europe. Forestry 65:237–251
Jensen FS (1999) Forest recreation in Denmark from the 1970s to the 1990s. The research series, vol 26. Danish Forest and Landscape Research Institute, Hoersholm
Jensen FS, Koch NE (2004) Twenty-five years of forest recreation research in Denmark and its influence on forest policy. Scand J For Res 19:93–102
Jenssen M, Hofmann G (1996) Der natürliche Entwicklungszyklus des Baltischen Perlgras-Buchenwaldes (Melico-Fagetum). Forstwirtschaft und Landschaftsökologie 30(3):114–124
Kaplan S (1992) Environmental preference in a knowledge-seeking, knowledge-using organism. In: Barkow JH, Cosmides L, Tooby J (eds) The adapted mind. Oxford University Press, Oxford, pp 581–598
Kaplan R (1995) Informational issues: a perspective on human needs and inclinations. In: Bradley GA (ed) Urban forest landscapes. Integrating multidisciplinary perspectives. University of Washington Press, Seattle, pp 60–70
Konijnendijk CC, Nilsson K, Randrup TB, Schipperijn J (eds) (2005) Urban forests and trees. Springer, Berlin
Kowarik I (2005) Wild urban woodlands: towards a conceptual framework. In: Kowarik I, Körner S (eds) Wild urban woodlands. Springer, Heidelberg, pp 1–32
Krott M (1998) Urban forestry: management within the focus of people and trees. Pp 9-19 In: Krott IM, Nilsson K (eds) Multiple-use of town forests in international comparison. Proceedings of the first European Forum on Urban Forestry, 5–7 May 1998, Wuppertal. IUFRO Working Group S.6.14.00, pp 9–19
Larsen JB (2000) From plantation management to nature-based silviculture – a Danish perspective. In: Third international congress on sustainability in time and space, Pro Silva Europe, p 187
Larsen JB (2005) Functional forests in multifunctional landscapes – restoring the adaptive capacity of landscapes with forests and trees. EFI-Proc 53:97–102
Larsen JB, Nielsen AB (2007) Nature-based forest management – where are we going? – Elaboration forest development types in and with practice. For Ecol Manage 238:107–117
Larsen JB, Emborg J, Rune F, Madsen P (2001) Skov og biodiversitet. Skovbrugsserien nr. 30, Skov & Landskab, Denmark (in Danish)
Leibundgut H (1959) Über Zweck und Methodik der Struktur- und Zuwachsanalyse von Urwäldern. Schweizerische Zeitschrift für Forstwesen 110:11–124
Lindenmayer DB, Franklin JF, Fischer J (2006) Conserving forest biodiversity: a checklist for forest and wildlife managers. Biol Conserv 129:511–518
Lindhagen A (1996) Forest recreation in Sweden. Four case studies using quantitative and qualitative methods. Dissertation, Swedish University of Agricultural Sciences, Department of Environmental Forestry, Uppsala
Meyer H, Neumann M (1981) Struktureller und entwicklungsdynamischer Vergleich der Fichten-Tannen-Büchen-Urwälder Rothwald/Niderôstereich und Corkova Uvala/Kroation. Forstwiss Centralblatt 100:111–132

Mueller-Dombois D (1987) Natural Dieback in Forests. Group of neighbouring trees may die in response to natural phenomena, as well as to stress induced by human activity. Bioscience 37:575–583

Nielsen AB, Jensen RB (2007) Some visual aspects of planting design and silviculture across contemporary forest management paradigms – perspectives for urban afforestation. Urban For Urban Green 6:143–158

Nielsen AB, Møller FG (2008) Is coppice a potential for urban forestry? The social perspective. Urban For Urban Green 7:129–138

Nielsen AB, Oustrup L, Stachlschmidt P, Nielsen JB (2005) Oplevelsesaspektet i naturnære skove. In: Larsen JBo (red) Naturnær skovdrift. Dansk Skovbrugs Tidsskrift, Dansk Skovforening, København, pp 361–385 (in Danish)

Nielsen AB, Olsen SB, Lundhede T (2007) An economic valuation of the recreational benefits associated with nature-based forest management practices. Landsc Urban Plan 80:63–71

Olsen IA, Wiegersma L (2005) NeighbourWoods. Comparative analysis of three urban woodlands in Denmark and the Netherlands. The Royal Veterinary and Agricultural University, Copenhagen, Denmark. Department of Economics and Natural Resources, Unit of Landscape

Olwig K (2002) Landscape, nature and the body politic: from Britain's renaissance to America's new world. University of Wisconsin Press, Madison

Peterken GF (1993) Woodland conservation and management, 2nd edn. Chapman and Hall, London

Rackham O (2003) Ancient woodland, its history, vegetation and uses in England. Castlepoint Press, Colvend

Ribe RG (1989) The aesthetics of forestry: what has empirical preference research taught us? J Environ Manag 13: 55–74

Rizell M, Gustavsson R (1998) Att alägga skogsbryn – modeller och referenser för anläggning och rekonstruktion. Stad & Land 160, MOVIUM, Swedish University of Agricultural Sciences

Rowntree RA (1995) Towards Ecosystem Management: shifts in the core and context of urban forest ecology. In: Bradley GA (ed) Urban forest landscapes. Integrating multidisciplinary perspectives. University of Washington Press, Seattle, pp 43–59

Rune F (1997) Decline of mires in four Danish state forests during the 19th and 20th century. Forskningsserien nr. 21. Forskningscentret for Skov og Landskab, Hørsholm

Schütz J-P (2006) Opportunities and strategies of bio rationalisation of forests tending within nature-based management. Studia Forestalia Slovenica 126:39–46

Sepahi A (2000) Nature as a model for large-scale planting design. Landsc Res 25:63–77

Smidt GF, Iremonger S, Kelly DL, O'Donoghue S, Mitchell FJ (2007) Enhancing vegetation diversity in glades, rides and roads in plantation forests. Biol Conserv 136:283–294

Sorte G (1989) Landskapet I det mångdimensionella ögat. Sammanfattningar av installations-föreläsningar. Institutionen för landskapsplanering, Swedish University of Agricultural Sciences (in Swedish)

Tregay R (1986) Design and ecology in the management of nature-like plantations. Ecology and design in landscape. In: Bradshaw AD, Goode DA, Thorp EHP (eds) Ecology and design in landscapes. Blackwells Scientific Publications, Oxford, pp 275–284

Tyrväinen L, Silvennoinen H, Kolehmainen O (2003) Ecological and aesthetic values in urban forest management. Urban For Urban Green 1:135–149

Tyrväinen L, Pauleit S, Seeland K, de Vries S (2005) Benefits and uses of urban forests and trees. In: Konijnendijk CC, Nilsson K, Randrup TB, Schipperijn J (eds) Urban forests and trees. Springer, Dordrecht, pp 81–114

United Nations (2004) World urbanization prospects: the 2003 revision. United Nations Publications, New York, 323 pp

Van Herzele A (2006) A forest for each city and town: story lines in the policy debate for urban forests in Flanders. Urban Stud 43:673–696

Chapter 10
Watershed-Scale Adaptive Management: A Social Science Perspective

Catherine Allan, Allan Curtis, and Bruce Shindler

10.1 Introduction: A Social Perspective

Watersheds are physical realities, but the perception and understanding of those physical realities are human, and in particular social, activities. It is societies that define watersheds as entities, and social processes that determine goals for, and management actions within, watersheds. Information on the physical and ecological stresses on watersheds is acted upon by environmental/ resource managers in ways shaped by their societies. To understand watersheds, therefore, it is necessary to understand the values and needs of the people associated with watersheds. In this chapter we draw on understandings developed through our combined social research experience to suggest strategies to achieve truly adaptive management of watersheds.

Catherine Allan and Allan Curtis have evaluated a number of watershed-scale adaptive management projects in Australia, and Bruce Shindler has had extensive involvement with the Adaptive Management Areas in the Forests of the Pacific North West. We are particularly interested in the social, cultural and institutional aspects of adaptive management because management is first and foremost a social activity. As social scientists we emphasize the importance of considering context, and so offer a summary of the world view in which this chapter sits. We acknowledge multiple ways of knowing the world around us, including local and indigenous knowledges. The notion of multiple forms of knowledge includes the relatively straightforward idea that assets can be viewed differently—one person's nature

C. Allan (✉) • A. Curtis
Institute for Land, Water and Society, Charles Sturt University, Albury, Australia
e-mail: callan@csu.edu.au; acurtis@csu.edu.au

B. Shindler
Department of Forest Resources, Oregon State University, Corvallis, OR, USA
e-mail: bruce.shindler@oregonstate.edu

J. Stanturf et al. (eds.), *Forest Landscape Restoration: Integrating Natural and Social Sciences*, World Forests 15, DOI 10.1007/978-94-007-5326-6_10,

walk is another person's source of lumber, for instance—but also acknowledges more profound differences in the ways in which people understand the world, the roles of humans in that world, and even the nature of knowledge and truth (for example, Roe 1996; Miller 1999). We therefore understand planning as a political endeavor. There are social consequences from every planning decision because there are always winners and losers in some sense. 'Successful' resource management programs of any sort must enjoy social acceptability in addition to their bio-physical possibility and economic feasibility. Public judgements about the appropriateness (acceptability) of management activities are based on more than just physical or scientific 'facts.' Judgments are made in response to a complex suite of factors, including knowledge of alternatives and their consequences, and levels of trust in decision makers (see for example Shindler et al. 2002; Howe et al. 2005).

As social scientists we are also interested in how the watershed concept has been, and continues to be 'constructed' to serve the changing needs of our societies. Social discourse shapes the ways in which watersheds are understood, valued and managed. Watersheds are an obvious unit for managing water, but currently watersheds are also considered appropriate as the frame for managing other environmental assets/natural resources, including forests. Although not a new phenomenon, using watersheds as planning units has become particularly popular in the last decade or two (Blomquist and Shlager 2005).

The current focus on watersheds appears to follow from a desire for an integrated approach to natural resource management; not only integration of different landscape features such as surface and groundwater, soil and vegetation and other discipline-based knowledge of these landscape elements, but also the integration of the efforts/resources of governments, private corporations and individuals, and the roles of managers/planners, researchers and landholders (Curtis and Lockwood 2000). Watershed organizations have been established as part of government initiatives to implement integrated natural resource management in both the US (where they are most often called watershed councils) and in Australia (where landcare groups perform many similar functions), although their organizational support structures are different in each country (Curtis et al. 2002). Current discourses of 'watersheds' imply that people within a particular watershed will share some sense of 'place', through either ownership or an emotional attachment to the land in its broadest definition (Kruger 2001). A watershed thus becomes a space that links people with their environment, and with each other. The watershed concept has been used as a means of scaling-up very small, locally focused environmental management activities in both Australia and the US (Ewing 1999; O'Neill 2005). Because watersheds can encourage and enable people to work together on larger-scale projects while maintaining their sense of connection with the environment, watersheds also represent an important venue in which learning can be promoted and the results of that learning applied.

Watershed management occurs within the context of prevailing societal norms and practices. For the past century, scientific management, a reductionist approach to understanding and controlling activity through rational planning, has dominated Western policy making regarding natural resource management (Smith 1997).

Rational planners work from the premise that the world is knowable and predictable, that the goals of management are clear, and that there is one 'right' answer that can be found through objective, technical enquiry (Rittel and Webber 1973). 'Rational' reductionist management of water, forests, agriculture and other natural resources worked well when the goals of management were narrowly defined (exploitation, harvesting, human use, human protection), and while the ecosystem consequences of this narrow focus could be ignored (Holling 1995; Holling and Meffe 1996). The environmental degradation that has resulted from this focus on exploitation in both the US and Australia is sobering, even when only a partial reckoning of that degradation is possible. For example, Australia's State of the Environment report notes that "*Comprehensive ecological information is available on at least 10% of mammal, bird and amphibian species, and partial ecological information is available on around 60% of known forest-dwelling vertebrate and vascular plant species. However, very limited information is available on forest-dwelling invertebrates, fungi, algae and lichens. A total of 1,287 forest-dwelling species are listed as vulnerable, endangered or threatened under the Environment Protection and Biodiversity Conservation Act 1999)*" (Commonwealth of Australia 2008).

Similarly, it is noted that around one third of the US's native animal species, including aquatic animals, are considered 'at risk' to some degree (The Heinz Center 2002). These and other degradation problems are likely to be exacerbated by the predicted impacts of climate change, highlighting just how little knowledge we have as a basis for rational planning of future action.

Of course, natural resources managers have not been idle. There have been many management responses to ecosystem collapses, and numerous attempts to reduce negative impacts on water, forests, soils and biodiversity, and restoration figures prominently in many of these. Often, however, these endeavors have been constrained by the growing recognition of complexity and uncertainty, along with some entrenched institutional habits of problem framing and response implementation. The attributes of current/traditional natural resource management organizations (including formal structures such as Federal and State government agencies) and institutions (including informal structures such as social movements and watershed citizen groups) exert a strong influence on the responses that can be made. Current natural resource/ environmental asset managers (and the societies which encourage their employment) have a preference for activity—they like to be doing, and to be seen to be doing. In other words, this style embodies the 'can do' spirit which characterizes those in the natural resource professions. This preference is supported by government programs and projects which emphasize milestones, and targets, and rapid accountability in the form of activity audits (Allan and Curtis 2005). Natural resource management organizations and institutions are also patently risk averse (Allan and Curtis 2003; Stankey et al. 2006a). Risk averse societies and their institutions and organizations fear doing, or being seen to do, something that turns out badly. This has a constraining effect on managers, making them unwilling, or even afraid, to experiment. The product of these pressures is a generation of agency staff who are always seeking to be active, but only in safe and predictable ways.

Against this background is a growing appreciation of the limits of predictive models when faced with high levels of complexity and uncertainty, coupled with the challenges of dealing with a world full of surprises, secondary- and tertiary-effects, and social and political volatility (Herrick and Sarewitz 2000). Over recent decades ecosystem collapses, combined with a broadening range of accepted human expectations and understandings, have led to management impasse where, despite the strong desire to be active, in many places it seems impossible to undertake any management at all (Stankey et al. 2003). Alternative management paradigms are needed to deal with high levels of complexity and uncertainty, and to accelerate the rate at which managerially relevant knowledge can be acquired. Adaptive management appears to promise to be one of those alternative paradigms.

10.2 Adaptive Management as a New Approach

Adaptive managers deliberately set out to learn from policy experiments to improve future practice. In a natural resource management context adaptive management involves learning from the outcomes—expected and unexpected—of implementing project activities. Adaptive management initially was conceived of and presented as a technical response to problems with ecological and social resilience (for example Holling 1978; Walters 1986; Walters and Holling 1990), but has increasingly become as much a social and civic undertaking as a technical one (Lee 1993; Gunderson et al. 1995). Adaptive management stands in sharp contrast to traditional reductionist scientific inquiry because of its emphasis on learning from management practice (Hillman et al. 2000). Adaptive management is also different from traditional incremental, bumbling along approaches because it is planned and purposeful, and has a focus on improving management through deliberate learning. While there are many different ways of interpreting and understanding adaptive management we suggest that watershed-scale adaptive management involves:

- management activities specifically designed to test hypotheses through ecosystem-scale, holistic experiments;
- active reflection on the outcomes of those management activities;
- provision of mechanisms for multi-disciplinary and multi-stakeholder involvement;
- an emphasis on collaborative or participatory social learning;
- provision of mechanisms for incorporating learning into planning and management; and
- development and maintenance of appropriate communication fora for all project participants.

In the US the theory of adaptive management is widely accepted, and it underpinned the development of the Northwest Forest plan (Stankey et al. 2006a, b). In Australia adaptive management has become formally embedded in watershed planning through the bilateral agreements between the Federal and State governments that underpin the National Action Plan for Salinity and Water Quality (NAP)

(Commonwealth of Australia 2003). The management of forest resources, invasive plants, pest animals, flow regimes in rivers, soil health and biodiversity maintenance are all supposedly guided by adaptive management principles; as an example adaptive management is applied to protect threatened species at risk from forestry operations in NSW (NSW Department of Primary Industries 2008). The pervasiveness and acceptability of (at least the rhetoric of) adaptive management can be gauged by a casual browser search of the World Wide Web, restricting the search to 'adaptive management' and 'environment' or 'natural resources.' This will yield hundreds of thousands of results. Yet despite all of this interest, catching adaptive management in practice, especially at watershed-scales, has proven difficult (Lee 1999; Allan and Curtis 2005; Stankey et al. 2006a, b). We suggest that the scarcity of operational adaptive management of natural resources is caused by social constraints, rather than technical difficulties.

10.3 The Challenges to Successful Implementation of Adaptive Management

Numerous constraints on undertaking adaptive management have been identified in the growing literature on adaptive management of natural resources. These constraints include risk aversion, inadequate protocols and inadequate resources.

Effective adaptive management embraces failures and shortcomings because they provide an opportunity to learn and change (Gunderson 1999). Embracing, or even acknowledging, failure or surprise remains antithetical to many managers and policy makers. Wildavsky and Dake (1990) showed that cultural biases are a more powerful predictive tool for risk perception than either knowledge or personality type, and we appear to be particularly risk averse societies at present. Across society, there is a growing inability to portray the concept of risk in a socially acceptable manner (Slovic et al. 2004). All actions, including 'no' action, involve risk, yet, increasingly, resource managers are called upon to avoid actions, policies, and programs that may lead to risk for species, conditions, and values. We suggest that aversion to failure also follows from the strong technological orientation developed in our cultures from over a century of reliance on the scientific method; errors are seen as the result of shortcomings, incompetence and poor planning, rather than an inevitable result of working in the face of complexity and uncertainty (for more on this see Stankey et al. 2006a, b). In this cultural environment there are very few incentives and rewards to encourage risk-taking, failure-accepting behaviour. There is also the corrosive belief that adaptive approaches could reveal that past policies and practices have been flawed; publicly acknowledging such past shortcomings is anathema to many. Admitting to past 'failures' also highlights the need for conventional ways of acting to change, and such changes often are strongly resisted in organizations. Finally, vested interests and powers are often content with the way things are and adaptive management is viewed with suspicion and alarm because it could lead to changes inimical to those interests.

That current and potential adaptive managers often feel constrained by institutional protocols and practices is suggested in the research literature focused on cultural conditions (Ison and Watson 2007; Jacobson et al. 2006) and institutional and legal practices (for example Thrower 2006) were echoed at practitioner workshops in Australia (Allan and Curtis 2003) and the US (Allan et al. 2008).

The participants of both the Australian and US workshops also stressed that inadequate resources are a fundamental constraint. There does not appear to be the people, money or the political will to allow projects to mature, or for purposeful reflection and learning to be undertaken.

To illustrate and explore these constraints further we draw briefly on two examples with which we have been involved as participants and evaluators; the development and implementation of the adaptive management plan for the forests of the US Pacific Northwest, and the implementation of 'Heartlands' in southeastern Australia.

Ten adaptive management areas (AMA) were established in 1994 in the forests of Washington, Oregon and northern California as sites for ecological, social and organizational learning. Each area included districts with mixed forest ownership, including federal lands, and most AMA were associated with communities affected by reduced timber harvests from federal lands (Stankey and Shindler 1997a, b). The AMA provide an innovative institutional structure for achieving the goals of the Northwest Forest Plan, and for improving shorter and longer term understanding of the region's biophysical and socioeconomic systems and the adaptive management process (Stankey et al. 2006a, b).

Operating between 2000 and 2003, Heartlands was a government initiated watershed-scale project which sought to design and implement landscape scale land-use change. Although on a much smaller scale than the AMA process described above, Heartlands had similar aims of learning about the landscape and how to restore its function. The project team defined Heartlands as an adaptive management project (CSIRO Heartlands Core Group 2000). Heartlands combined implementation of on-farm land management works with scientific enquiry into farm forestry, catchment hydrology and biodiversity. One of the Heartlands sites was the Billabong watershed, located in southwestern slopes of the state of New South Wales. The Billabong is a foothills environment that currently supports forestry and grazing enterprises in the east and predominantly cropping enterprises in the west of the watershed. Over a 2-year period on-ground works under the Heartlands banner were funded by Federal and State programs, while the Murray Darling Basin Commission provided money for research. The Billabong Operations Group (BOG) comprising local landcare staff and members, CSIRO scientists, landholders, and representatives of State natural resource management agencies was created to steer the project, manage the funds, co-ordinate implementation and report on activities and findings.

Project management involves risks to both the resource and to management, and aversion to each type of risk can impact on different aspects of adaptive management. In one of the AMA in the Pacific Northwest a proposed evaluation of alternative forestry management prescriptions for enhancing old growth conditions along the riparian zone was opposed because the researcher was unable to give fishery biologists and regulators a guarantee that the experiment would not jeopardize salmon. To put that another way, the experiment would only be able to take place if the outcome was sufficiently known to be able to make guarantees, and if that was the case, the experiment would not need to be conducted (Stankey et al. 2003). In the Heartlands case aversion to the risk of losing future funding constrained the opportunities for sharing lessons with the policy group in the lead agency—lessons which could have been perceived as failures were carefully managed. For instance, in 2002, in response to severe drought, many landholders chose not to plant the tubestock trees and shrubs supplied through the project in the autumn period, holding off this work until sufficient rain had fallen for the plants to survive. However, the report to the funding body implied that planting had occurred in autumn to satisfy the funding agency's requirements that the funds be expended and works completed on time. This little piece of information management resulted in achieving dual goals of living plants and satisfied funders, but there was no communication of the important lesson about the need for flexible planting times (Allan 2004).

Adaptive management requires broad based participation and involvement. Research five years after the establishment of the AMA suggests that the internal operation of management agencies is a major stumbling block for successful public engagement in adaptive management. In this case, there were very low levels of organizational support for personnel in adaptive management functions (Shindler et al. 2003).

The Heartlands project provides a clear example of the importance of financial commitment for the medium- to longer-term. Heartland was launched as a grand vision that was to have involved medium- to long-term evaluation of implemented activities. However, initial funding for the project was only for three years, guaranteed one year at a time. Shifts in priorities for environmental management led to funding for the project being stopped in 2003. On ground changes had occurred as a result of project activities, but the opportunities for learning from them and the monitoring regimes established were greatly reduced by the sudden end to the project. Many of the local participants were disappointed, and felt let down by the way the project was concluded (Allan 2009).

10.4 Building Effective Adaptive Management Organisations and Processes

Adaptive management is proposed for many different resource sectors, at a variety of scales and a range of social contexts, so it is inevitable that there is a variety of ways that effective adaptive management can be accomplished. Indeed, one of the key lessons identified in a recent book in which adaptive managers reflected on their practice was the importance of recognizing and understanding the context in which the adaptive management develops (Allan and Stankey 2009). However, we propose some general strategies for adaptive managers to consider.

Strategy 1- Define adaptive management, focus and be purposeful. As a pre-requisite for other actions, watershed-scale adaptive management must be recognized as a radical departure from established ways of managing natural resources. Adaptive management is not 'business as usual', nor should it be seen as an excuse to muddle through management problems.

Grint (2005) suggests that leaders create the conditions for managers by defining the context in which appropriate management actions can occur. There is a role for effective leadership to create the institutional conditions that will enable and encourage experimental and reflective management. Sound leadership, not just in a hierarchical sense, but throughout an organization, is required to support this radical departure in thinking and practice (Shindler et al. 2002). In other words, adaptive management needs champions who have (or who have been given) the time, resources, capacities and power to influence the ways in which policy is devised and its tools are implemented. When such leadership is in place purposeful activities can be developed and implemented.

Strategy 2- Encourage and support evaluation. Evaluation, an activity that involves learning rather than simple auditing, is central to the adaptive management cycle. There is an excellent body of work on effective program evaluation (for example Cook and Shadish 1986; Guba and Lincoln 1989; Rossi and Freeman 1993; Weiss 1997). However, to be able to apply these or other evaluation approaches space must be actively created in a project or program to allow genuine reflection on processes and outcomes. Creating such space may require a serious reassessment of institutional rewards and punishments. There must be an acceptance of the limits of knowledge and the possibility of errors and mistakes. Even the most rigid and conservative *status quo* management can produce major errors (i.e., there is no such thing as a 'no action' alternative; no action *is* an action). Acceptance of evaluation as an integral part of implementation needs to extend beyond scientists and managers to include citizens, policymakers, and politicians. Adaptive management relies on acknowledging that because we are dealing with uncertainty and complexity, it often will prove necessary to pause, reflect and maybe even start over, as new knowledge and understanding reveal that the intended course is unsuitable or undesirable.

Strategy 3- Collaborate and integrate. For multi-disciplinary, collaborative and/or participatory approaches to operate effectively there must be recognition and acceptance of multiple ways of knowing and understanding the world. Again, there is a wealth of theory and experience available to guide managers who are willing to use participatory approaches, ranging from collaboration (for example Allen et al. 2001; Poncelet 2004) through to full participatory methods (for example Spencer 1989) and social learning (Ison and Watson 2007; Roling et al. 2001; Schusler et al. 2003), as well as useful warnings about potential disbenefits (for example Kapoor 2001; Rahnema 1997; Swanson 2001). However, these approaches cannot be simply lifted from a text and applied; ideas surrounding the creation and legitimacy of the knowledge that underpins management decisions need to be discussed, clarified and acted on in the areas in question, especially when integration of scientific and other information is proposed. For example, there are legitimate concerns with the question of local capacity in terms of the ability to interpret scientific work; if nothing else the work and language of scientists is unfamiliar to many non-scientists. There are also problems related to belief systems and assumptions about scientific inquiry; scientists (and many non-scientists) might find it difficult to accept that local/experiential knowledge is both legitimate and useful. All people who create and /or use knowledge must be diligent in efforts to achieve real communication. Researchers and managers alike must work together at the watershed-scale to bridge the gaps between theory and practice, and between social and technical understanding of communities dependent on watersheds. The challenge is to use their combined knowledge and skills to engage potential supporters among administrative, legal, and political actors to build commitment to adaptive management. Underpinning all of this is the need for well informed and committed leaders.

10.5 Conclusion

Adaptive management uses greater resources than conventional management, and is not necessarily needed in every situation. Adaptive management is useful when there is uncertainty about the nature of the resource base and the outcomes of potential management activities, combined with multiple and/or changing societal goals for the resource base. Once a decision is made to use adaptive management, however, it should be undertaken in a purposeful and conscious manner. Anyone planning to use adaptive management needs to consider the cultural contexts of the project. Firstly, it is important to understand which activities are most valued and rewarded by the wider society, as some may be conducive to adaptive management (for example questioning, reflecting, and embracing complexity) while others may constrain its use (for example achieving milestones, embracing simple solutions to manage complexity). Secondly the cultures within governments and their bureaucracies should be understood, with a commitment to modifying them if needed. Commitment to supporting the essential qualities of adaptive management within these organizations is necessary to ensure that adequate resources are available for adaptive management; this includes money, but just as importantly, time and intellectual resources. Planners and managers require educational, administrative, and political support as they seek to understand when and how to implement adaptive management.

Adaptive management encourages scrutiny of prevailing social and organizational norms and this is unlikely to occur without a change in the culture of natural resource management and research. Formal links between monitoring, evaluation and learning and policy development could be developed through formalized documentation protocols or mandatory learning summaries, much like those used effectively on the Applegate AMA in Oregon to capture lessons learned from each project (Shindler et al. 1999). New structures may be required to reduce the current barriers between 'researchers' and 'implementers'. Different power sharing structures may also need to be developed and maintained, in particular allowing power to be shared between technical and non-technical experts. Finally, new educational, training and skill development structures are required, both within management institutions and formal education venues.

Once suitable structures are developed, there will be a need for appropriate, possibly new, institutional processes.

We suggest the following processes for the facilitation of adaptive management:

- Training and support for staff and other participants in watershed organizations. In Australia this may include support for staff in the new Catchment Management organizations and for Landcare participants. In the US, there is a need to make adaptive management an essential element in natural resource planning at the national, regional, and state level;
- Reflective practice should be part of natural resource management related tertiary learning and training;
- Rewards in performance management reviews for taking risks, and for taking time to learn from activities;
- A genuine commitment to evaluation as well as to mere project auditing;
- A commitment to multi-party monitoring and evaluation;
- Establishment of appropriate community fora to facilitate social learning;
- Particularly in the US, there is a need for some form of relief from statutory prescriptions that act to limit experimentation, particularly under conditions of high uncertainty; and
- In both countries, there is a need to examine critically the educational curricula of natural resource management programs to encourage critical thinking skills to support professionals as we move away from a high dependence on rule-based decision making.

Is adaptive management simply too hard for us humans to do? We suggest that adaptive management will not occur spontaneously, and that it will require the efforts of leaders and others to overcome entrenched social constraints and institutional structures. Mobilizing resources and energy for adaptive management will not always be sensible or appropriate. When it is sensible and appropriate, however, genuine commitment to adaptive management should benefit both physical watersheds and the people who depend on them.

Acknowledgements We acknowledge the contribution made by George Stankey to the American Water Resources Association Journal paper on which this chapter is based.

References

Allan C (2004) Improving the outcomes of adaptive management at the regional scale. PhD thesis, Charles Sturt University, Albury, New South Wales

Allan C (2009) Reviewing adaptive management through a wicked lens. In: Lane M, Robinson C (eds) Contested country. CSIRO Publishing, Brisbane

Allan C, Curtis A (2003) Regional scale adaptive management: lessons from the North East Salinity Strategy. Austr J Environ Manag 10:76–84

Allan C, Curtis A (2005) Nipped in the bud: why regional scale adaptive management is not blooming. Environ Manag 36:414–425

Allan C, Stankey GH (eds) (2009) Adaptive environmental management: a practitioner's guide. Springer, Amsterdam
Allan C, Curtis A, Stankey G, Shindler B (2008) Adaptive management and watersheds: a social science perspective. J Am Water Resour Assoc 44:166–174
Allen W, Bosch O, Kilvington M, Oliver J, Gilbert M (2001) Benefits of collaborative learning for environmental management: applying the integrated systems for knowledge management approach to support animal pest control. Environ Manag 27:215–223
Blomquist W, Shlager E (2005) Political pitfalls of integrated watershed management. Soc Nat Resour 18:101–117
Commonwealth of Australia (2003) Agreement between the Commonwealth of Australia and New South Wales, 2003. http://www.napswq.gov.au/downloads/pdf/nsw-bilateral.pdf. Accessed Jan 2004
Commonwealth of Australia (2008) Australia's State of the forests report 2008. http://adl.brs.gov.au/forestsaustralia/publications/execsummary.html. Accessed Jan 2009
Cook TD, Shadish WR (1986) Program evaluation: the worldly science. Annu Rev Psychol 37:193–232
CSIRO Heartlands Core Group (2000) Draft heartlands five year plan for 2001–2005: CSIRO/MDBC
Curtis A, Lockwood M (2000) Landcare and catchment management in Australia: lessons for state-sponsored community participation [review]. Soc Nat Resour 13:61–73
Curtis A, Shindler B, Wright A (2002) Sustaining local watershed initiatives: lessons from Landcare and Watershed Councils. J Am Water Resour Assoc 38:1207–1216
Ewing S (1999) Landcare and community-led watershed management in Victoria, Australia. J Am Water Resour Assoc 35:663–673
Grint K (2005) Problems, problems, problems: the social construction of 'leadership'. Hum Relat 58:1467–1494
Guba EG, Lincoln YS (1989) Fourth generation evaluation. Sage, Thousand Oaks
Gunderson L (1999) Resilience, flexibility and adaptive management – antidotes for spurious certitude? Conserv Ecol Online 3(1). http://www.consecol.org/vol3/iss1/art7
Gunderson LH, Holling CS, Light SS (1995) Barriers & bridges to the renewal of ecosystems and institutions. Columbia University Press, New York
Herrick C, Sarewitz D (2000) Ex post evaluation: a more effective role for scientific assessments in environmental policy. Sci Technol Hum Value 25:309–332
Hillman TJ, Koehn JD, Mitchell D, Thompson D, Sobels JD, Woodside D (2000) The Murrumbidgee: assessing the "health" of a working river. Irrigated Agribusiness Taskforce and Department of Land and Water Conservation, Canberra
Holling CS (ed) (1978) Adaptive environmental management and assessment. Wiley, Chichester
Holling CS (1995) What barriers? what bridges? In: Gunderson LH, Holling CS, Light SS (eds) Barriers and bridges to the renewal of ecosystems and institutions. Columbia University Press, New York
Holling CS, Meffe GK (1996) Command and control and the pathology of natural resource management. Conserv Biol 10:328–337
Howe GT, Shindler B, Cashore B, Hansen E, Lach D, Armstrong W (2005) Public influences on plantation forestry. J For 103:90–94
Ison R, Watson D (2007) Illuminating the possibilities for social learning in the management of Scotland's water. Ecol Soc 12:1–21
Jacobson SK, Morris JK, Sanders JS, Wiley EN, Brooks M, Bennetts RE et al (2006) Understanding barriers to implementation of an adaptive land management program. Conserv Biol 20: 1516–1527
Kapoor I (2001) Towards participatory environmental management? J Environ Manag 63:269–279
Kruger L (2001) What is essential is invisible to the eye: understanding the role of place and social learning in achieving sustainable landscapes. In: Sheppard SRJ, Harshaw HW (eds) Forest and landscapes: linking ecology, sustainability and aesthetics. CABI Publishing, Wallingford

Lee KN (1993) Compass and gyroscope: integrating science and politics for the environment. Island Press, Washington, DC
Lee KN (1999) Appraising adaptive management. Conserv Ecol online 3(2). (http://www.consecol.org/vol3/iss2/art3)
Miller A (1999) Environmental problem solving: psychosocial barriers to adaptive change. Springer, New York
NSW Department of Primary Industries (2008) Forests NSW seeing report 2007–2008. Forests NSW, Sydney
O'Neill KM (2005) Can watershed management unite town and country? Soc Nat Resour 18:241–253
Poncelet EC (2004) Partnering for the environment: multistakeholder collaboration in a changing world. Rowman and Littlefield Publishers, Oxford
Rahnema M (1997) Participation. In: Sachs W (ed) The development dictionary: a guide to knowledge as power. Orient Longman, Hyderabad
Rittel HWJ, Webber MM (1973) Dilemmas in a general theory of planning. Policy Sci 4:155–169
Roe E (1996) Why ecosystem management can't work without social science: an example from the Californian Northern Spotted Owl controversy. Environ Manag 20:667–674
Roling NG, Jiggins J, Geisler CC, Schelhas J, Wollenberg E (2001) Agents in adaptive collaborative management: The logic of collective cognition. In: Buck LE (ed) Biological diversity: balancing interests through adaptive collaborative management. CRC Press , Boca Raton, Florida, pp 145–170
Rossi P, Freeman H (1993) Evaluation: a systematic approach. Sage, Thousand Oaks
Schusler TM, Decker DJ, Pfeffer MJ (2003) Social learning for collaborative natural resource management. Soc Nat Resour 16:309–326
Shindler B (2000) Landscape-level management: it's all about context. J For 98:10–14
Shindler B, Aldred-Cheek K, Stankey GH (1999) Monitoring and evaluating citizen-agency interactions: a framework developed for adaptive management. General Technical Report PNW-GTR-452. U.S. Department of Agriculture, Forest Service, Pacific Northwest Research Station, Portland
Shindler B, Brunson M, Stankey GH (2002) Social acceptability of forest conditions and management practices: a problem analysis. General Technical Report PNW-GTR-537. U.S. Department of Agriculture, Forest Service, Pacific Northwest Research Station, Portland
Shindler BA, Beckley TM, Finley C (eds) (2003) Two paths through the forest: the sustainability challenge in Canada and the United states. Oregon State University Press, Corvallis
Slovic P, Finucane M, Peters E, MacGregor DG (2004) Risk as analysis and risk as feelings: Some thoughts about affect, reason, risk, and rationality. Risk Anal 24:311–322
Smith GR (1997) Making decisions in a complex and dynamic world. In: Kohm KA, Franklin JF (eds) Creating a forestry for the 21st century: the science of ecosystem management. Island Press, Washington, DC
Spencer LJ (1989). Winning through Participation. Dubuque: Kendall/Hunt Publishing Company
Stankey GH, Shindler B (1997a) Adaptive management areas: achieving the promise, avoiding the peril. General Technical Report. PNW-GTR-654. U.S. Department of Agriculture, Pacific Northwest Research Station, Portland
Stankey GH, Shindler B (1997b) Adaptive management areas: achieving the promise avoiding the Peril. United States Department of Agriculture Forests Service Pacific Northwest Research Station, Portland
Stankey GH, Bormann B, Ryan C, Shindler B, Sturtevant V, Clark R, Philpot C (2003) Adaptive management and the Northwest Forest Plan: rhetoric and reality. J For 101:40–46
Stankey GH, Clark RN, Bormann BT (eds) (2006a) Learning to manage a complex ecosystem: adaptive management and the Northwest Forest Plan. Research Paper PNW-RP-567. U.S. Department of Agriculture, Forest Service, Pacific Northwest Research Station, Portland
Stankey GH, Clark RN, Bormann BT (2006b) Learning to manage a complex ecosystem: adaptive management and the Northwest Forest Plan. Department of Agriculture Forest Service, Portland

Swanson LE (2001) Rural policy and direct local participation: democracy, inclusiveness, collective agency, and locality based policy. Rural Sociol 1:1–2
The Heinz Center (2002) The state of the nation's ecosystems. Cambridge University Press, Cambridge
Thrower J (2006) Adaptive management and NEPA: how a nonequilibrium view of ecosystems mandates flexible regulation. Ecol Law Q 33:871–895
Walters C (1986) Adaptive management of renewable resources. Macmillan, New York
Walters CJ, Holling CS (1990) Large-scale management experiments and learning by doing. Ecology 71:2060–2068
Weiss CH (1997) How can theory-based evaluation make greater headway? Eval Rev 21:501–524
Wildavsky A, Dake K (1990) Theories of risk perception: who fears what and why? Dædalus 19:41–60

Chapter 11
The Economics of Restoration

Kerrie A. Wilson, Megan Lulow, Jutta Burger, and Marissa F. McBride

11.1 Economics and the Environment

The main premise of economics is simply the efficient allocation of scarce resources. Scarcity occurs when the needs and wants of an individual (or that of a group) exceed the resources available to satisfy them. Because of scarcity, competition may arise and the available resources have to be rationed (either through price adjustment or through management of production, distribution, exchange and consumption of goods and services). As a consequence of this rationing, choices and allocation decisions must be made, and trade-offs are inevitable.

Macroeconomics is concerned with how the economy works as a whole, which is generally measured as the sum total of economic activity, dealing with the issues of growth, inflation and unemployment, and the effects of government actions (such as changing taxation levels or interest rates) on these factors. Macroeconomic policy occurs at two scales – at the scale of country-level fiscal and monetary policy and at the international scale. In its simplest form, macro-economic policy at a country-level is aimed toward achieving the maximum inflation-acceptable (usually 3% per annum) growth rate of real gross domestic product (GDP). Economists are therefore, predominantly concerned with the level of production of an economy at a

K.A. Wilson (✉)
School of Biological Sciences, The University of Queensland,
St. Lucia, Brisbane, QLD 4072, Australia
e-mail: k.wilson2@uq.edu.au

M. Lulow • J. Burger
Irvine Ranch Conservancy, 4727 Portola Parkway, Irvine, CA 92620-1914, USA

M.F. McBride
School of Botany, University of Melbourne,
Parkville, VIC 3010, Australia

J. Stanturf et al. (eds.), *Forest Landscape Restoration: Integrating Natural and Social Sciences*, World Forests 15, DOI 10.1007/978-94-007-5326-6_11,

whole, and the indices that estimate its health, such as the economic growth rate, the unemployment rate, and the inflation rate (the rate at which the average price of the goods in the economy increases over time).

Macroeconomics affects biodiversity conservation and natural resource management activities, such as landscape restoration, through the decisions made to invest in such activities (versus other societal priorities such as health and defence) and the development of policies that influence this macro-allocation. For example, the Organization for Economic Co-Operation and Development (OECD) (2007) reported that for member countries for the years 1990–2004 environmental expenditure for pollution abatement and control was in the range of 0.2–2.7% of national GDP and public research and development (R&D) financing for environmental protection was in range of 0.1–4.8% of the total R&D budget allocations (http://www.oecd.org/dataoecd/37/45/38230860.pdf). Local governments in Australia receive over AUD$1.5 billion in revenue for natural resource management activities, which equates to 8% of their total revenue. These funds are then allocated to issues as diverse as the management of parks and reserves, to that of water supply (Australian Bureau of Statistics 2004). Macroeconomic decisions associated with restoration are likely to entail an analysis of the costs and benefits to the economy, society and the environment of the investment and an evaluation of the international obligations related to the decision (such as the Convention on Biological Diversity or the Kyoto Protocol).

Understanding the key components of allocations at a macro-level is useful for estimating the benefits of expenditures. In macroeconomic terms, goods and services are considered to be delivered through resources, and economists generally divide resources into three types: land, labour, and capital. Macroeconomists typically view the environment as capital. Two types of macro-level activities involving natural capital can be distinguished – those that increase the total volume of the goods and services delivered and those that maintain the existing flow of goods and services. Ecologists and natural scientists typically refer to ecosystem goods and services. Ecosystem goods and services are the direct or indirect benefits that humans obtain from ecosystems and can be categorized as provisioning (e.g. of food, medicines, firewood), regulating (e.g. of climate and floods), supporting (e.g. pollination) and cultural (e.g. of religious significance) (Costanza et al. 1997; Millennium Ecosystem Assessment 2005)

Property rights or tenure system arrangements represent an important consideration in the allocation of resources to ensure the correct functioning of markets. Such arrangements may entail rights and privileges to use a particular resource, or policy to regulate or control access to a resource. In theory, the owner of a resource with a well-defined right to that resource has an incentive to use the resource efficiently in order to minimise the decline in its value. Property-rights are therefore a key issue for marketable goods obtained from the environment, particularly in a free-market economy. The key characteristics of property rights to produce efficient and effective allocations of resources are:

1. Clearly defined – to remove ambiguity about the asset, owners, and penalties for illegitimate use,

2. Exclusive – all benefits and cost as a result of owning and using the resources should be accrued to the owner,
3. Transferable – all property rights should be transferable, and
4. Enforceable – all property rights should be secure.

The extent to which environmental goods and services are accounted for in national accounting systems is an important consideration. Environmental economists argue that the scarcities of natural resources and the externalities that arise through economic activity be accounted for to avoid market failures. An externality is an unintended and uncompensated loss or gain in the welfare of one party resulting from an activity by another party (Daly and Farley 2010). Contemporary economic policy with inadequate cost-internalisation (incorporation of negative external effects) is generally viewed as unsustainable and contributing to the loss of biodiversity and environmental degradation (Lawn 2008). When this occurs the damage to natural capital is not accounted for nor is the costs of repair or maintenance. A possible solution is that the use and misuse of natural resources and the environment should be measured and accounted for at a national scale (i.e. Green Accounting). Under such a scenario, the value of resource use would be accounted for in the determination of a nation's GDP. The basic argument for accounting for the environment in the determination of national accounts is that by not doing so and treating such services as free goods, the value of the natural environment will remain at zero and provide no incentive for its preservation either now or in the future. Achieving compatibility between economic decisions and the environment is the basis of the concept of sustainability.

Microeconomics is a branch of economics that studies how individuals or organisations make decisions to allocate limited resources and typically concerns how one part of the economy works. It is often viewed that an alignment with sustainable development and environmental protection goals will occur at the microeconomic level. That is, the increased scarcity of natural resources will result in high resource prices (particularly for those which have a market value and the rates of regeneration and depletion are quantified), which will result in greater efficiency of resource use and thereby lessen demands on natural resources. From a restoration perspective, the role of microeconomics might be to identify the best signals (e.g. prices, incentives) to encourage landholders to restore their land. Here the role of ecologists is to identify the point where ecosystem goods and services are threatened (i.e. ecological thresholds) and the preferred approach to management, protection and/or amelioration of the ecological system.

Environmental degradation (and restoration) often affects a large area and can entail substantial off-site impacts (e.g. land clearing and over-irrigation in a catchment may result in salinisation of the immediate catchment and also impact downstream catchments). A distinction is often made between private benefits and public benefits. Private benefits refer to the benefits that are accrued to the individual landholder or organisation. Public benefits are the benefits accrued to society and not just the individual making the decision. Any particular project or decision can have a combination of both direct and indirect private and public benefits that are both positive and negative and therefore entail externalities (benefits and costs) that are not reflected in market transactions and not borne by the decision-maker (Mercer 2004). Many ecosystem goods and services possess the characteristics of being non-exclusive and non-rival, meaning that it is impossible to exclude the people that do not bear the costs of providing these and that they offer collective (private and public) benefits (e.g. clean air, aesthetic values) (Ostrom et al. 1999). Such characteristics can lead to free-rider problems and undervaluation in a free-market economy in which it is recognised that market based methods alone will not facilitate their provision.

11.2 The Economics of Forest Landscape Restoration

While our general preference might be to prevent further habitat loss, in many areas there is a need to reverse the negative impacts of habitat clearing and degradation (Carroll et al. 2004; Brooks et al. 2002; Tilman et al. 1994). Globally it is estimated that between 40% and 50% of original forest cover has been removed or degraded, and in tropical regions alone it is estimated that 850 million hectares of previously forested land is degraded (Bradshaw et al. 2009).

Restoration refers to the process of assisting the recovery of an ecosystem that has been degraded, damaged, or destroyed (Mansourian 2005; Gann and Lamb 2006) and may entail re-establishing the composition, structure and function of the pre-degraded ecosystem (Hobbs and Norton 1996). Restoration projects typically occur at a site or property scale, however, the benefits delivered and costs incurred are likely to occur at a landscape scale. Forest landscape restoration refers explicitly to the restoration of land at a landscape scale and aims to regain ecological integrity and enhance human-wellbeing in deforested or degraded forest landscapes (see Lamb et al. Chap. 1).

The objective of restoration projects can be diverse, aimed for example, at re-establishing ecosystem services (e.g. to reduce the impact of salinisation by restoring the hydrological balance, see Harper et al. Chap. 14), recreating native habitat (see Davis et al. Chap. 15), and improving livelihoods and human well-being (see Rosengren Chap. 17). As a consequence, forest landscape restoration can be essential for providing a suite of ecosystem goods and services. Some of these can be valued including the carbon sequestered (measured through carbon market prices); improvements in water quality (measured through avoided costs of treatment plants); and amenity value (measured through hedonic price models).

The funding for restoration projects is often provided by governments through regionally-based natural resource management organisations. The Australian government for example, invested AUD$16.5 million in revegetating approximately 650 ha of rainforest in north Queensland, which does not account for the opportunity cost of foregone protection on the revegetated land. A favourable economic and policy setting is also required to strengthen the capacity of restoration delivery agents, R&D institutions, and reform incentives that result in land degradation or restoration projects that fail to deliver desired environmental benefits at the desired scale.

Table 11.1 Relative costs of restoration, and indicative rates of recovery of biodiversity and the ecosystem service benefits of different approaches to restoration

Method	Cost	Rate of recovery of biodiversity	Ecosystem services benefit
Prime focus on biodiversity			
Passive restoration	Low	Slow	High
Enrichment planting	Low-medium	Slow-medium	High
Direct seeding	Low-medium	Medium	High
Scattered plantings	Low	Slow	Medium
Close plantings of a few species	Medium	Medium	High
Intensive planting after mining	High	Fast	High
Prime focus on productivity and biodiversity			
Managing secondary forests	Low-medium	Medium	High
Enrichment plantings	Low-medium	Medium	Medium-high
Agroforestry	Medium-high	Medium	Medium-high
Monoculture plantations with buffers	High	Slow	Medium
Mosaics of monocultures	High	Slow	Low-medium
Mixed species plantations	High	Slow	Medium
Enhanced understorey development	Low	Slow	Medium-high

Adapted from Lamb and Gilmour (2003)
The ranks allocated are examples only as the costs and benefits would depend on the initial conditions of the land

11.3 Valuing Forest Landscape Restoration

11.3.1 Accounting for Restoration Costs

The cost of a restoration project is determined by the cost of delivery, implementation, and maintenance of the restoration approach employed (Table 11.1), and the opportunity costs of land no longer available for other forms of productive or income-producing uses (Harrison et al. 2001; Bryan et al. 2009). Delivery costs relate to the difficulty in restoring an area and achieving biodiversity and ecological services benefits. Implementation and maintenance costs vary by the approach to restoration taken, with passive restoration (e.g. through removal of grazing stock and allowing for natural recolonisation) potentially being inexpensive (depending on the opportunity costs) and intensive planting of diverse species over large areas potentially being more costly. Different restoration approaches also take different amounts of time to achieve their biodiversity and ecosystem service goals, and often the biodiversity and ecological services benefits are unlinked.

The costs and benefits of restoration are unlikely to be evenly distributed, but instead are likely to vary spatially. Such variation is important for considerations of equity and necessitates a need for taking a landscape-scale perspective when planning site-based restoration projects. The costs and benefits are also likely to vary through time and decisions made now will also have a consequence for decisions that are able to be made in the future. Present value cost calculations (known as discounting) explicitly incorporates the time value of any amount of money X, which is calculated as $\frac{X}{(1+r)^n}$, where r is the discount rate and n is the time period of interest. Discounting accounts for the fact that while restoration costs are incurred immediately, the benefits are unlikely to be realised for several years. The appropriate discount rate should reflect the opportunity cost of resources not being available in the short-term and is likely to vary with an individual's needs, their personal time preferences, and their available resources. For example, a person who is not meeting their basic needs, may place higher value on receiving benefits from goods or services in the immediate future, and place a lower value on receiving these benefits later (translating to a high discount rate). A more affluent person who has less concern for resourcing their immediate needs may have an enhanced capacity to give consideration for the needs of future generations and therefore may chose a lower discount rate.

The opportunity cost of restoration is the difference in value of the land in its highest and best use and its value following restoration. It can strongly affect a landholder's land use decisions. In some cases compensation may be required to avoid perverse ecological and social outcomes arising from foregone economic opportunities, but opportunity costs can be minimised if restoration is targeted to marginal or degraded lands where alternative land uses or capability are limited or commercial uses are not sought (Bryan et al. 2009). To ideally calculate the opportunity costs of restoration, one needs to identify all feasible alternative activities and estimate the net income of each. However, areas with low opportunity costs (e.g. mountainous rangelands, areas suffering from desertification), may coincide with greater restoration costs as a result of difficult accessibility or a requirement for more intensive restoration approaches to be implemented.

11.3.2 Accounting for the Benefits of Restoration

Given the landscape-scale focus, forest landscape restoration has the potential to deliver a diverse array of goods and services to a wide range of stakeholders (see Box 11.1). While some ecosystem goods and services can be provided from highly transformed landscapes (e.g. large scale food or timber production through planting monoculture crops or stands), others are best delivered by more diverse and natural as well as cultural, novel or successional ecosystems (e.g. good quality water supplies).

Forest restoration can produce a mix of ecosystem goods and services that have both private and public benefits. For example, while the revegetation of riparian vegetation might be costly to a landholder, the public benefits could be great, due to improved water quality and habitat connectivity for wildlife. In comparison, the benefits to the landholder of converting grazing land to a Eucalypt plantation may exceed the public benefits due to reduced water flows in the catchment. In general,

Box 11.1 Forest Landscape Restoration in the Shinyanga Region of Tanzania

Extensive clearing of Miombo and Acacia woodlands in the Shinyanga region of Tanzania for agricultural development led to a deterioration of ecosystem goods and services provided by the woodlands, such as fuelwood and fodder. In 1986 the government of Tanzania established a program of reintroduction and restoration of traditional managed woodland and forage enclosures (known locally as *Ngitili*). The reduced grazing pressure and natural regeneration of trees and grasslands within these enclosures resulted in the restoration of between 300,000 and 500,000 ha of land by the year 2000. The human-wellbeing and livelihood outcomes of this restoration included up to 6 hours saved per day by households in collecting fuelwood, thatch or fodder, the *Ngitili* were reported to have a monthly value per person of US$14 (with the rural average monthly consumption in Tanzania equating to US$8.5 per person) and 14% of local households were reported to collect medicinal herbs from the restored lands.
http://www.uicn.org/about/work/programmes/forest/fp_resources/fp_resources_publications/fp_resources_newsletters/?611/arborvitae-Issue-28

the benefits of restoration are received by the community at large (e.g. habitat provision, carbon sequestration, reduction in flooding damage) and the opportunity costs are typically borne by the landholder. The extent to which ecosystem goods and services are public goods determines the need for additional incentives and mechanisms for restoration (see section below on mechanism and incentives Section 11.5) and the extent to which individuals or organisations may free-ride on the decisions of others (Pannell 2008).

The value of a resource can be split into three main components (1) the use value reflecting the direct (e.g. timber or non-timber forest products, recreation) or indirect use (e.g. water purification); (2) the option value reflecting the value people place on the ability to use an ecosystem good or service in the future; and (3) the existence value, the value attributed to the mere existence of a good and reflects the circumstance that people might be willing to pay for improving or preserving resources that they will never use. Existence values are therefore typically the least tangible.

In the conservation and natural resource management literature, much attention has been given to valuing ecosystem goods and services, recognizing that ecosystem services are best communicated to policy makers and managers in the currency of their concern (Daily 1997; Nelson et al. 2009). Some goods and services that arise from forested landscapes can be directly valued in money terms if they are traded in markets (such as timber and non-timber forest products), or if there is a market that regulates the supply and demand for the goods or service (e.g. voluntary and regulatory markets for carbon and water). Not all goods and services are directly marketable however, and therefore valuing these in monetary terms is complex and can involve non-market valuation techniques. There exists a variety of non-market valuation tools, each with their advantages and challenges associated with their application.

For such goods and services indirect valuations can be made by measuring the marketable resources that are sacrificed to obtain or access the goods or services. For example, the travel cost method of valuation is used to value sites that have recreational potential and inferring value by *revealed preferences* – i.e. how much time visitors spend getting to a site and how many times they visit the site.

The contingent valuation method is an alternative valuation approach and involves asking respondents what they will be willing to pay for an environmental benefit (e.g. preservation or avoided damage) based on a hypothetical scenario of the terms under which it will be provided (Mitchell and Carson 1989). For example, using this approach, respondents might be asked what value they would place on an environmental change or whether they would be willing to pay $X to prevent the change or instigate an action. The name of the contingent valuation approach is derived from its nature – the willingness to pay response elicited is contingent upon the initial information provided and therefore the way the information is framed and who responds is a key determinant of the reliability of this method. Another limitation of this valuation approach is that there is often inconsistency between hypothetical responses (*stated preferences*) and actual commitments (*revealed preferences*) (Wills 2006). Costanza et al. (1997) also reported that the values of ecosystem goods and services that they elicited were conservative due to the potential for irreversibilities or thresholds in ecological systems, and failure to account for the infrastructure value of ecosystems.

The replacement cost method of valuation reflects the cost if the naturally occurring service did not exist (e.g. for flood control or water purification) and therefore to some extent accounts for the infrastructure value of ecosystems.

For those goods and services for which there is market failure there can be an attempt to measure and value the goods and services and create a market for them. Alternatively, or more commonly in combination, property rights can be allocated (e.g. goods and services can be privatised) so that there is a vested interest in adopting strategies to maximise the value of the asset. Valuation however remains the most common method adopted by governments to ensure the sustainable management of ecosystem goods and services. While we have focussed on economic methods of valuation, it is important to note that methods for valuation also originate from ecology, psychology, and philosophy.

11.4 Evaluating and Prioritising Restoration Projects

Generally speaking, the funds available for natural resource management are small relative to the required funding to achieve everything that we would like to do everywhere all at once (James et al. 2001). Restoration for example, is often a costly, time-intensive process that typically takes many years to achieve the desired outcomes. However, increasingly funding agencies are seeking transparency in funding allocation decisions and in the evaluation of outcomes in order to confirm that an expenditure has been effective and appropriate (Ferraro and Pattanayak 2006).

With multiple sites requiring restoration, and a variety of restoration approaches that could be implemented, the effective and efficient allocation of resources for restoration becomes even more pivotal. The scarcity of resources combined with the spatial and temporal heterogeneity in the costs and benefits of restoration drives a need to prioritise the allocation of restoration funds. At a micro-level (project planning) there is a need for transparent procedures that help with minimising misallocation of funds, minimising inequities, distributing expenditures most effectively, and maximising success for a given investment.

Economics provides tools for decision making, once a clearly specified objective has been stated and an objective function developed in order to compare alternative strategies or outcomes has been identified. The objectives of restoration projects can be as diverse as reducing salinisation, improving water quality, improving habitat for fauna, producing timber and firewood, producing food, sequestering carbon, or stabilising land. Forests also have substantial cultural values for local communities. Given these multiple objectives, forest landscape restoration will often entail multiple stakeholders with differing and potentially competing needs. When there are multiple objectives for a restoration project there is both the potential for synergy and conflict (Higgins et al. 2008).

Given such context, the identification of socially, environmentally, and economically acceptable options for forest landscape restoration through an explicit analysis of trade-offs and participatory processes are likely to be particularly important (Polasky et al. 2005). In the previous section methods were described to value the costs and benefits of restoration, in this section we discuss the most commonly used approaches to evaluate the priority of restoration projects, and how this information may be used in spatial and temporal planning of restoration activities.

11.4.1 Cost-Effectiveness Analysis

Cost-effectiveness analysis allows the determination of the least-cost means of achieving an objective. This may not be the most efficient allocation, since the predetermined objective may not be the most efficient and effective, but the least-cost means to achieving the objective will be identified. Common guidance for setting objectives is to make them SMART (i.e. Specific, Measurable, Achievable, Relevant and Time-bound) (Tear et al. 2005; Wilson et al. 2009). There are several examples of restoration activities within a single site, or for a handful of sites, being prioritised on the basis of cost-effectiveness analyses (Dymond et al. 2008; Macmillan et al. 1998; Petty and Thorne 2005; Hyman and Leibowitz 2000; Goldstein et al. 2009). Pannell et al. (2009) have developed a general tool for planning and prioritising public investments in natural resource management, where the aim is to protect or enhance specific assets in a cost-effective manner. The process entails identifying a measureable goal for the assets of concern, on-ground actions that will achieve the goal, risks to the success of the projects to deliver the actions, and their associated costs. A cost-effectiveness index is then used to compare alternative projects and to guide the selection of associated policy mechanisms and instruments.

In its simplest sense, cost-effectiveness analyses can provide a static evaluation of restoration priorities but fail to account for spatial dependencies between sites and not provide a temporal sequencing of restoration projects. There is however a growing body of theory and associated tools for prioritisation in the context of spatial and temporal dynamics (Wilson et al. 2009; Moilanen et al. 2009; Bryan and Crossman 2008; Crossman and Bryan 2009; Wilson et al. 2011). Spatially and temporally explicit cost-effectiveness analysis can ensure that restoration projects are designed and implemented at a landscape scale, rather than on a piece-by-piece basis and can allow other important factors such as the likelihood of success (e.g. willingness of local communities to participate) and stochastic events (e.g. fire) to be accounted for (for an example, see Box 11.2).

Box 11.2 Restoration Planning on the Irvine Ranch Natural Landmark in Southern California (Wilson et al. 2011; McBride et al. 2010)

The Irvine Ranch Natural Landmark is a collection of permanently protected wildlands and parks located near the Santa Ana Mountains in Southern California. The circa 17,600 ha of wildlands and parks contain some of the largest remaining stands of coastal sage scrub, oak-sycamore woodland, native grassland, and chaparral vegetation types left in southern California. However, much of it has been degraded by agriculture, intensive grazing, woodland clearance, adjacent development, invasive species, and too-frequent fire and is in need of restoration. The Irvine Ranch Conservancy, a local non-profit conservation organisation, is managing most of the protected portions of the Natural Landmark. Existing levels of degradation and vulnerabilities to further disturbances from fire and invasive species, mean they face the important problem of determining how best to target funds available to ensure the recovery and preservation of the significant ecological values of the area. A prioritisation model has been developed to prioritise their future habitat restoration activities with a limited budget. The objective of the restoration program is to maximise the overall area restored and the habitat of species of special concern, while also enhancing the resilience at both population and landscape scales.

A total of 923 sites have been identified as potential candidates for restoration action, and an annual budget of $700,000 is targeted each year over a 20 year time period for restoration. The contribution of restoration to enhancing resilience is measured in terms of the achievement of a set of additional criteria that target particular indicators or processes, such as a decrease in the risk of fire ignition on a landscape scale, and representation and recruitment among functional groups important to succession on a community scale. At the Irvine Ranch Natural Landmark the cost of restoration at a site is dependent on the

(continued)

Box 11.2 (continued)

desired habitat type, the restoration technique to be employed, and the area, slope, and accessibility of each site. The likelihood of success, and therefore the delivery of the expected benefit, varies depending on the restoration action, degradation state, and desired habitat type. It also varies with the slope of the site, its aspect, and the condition of neighbouring sites. Sites undergoing restoration on the Irvine Ranch are also vulnerable to fire, on average, once every 15 years.

A dynamic simulation approach has been used to determine a close-to-optimal restoration schedule – i.e. the combination of restoration sites and activities and the schedule for their implementation that will be the most cost-effective by providing the greatest and most resilient improvement in habitat coverage given a fixed budget and operational constraints. The approach is spatially and temporally-explicit and accounts for the likelihood of restoration success, the probability of a major catastrophic fire event, the benefit of spatial connectivity, and the relative contribution of restoring a given site towards enhancing ecologic resilience of the broader ecosystem.

Mapping tools such as Geographical Information Systems (GIS) are likely to be of value when planning restoration at a landscape scale, particularly if the goal is to deliver a mosaic of land uses, as is the case with forest landscape restoration. Crossman and Bryan (2009) for example identified spatially-explicit priority locations for ecological restoration to cost-effectively restore natural capital in the Lower Murray region of southern Australia. Locations for ecological restoration were selected based on their ability to improve biodiversity, mitigate dryland salinity and soil erosion, sequester carbon, and do so in a cost-effective manner (i.e. have least impact on farm income). In order to achieve restoration targets, Crossman and Byran found trade-offs to exist between the delivery of benefits and average farm profitability. This is an example of conflicts arising between multiple objectives.

11.4.2 Cost-Utility Analysis

Cost-utility analysis represents an extension of cost-effectiveness analysis, which places a value on a better outcome. Cost-utility analysis considers multiple objectives and aggregates them into a single utility function. Utility curves can be simple or complex and can potentially involve ecological threshold effects. Indifference curves can be used to estimate the marginal rate of substitution of one objective for another and multi-criteria decision analysis can be used to elicit weightings for different objectives (Maron and Cockfield 2008).

Multi-criteria evaluation is a commonly used approach to evaluate restoration projects and is often driven by expert opinion and ranking systems (Bryan and Crossman 2008; Cipollini et al. 2005; Twedt et al. 2006). With such approaches the potential restoration sites are scored against multiple social, ecological, and economic criteria and priorities are evaluated by ranking sites based on a sum of the weighted scores for each criterion. Preferences from stakeholders can be incorporated via specification of the relative weights given to each criterion.

While commonly employed, weighted scoring systems tend to suffer from a lack of transparency and repeatability. Choice of ranking criteria is often arbitrary and not well defined, and as a result the assignment of weights to the different criteria tends to be value-based, although there are a multitude of quantitative and semi-quantitative methods to assist with their assignment (e.g. the Analytic Hierarchy Process (Saaty 1980) where weights are derived though paired comparison of criteria).

11.4.3 Cost Benefit Analysis

A fundamental economic tool is to evaluate the costs and benefits of different projects (Hanley and Spash 1994). The measurement of benefits and costs is typically from an anthropocentric perspective, measured in dollar terms and valued according to the effects on humanity. We have already learnt that this will be difficult when the benefits are non-marketable and/or intangible, or where the assignment of rights is not clear, and therefore, there is the potential for issues of inequity to arise. Cost Benefit Analyses have been criticised for inadequately accounting for benefits of ecosystem goods and services by focusing only on market values or economic benefits, and for failing to account for the costs associated with the irreversible loss of ecosystem goods and services (Hunt 2008). In the simplest case, if the benefits exceed the costs then the project is favourable, that is, if the restoration project should proceed. Naturally if the information on costs and benefits are uncertain (that is the confidence interval on the estimate of costs overlaps that of the benefit) then a decision based on such an evaluation cannot be made with certainty, and a sensitivity analysis to explore the robustness of decisions to key uncertainties should be undertaken. In a restoration context, we might determine what the efficient amount of restoration would be to maximise the net benefits.

11.5 Economic Instruments and Policy Mechanisms to Facilitate Forest Landscape Restoration

We have learnt that forest landscape restoration may entail significant opportunity costs, and under such circumstances a range of economic instruments and policy mechanisms can be used to encourage participation. This is particularly so on private lands, where there is the potential for mismatch between the public and private benefits and

Table 11.2 Economic instruments for facilitating forest landscape restoration.

Instrument category	Examples
Positive incentives	Subsidies, grants, tax or rate concessions, payments for ecosystem services, and reduced or removed subsidies for competing land uses
Negative incentives	Regulation, taxes
Extension	Technology transfer, education, awareness raising, pilot programs
Technology development	Research and development programs, infrastructure support

costs. These instruments include a range of positive and negative incentives, extension activities, and technology development (Table 11.2). Such instruments can be used to internalise externalities and be imposed in either a mandatory or voluntary manner (Mercer 2004). Pannell et al. (2006) identify that voluntary adoption of land management practices by landholders is likely to occur when they have a high relative advantage and when they are readily trialable. Impacts on profitability, asset value and lifestyle have also been identified as motivators for on-farm vegetation management (Mallawaarachchi and Szakiel 2007). The general key is that the landowner must perceive some gain or at least no net loss from participating (Mercer 2004).

A challenge with implementing such approaches is determining the appropriate instruments, and the appropriate level of subsidy or incentive required (Mercer 2004). Pannell (2008) illustrated that the choice of instrument can be determined by the relative levels of private and public benefits. For example, positive incentives might be employed where public benefits are highly positive and private benefits are close to zero. Negative incentives might be employed where public benefits are highly negative and private benefits are slightly positive. Extension on the other hand might be favoured where both benefits are positive. Another challenge is targeting the sites and individuals that are most crucial (both a high likelihood of participation and also high possible benefits).

There are a range of support mechanisms and policies that might be associated with the delivery of such instruments, such as conservation easements and market-based instruments. Conservation easements are a legally binding agreement between a government agency or land trust organisation and landowners. The landowner foregoes the right to use the land in specific ways (or agrees to manage the land in particular ways), in exchange for tax benefits or other monetary compensation (Mercer 2004). Conservation easements provide a mechanism to deliver positive incentives, in a way that has the potential to be long-lasting, particularly if the easement is attached to the title of the land.

Market-based mechanisms are economic policy instruments that can be applied where there has been market failure, that is, where a change in the way resources are managed would enhance efficiency but this is unlikely to arise spontaneously from the market, deeming government intervention necessary. Conservation tenders or reverse auctions are an example of a market-based mechanism, where amenable landowners offer environmental services from their land in return for financial support for delivering improved land management practices (Stoneham et al. 2003;

Hajkowicz et al. 2007; Connor et al. 2008). Agencies rank and select bids for funding based on some measure of cost-effectiveness. Like conservation easements, such an approach requires monitoring and enforcement to ensure desired outcomes are delivered.

Schemes for trading in carbon dioxide emissions reductions have relevance for restoration activities and investment in carbon sequestration may present an opportunity to finance future restoration activities (Harper et al. 2007). The development of regulatory carbon sequestration markets depends on the establishment of national emissions policies and targets and a statutory basis for the ownership and protection of carbon rights is required in order to facilitate trading. The profitability of restoring land in response to opportunities presented by markets for carbon sequestration depends on the price of carbon, on the type of land restored (particularly in the context of opportunity costs), and on the species used (Hunt 2008). The biodiversity outcomes of such programs are also determined by the species used (Caparros and Jacquemont 2003). Harper et al. (2007) estimate that 2200 Mt CO_2-e could be sequestrated by reforesting 16.8 million hectares of cleared farmland in Australia, but with carbon sinks only likely to become profitable when carbon prices exceed AUD$15/t CO_2-e.

Mitigation and biodiversity/vegetation offsets can also play a role in facilitating forest landscape restoration, particularly if implemented at a landscape scale and if outcomes are monitored. Such policies support the creation, enhancement or restoration of habitats in response to an action that negatively impacts an ecosystem. “Banks” can be established which constructs a market in which restored ecosystem values are quantified as credits and later purchased by developers or landowners when compensation is required for loss of values elsewhere.

11.6 Conclusions

Forest landscape restoration can provide a diverse array of ecosystem goods and services – timber for housing and fuel, paper from wood fibre, sequestration of carbon dioxide, habitat for flora and fauna, and improvements to the quantity and quality of water supply. While economic tools exist to value the goods and services that arise from restoring forest landscapes, not all goods and services are amendable to market solutions, either because there is no market and/or a way to quantify the goods and services. It is clear from analyses of the value of ecosystem goods and services that they do however make an important contribution to national incomes and human wellbeing, providing a strong case for modifying systems of national accounting at a macro-level to better reflect the non-monetary (less directly quantifiable) value of ecosystem goods and services. We have also learnt that forest landscape restoration may entail significant opportunity costs to the landholder, and under such circumstances a range of economic instruments and policy mechanisms can be used to encourage participation. A variety of economic tools exist for evaluating restoration projects and prioritising the allocation of limited funds between projects. A potentially useful framework for prioritising investments in restoration is identify and target those activities that will restore highly valuable goods and services that are otherwise scarce or threatened, and for which social, economic and environmental benefits can be achieved at a reasonable cost (Pannell et al. 2009; Hobbs and Kristjanson 2003).

11.6.1 Management Implications

- The costs of restoration may entail operational costs and that of foregone opportunities.
- The benefits of restoration may reflect biodiversity outcomes or the provision of ecosystem goods and services, some of which can be valued.
- The economic aims of restoration (for example, to produce timber or restore watersheds) may be at odds with other aims of restoration (such as to enhance biodiversity). Trade-offs will likely exist and identifying and targeting synergies between stakeholder objectives can be important.
- There are a variety of approaches to measure the costs and benefits of restoration projects each with their strengths and limitations. Taking an economic perspective encourages valuable analysis of the efficiency of restoration actions.
- There are a range of economic instruments and policy mechanisms available to encourage participation in restoration, particularly on private lands where there is potential for mismatch between the public and private benefits and costs.
- Restoration prioritisations can establish and define unique and fairly complex utility functions, thereby allowing planning to be driven by a suite of benefits and thresholds to benefits, tailored to a region or organisation.

Literature Cited

Australian Bureau of Statistics (2004) Environment expenditure. Local Government, Australia, 2002–2003 Australian Government, Canberra

Bradshaw CJA, Sodhi NS, Brook BW (2009) Tropical turmoil: a biodiversity tragedy in progress. Front Ecol Environ 7(2):79–87

Brooks TM, Mittermeier RA, Mittermeier CG et al (2002) Habitat loss and extinction in the hotspots of biodiversity. Conserv Biol 16(4):909–923

Bryan B, Crossman ND (2008) Systematic regional planning for multiple objective natural resource management. J Environ Manage 88:1175–1189

Bryan BA, King D, Ward JR (2009) Modelling and mapping agricultural opportunity costs to guide landscape planning for natural resource management. Ecol Indicators 11:199–208

Caparros A, Jacquemont F (2003) Conflicts between biodiversity and carbon sequestration programs: economic and legal implications. Ecol Econ 46:143–157

Carroll C, Noss RF, Paquet PC et al (2004) Extinction debt of protected areas in developing landscapes. Conserv Biol 18(4):1110–1120

Cipollini KA, Maruyama AL, Zimmerman CL (2005) Planning for restoration: a decision analysis approach to prioritisation. Restor Ecol 13(3):460–470

Connor JD, Ward JR, Bryan B (2008) Exploring the cost effectiveness of land conservation auctions and payment policies. Aust J Agric Res Econ 51:303–319
Costanza R, d'Arge R, de Groot R et al (1997) The value of the world's ecosystem services and natural capital. Nature 387:253–260
Crossman ND, Bryan BA (2009) Identifying cost-effective hotspots for restoring natural capital and enhancing landscape multifunctionality. Ecol Econ 68:654–668
Daily G (1997) Nature's services: societal dependence on natural ecosystems. Island Press, Washington, DC
Daly HE, Farley J (2010) Ecological economics: principles and applications, 2nd edn. Island Press, Washington, DC
Dymond JR, Ausseil AE, Overton J (2008) A landscape approach for estimating the conservation value of sites and site-based projects, with examples from New Zealand. Ecol Econ 66:275–281
Ferraro PJ, Pattanayak S (2006) Money for nothing? A call for empirical evaluation of biodiversity conservation investments. PLoS Biol 4(105)
Gann GD, Lamb D (2006) Ecological restoration: a means of conserving biodiversity and sustaining livelihoods (version 1.1). Society for Ecological Restoration International and IUCN, Tucson/Gland
Goldstein JH, Pejchar L, Daily GC (2009) Using return-on-investment to guide restoration: a case study from Hawaii. Conserv Lett 1(5):236–243
Hajkowicz SA, Higgins A, Williams K et al (2007) Optimisation and the selection of conservation contracts. Aust J Agric Res Econ 51:39–56
Hanley N, Spash C (1994) Cost benefit analysis and the environment. Edward Elgar Publishing Company, Brookfield
Harper RJ, Beck AC, Ritson P et al (2007) The potential of greenhouse sinks to underwrite improved land management. Ecol Eng 29:329–341
Harrison SR, Stern NH, Boydell S et al (2001) The valuation impact of reforestation on degraded farm land. In: Harrison SR, Herbohn JL (eds) Sustainable farm forestry in the tropics: social and economic analysis and policy. Edward Elgar Publishing Company, Cheltenham, pp 133–146
Higgins AJ, Hajkowicz S, Bui E (2008) A multi-objective model for environmental investment decision making. Comput Oper Res 35:253–266
Hobbs RJ, Kristjanson LJ (2003) Triage: how do we prioritize health care for landscapes? Ecol Manage Restor 4(1):39–45
Hobbs RJ, Norton DA (1996) Towards a conceptual framework for restoration ecology. Restor Ecol 4(2):93–110
Hunt C (2008) Economy and ecology of emerging markets and credits for bio-sequestered carbon on private land in tropical Australia. Ecol Econ 66:309–318
Hyman JB, Leibowitz SG (2000) A general framework for prioritizing land units for ecological restoration and protection. Environ Manage 25(1):23–35
James A, Gaston K, Balmford A (2001) Can we afford to conserve biodiversity? Bioscience 51:43–52
Lamb D, Gilmour D (2003) Rehabilitation and restoration of degraded forests. IUCN/WWF, Gland
Lawn P (2008) Macroeconomic policy, growth and biodiversity conservation. Conserv Biol 22(6):1418–1423
Macmillan DC, Harley D, Morisson R (1998) Cost-effectiveness analysis of woodland ecosystem restoration. Ecol Econ 27:313–324
Mallawaarachchi T, Szakiel S (2007) Nonbroadscale land clearing in Southern Australia: economic issues in managing native vegetation on farm land. ABARE research report 07.2. Prepared for the Natural Resource Management Division. Australian Government Department of Agriculture, Fisheries and Forestry, Canberra
Mansourian S (2005) Overview of forest restoration strategies and terms. In: Mansourian S, Vallauri D, Dudley D (eds) Forest restoration in landscapes. Beyond planting trees. Springer, New York, pp 8–13

Maron M, Cockfield G (2008) Managing trade-offs in landscape restoration and revegetation projects. Ecol Econ 18(8):2041–2049

McBride MF, Wilson KA, Burger J et al (2010) Mathematical problem definition for ecological restoration planning. Ecol Model 221(19):2243–2250

Mercer DE (2004) Policies for encouraging forest restoration. In: Stanturf JA, Madson P (eds) Restoration of boreal and temperate forests. CRC Press, Boca Raton

Millennium Ecosystem Assessment (2005) Ecosystems and human wellbeing: synthesis report. Island Press, Washington, DC

Mitchell RC, Carson RT (1989) Using surveys to value public goods: the contingent valuation method. Resources for the Future, Washington, DC

Moilanen A, Wilson KA, Possingham HP (2009) Spatial conservation prioritisation: quantitative methods and computational tools. Oxford University Press, Oxford

Nelson E, Mendoza G, Regetz J et al (2009) Modeling multiple ecosystem services, biodiversity conservation, commodity production, and tradeoffs at landscape scales. Front Ecol Environ 7(1):4–11

Ostrom E, Burger J, Field CB et al (1999) Revisiting the commons: local lessons, global challenges. Science 284:278–282

Pannell DJ (2008) Public benefits, private benefits, and policy intervention for land-use change for environmental benefits. Land Econ 84(2):225–240

Pannell DJ, Marshall GR, Barr N et al (2006) Understanding and promoting adoption of conservation practices by rural landholders. Aust J Exp Agric 46:1407–1424

Pannell DL, Roberts AM, Alexander J et al (2009) INFFER (Investment framework for environmental resources). INFFER working paper 0901. University of Western Australia, Perth

Petty JT, Thorne D (2005) An ecologically based approach to identifying restoration priorities in an acid-impacted watershed. Restor Ecol 13:348–357

Polasky S, Nelson E, Londsdorf E et al (2005) Conserving species in a working landscape: land use with biological and economic objectives. Ecol Appl 15(4):1387–1401

Saaty TL (1980) The analytical hierarchy process. McGraw-Hill, New York

Stoneham G, Chaudhri V, Ha A et al (2003) Auctions for conservation contracts: an empirical examination of Victoria's BushTender trial. Aust J Agric Restor Econ 47(4):477–500

Tear TH, Kareiva PK, Angermeier PL et al (2005) How much is enough? The recurrent problem of setting measurable objectives in conservation. Bioscience 55(10):835–849

Tilman D, May RM, Lehman CL et al (1994) Habitat destruction and the extinction debt. Nature 371:65–66

Twedt DJ, Uihlein WB, Elliot AB (2006) A spatially explicit decision support model for restoration of forest bird habitat. Conserv Biol 20(1):100–110

Wills I (2006) Economics and the environment: a signalling and incentives approach, 2nd edn. Allen and Unwin, Crows Nest

Wilson KA, Carwardine J, Possingham HP (2009) Setting conservation priorities. Ann N Y Acad Sci Year Ecol Conserv Biol 1162:237–264

Wilson KA, Lulow M, Burger J et al (2011) Optimal restoration: accounting for space, time, and uncertainty. J Appl Ecol 48(3):715–725

Chapter 12
Wild Ennerdale: A Cultural Landscape

Ian Convery and Tom Dutson

12.1 Introduction

The concept of cultural landscape within the context of the Northern UK uplands has been outlined by Convery and Dutson (2006) and Dutson and Convery (2007), who emphasize the complex dynamic that exists between people and place, and explore how elements in the cultural landscape of upland Northern England might contribute to community sustainability. As both Kirby (2003) and Blackshall et al. (2001) note, the countryside of England is very much a cultural landscape, a product of human management of one form or another. Thus whilst upland areas in England contain most of the remaining semi-natural habitats in the country, which often contain vegetation communities that are similar to natural communities in structure and function (Blackshall et al. 2001) their distribution is the product of thousands of years of human activity (Fielding and Haworth 1999; Carver and Samson 2004).

Yet despite the almost total lack of wilderness in England, there has been an increase in interest in the concept of 'wild land' and 'rewilding'[1] over recent years (Green 1995; Fenton 1996; Fisher 2003). The concept of wild land or wilderness has been used to good effect, both nationally and internationally, for conservation management (Habron 1998) and for tourism (Hall and Page 2002), though developing wild areas presents significant challenges for policy makers and practitioners alike (Höchtl et al. 2005; Waitt et al. 2003; Jerram 2004) and is frequently contentious.

[1] Editors' Note: The concept of wilding as applied to a cultural landscape does not comfortably fit the notion of "ecological restoration" in the sense of minimum human intervention. But it does represent an attempt to restore to a previous, pre-industrial state that is considered more natural. In our view, this only illustrates that restoration is a social choice informed but not determined by ecological reality.

I. Convery (✉) • T. Dutson
National School of Forestry, University of Cumbria, Cumbria, UK
e-mail: ian.convery@cumbria.ac.uk

J. Stanturf et al. (eds.), *Forest Landscape Restoration: Integrating Natural and Social Sciences*, World Forests 15, DOI 10.1007/978-94-007-5326-6_12,

For example, Höchtl et al. (2005), reporting on the decline in land-use in the Val Grande National Park, Italy, note how the main impacts on the inhabitants are psychological and economic in nature. They state that in the surrounding villages the effects of abandonment are viewed in a very negative light. However, visitors to the area judge the consequences of land abandonment differently. While they regard the resulting landscape's wildness positively, they also regret the cultural losses suffered by rural communities.

In the UK, the post-productivist period of agriculture has been typified by wide reaching changes in land use policy and support structures. As Green (1995) indicates the overcapacity in agriculture and changes to Common Agricultural Policy (CAP) subsidies, together with the poor market in timber are driving a new approach to our landscape, dominated by an ethos of environmental management and community access. In the forestry sector, the UK Forestry Commission aims to increase access to woodland, improve the quality of information about access, enhance the nation's forest estate and promote better understanding.

The Forestry Strategy (Forestry Commission 1998) also emphasizes 'forestry's role in the wider countryside including its contribution to the rural economy … There will be a focus both on the role of new woodlands and on how existing woodlands can be managed to deliver more benefits to local economies, including the need to 'work in partnership to deliver … objectives'. Significantly for this study, the strategy recognizes that new initiatives should be sensitive to stakeholder interests and history as well as being economically, culturally and environmentally sustainable in the longer term.

The North West Regional Forestry Framework Partnership (2005) highlights a holistic and participatory approach to forest management in England. Under Action Area Six (*supporting and resourcing the sector*) the partnership highlight the need for an integrated landscape scale approach to woodland and forestry development (Priority b) delivered through broad, cross-sector Partnerships (Priority c).

This chapter focuses on Wild Ennerdale (WE), a 'wilding' initiative in a relatively remote valley in the Western Lake District, Cumbria, North West England (Fig. 12.1). The Ennerdale valley is 14.5 km long, 5.6 km wide (at its widest) and extends to an area of 4,711 ha. The valley narrows from west to east and is surrounded by some of the Lake Districts highest summits: Green Gable, Great Gable, Pillar, Kirk Fell & Steeple.

Habitats range from the agricultural land and riparian zones of the valley bottom, through the coniferous and broadleaf woodland of the lower and middle slopes, to the heather moorland and rock of the upper slopes. The highest ridges and summits are characterized by a low growing montane heath. The valley is important for conservation, with over 40% of the WE area designated as Site of Special Scientific Interest (SSSI) and Special Area of Conservation (SAC), and also contains a number of Biodiversity Action Plan (BAP) habitats and species. These designated represent areas of value for nature conservation, geology and archaeology. According to Wild Ennerdale Partnership (WEP) these sites are designated for their regional, national or international importance (WEP 2006). In terms of human settlement, the identified archaeological remains at Ennerdale span a time-scale from the Bronze Age to the Pre-medieval period (Oxford Archaeology Unit 2003).

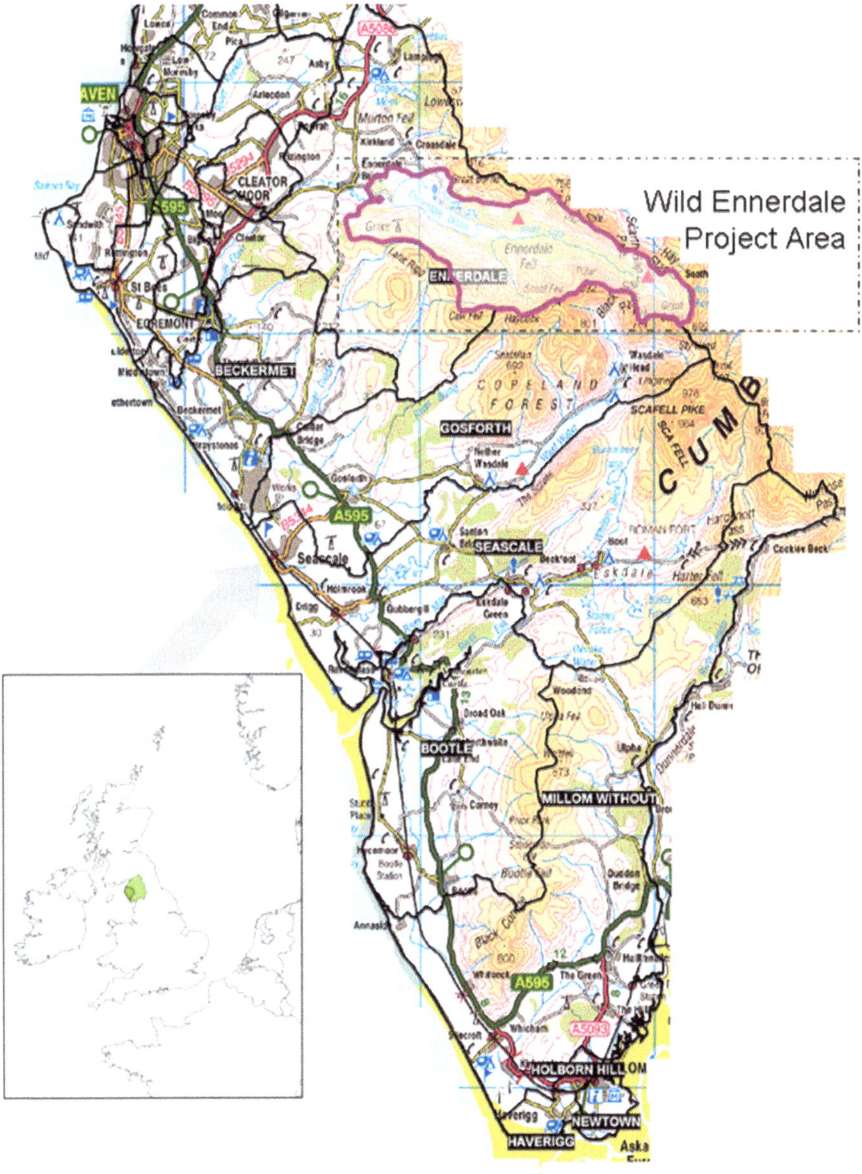

Fig. 12.1 Location of Wild Ennerdale. *Inset map* indicates location of Cumbria (*lighter shading*) & West Cumbria (*darker shading*) within UK (Adapted from Cumbria County Council 2006)

Ennerdale Ward[2] is sparsely populated and is serviced by relatively poor road and rail links. Further west from Ennerdale, along the lower lying coastal strip, are urban communities built on a tradition of manufacturing industries (coal mining & ship building) which include the towns of Cleator Moor, Egremont and Whitehaven. Accordingly, manufacturing is the dominant employment sector in the ward (accounting for the employment of 100 people, most of whom work outside the ward), followed by Health and Social Work (76 people). Agriculture and forestry employs 50 people (WEP 2006).

The valley has been managed as coniferous plantation forest since the 1920s by the Forestry Commission. In the early twentieth Century, UK timber stocks were so depleted by the demands of the First World War that the Forestry Commission (which was established in 1919) was given a good deal of freedom to acquire and plant land. During the 1920s, the Forestry Commission acquired part of Ennerdale valley (some 3,642 ha) as part of this emerging national strategy. Planting, of mainly high yielding exotic species such as Sitka spruce, began in Ennerdale in March 1925 and continued through the twentieth Century (WEP 2006). The afforestation caused a general outcry at the time and was one of the issues that prompted the establishment of the Lake District National Park (Oxford Archaeology Unit 2003).

Figures 12.2 and 12.3 show aspects of this forest plantation history. Figure 12.2 indicates remaining areas of plantation forestry, areas of clear fell and improved pasture. Figure 12.3 also indicates areas of plantation forestry, along with areas that have been cleared of spruce and some evidence of spruce regeneration.

WEP was established in 2002 between the three main landowners in the valley: The Forestry Commission,[3] National Trust[4] and United Utilities.[5] This was in part a response to the post Foot & Mouth rural recovery agenda (Cumbria was the most affected county in the UK, with over 95% of cases), ongoing agricultural reform,

[2] Electoral wards are the base unit of UK administrative geography.

[3] The Forestry Commission is the government department responsible for the protection and expansion of Britain's forests and woodlands. Founded in 1919, they are the largest land manager in Britain, with an estate covering some 258,000 ha.

[4] The National Trust is a charity established in 1895 by three Victorian philanthropists – Miss Octavia Hill, Sir Robert Hunter and Canon Hardwicke Rawnsley. Concerned about the impact of uncontrolled development and industrialisation, they set up the Trust to act as a guardian for the nation in the acquisition and protection of threatened coastline, countryside and buildings. The Trust now protects over 300 historic houses and gardens, 49 industrial monuments and mills, and a range of open air properties including forests, woods, fens, beaches, downs, moorland, islands, archaeological remains and nature reserves, all of which are open to the public. The Trust has 3.5 million members and 52,000 volunteers. During 2007, 15 million people visited Trust pay for entry properties, while an estimated 50 million visited open air properties (National Trust 2008).

[5] United Utilities (UU) is the UK's largest listed water company. They are a FTSE 100 company with a turnover of £2 billion. Alongside owning and operating the water network in north west England, UU are also a major landowner in the region of England, with 58,000 ha of catchment land. Nearly half of the land is in three National Parks, and nearly one third is designated as Sites of Special Scientific Interest (SSSIs).

Fig. 12.2 Plantation forestry in Ennerdale Valley (Adapted from WEP 2006)

Fig. 12.3 Ennerdale landscape post-spruce removal (Adapted from WEP 2006)

changing trends in UK forestry and a growing interest generally regarding the concept of 'wild land' in Britain (Convery et al. 2005). According to The Wild Ennerdale Stewardship Plan (WESP) the partnership developed a vision *'to allow the evolution of Ennerdale as a wild valley for the benefit of people, relying more on natural processes to shape its landscape and ecology'* (WEP 2006).

The WESP asks what will Ennerdale look like in 50, 100 or 200 years time? *'The reality is we don't really know. Based on our hopes and aspirations, we can however make broad assumptions; that we will have a series of naturally evolving and interacting ecosystems across the valley that are far more robust in the face of stresses such as climate change. Farming and forestry will continue to have a role in the valley, but with the aim of maximizing ecology and landscape value... we cannot predict exactly how bio-diversity may develop as natural processes take greater hold'* (WEP 2006). In short, the process of change in Ennerdale will be slow, with no fixed end point. Indeed, the process of change may well prove as important as any end result (Oxford Archaeology Unit 2003).

The chapter starts by discussing the farming landscape of Cumbria post Foot and Mouth Disease before then examining the relationship between the farming community and WE. Finally, it discusses WE within the broader cultural landscape of Ennerdale valley and the future role of the farming community. This chapter is based in part on a study undertaken by the National School of Forestry, University of Central Lancashire, which received funding from Cumbria Rural Enterprise Agency.[6]

12.2 Methodology

Following an initial orientation visit, a desk-based review was completed. Respondents from the relatively small farming population of Ennerdale were purposefully recruited (all farmers within the WE project area were interviewed). For some respondents, agricultural links to the valley span several generations. Seven semi-structured interviews (six with farmers and one with a tourism provider) and one group meeting (including various sectors[7] of the Ennerdale community) were completed.

Amongst other things, respondents were asked about the opportunities and constraints for farm business development; their level of involvement with WE; opportunities for collaboration linked to WE; their thoughts regarding wilding the Ennerdale valley and the future of farming in the valley. Interviews and group meetings were taped and transcribed and data were analyzed using the grounded theory – constant comparison method, where each item is compared with the rest of the data to establish and refine analytical categories (Pope et al. 2000). All interview transcripts have been anonymised and a coding system has been used to differentiate between respondents (for example, I01 indicates interview 1, I02 indicates interview 2 and so on).

---

[6] The views expressed in the study are those of the authors and not necessarily those of Cumbria Rural Enterprise Agency.

[7] Twenty-five people attended, this included representatives of the Parish council, local business owners and tourism providers, local residents and 'incomers' to the valley. This was an open meeting, all members of the community were invited and it had been publicized locally by the Parish Council.

12.3 Findings

12.3.1 Agricultural Change Post Foot and Mouth Disease

The 2001 Foot and Mouth Disease (FMD) disaster is a watershed period in recent UK farming history. As Convery et al. (2008) indicate, FMD created deep fissures in the lifescapes of Cumbria, so that much of the taken-for granted world, identity and sense of meaning within the farming community changed. Prior to FMD, the last decade had been very difficult for UK agriculture in general and hill farming in particular (Franks et al. 2003; Lowe et al. 2001; Curry 2002; Royal Society 2002; Ministry of Agriculture Fisheries & Food (MAFF) 1999), to the extent that by the mid-1990s, 'much of the profitability has drained from the industry' (Royal Society 2002). By the time of the FMD epidemic, farm incomes were 'on the floor' (Curry 2002).

More broadly, UK (and European Union) agriculture over the last 30 years or so have been characterized by a move from a production oriented countryside to a consumption oriented countryside (Marsden 1999), and an increasing emphasis on the provision of public goods by farmers.[8] This is perhaps particularly true of upland[9] areas, where 'multifunctionality' is often used as a lobbying point to secure funding for agriculture at a European level (Potter and Tilzey 2005). Commentators (e.g. Mansfield 2008; Burton et al. 2008) identify that there is strong public demand for upland landscape and biodiversity, whilst the returns from upland agricultural outputs are unreliable and rarely sufficient to generate profit. Indeed, Ennerdale valley is classified as 'severely disadvantaged' under the European Union Less Favored Areas (LFA) program, which means that farmers in the area qualify for additional support to compensate for permanent natural handicaps (e.g. short growing season, steep slope, poor soils & high rainfall). The LFAs are predominantly in the northern and south western areas of England and in areas of the Welsh Borders.

Recent debates around stocking levels and the introduction of programs like the English Nature Sheep and Wildlife Enhancement Scheme (SWES), the 2007–2013 Rural Development Plan for England (RDPE), changes in the structure of Department for Environment Food and Rural Affairs (DEFRA) agri-environment schemes and the introduction of the Single Farm Payment (SFP) have created further uncertainty within the farming sector (Conyers et al. 2008).[10]

---

[8] We use the term public goods to refer broadly to resources from which all may benefit, regardless of whether they have helped provide the good. Public goods are also distinguished by the fact that they are non-rival in that one person's use of the good does not diminish its availability to another person (Kollock 1998).

[9] There is no statutory UK definition for the uplands. In an agricultural context the term is generally used to refer to areas above the upper limits of enclosed farmland containing dry and wet dwarf shrub heath species and rough grassland, where management is predominantly through sheep grazing.

[10] The recent decision by DEFRA to roll the existing Hill Farm Allowance (HFA) over for a further 3 years is, however, being viewed as a positive move by the National Farmers Union. NFU uplands spokesman Will Cockbain (2006) states that 'the fact that in 2010 we will move to an uplands entry level scheme is also important as it means all farmers in the uplands will be eligible and can be rewarded for the hugely important role they play in the delivery of public goods.'

Slee et al. (2009) provide a concise overview of the current situation facing people and communities in the uplands.[11] Highlighting the importance of upland ecosystem goods and services for the rest of society, they state that land use and environment in the uplands are not uniquely multifunctional, but are exceptionally multifunctional; and the benefits to society arising from that multifunctionality are undoubtedly compromised, especially by agricultural change, but also by many other pressures. As the expectations that society have of the uplands increase, additional input is needed to support upland communities in delivering what is required. The recent turbulence in the rural economy and the economy as a whole caused by global events, from the rising food and energy costs of 2007 to the rapid and largely unforeseen major economic downturn in the last year are uncertain but potentially profound, especially if social and environmental tipping points are reached.

An equally stark picture of upland land use is also painted by Smyth (2006), who argues that regardless of policy, socio-economic forces are what drive change. She predicts an uplands future which is likely to include the abandonment of sheep farming across marginal areas; the colonization of attractive areas by lifestyle farming; non-governmental organizations buying run-down farms with high biodiversity or landscape value and activity tourism thriving in a variety of places.

The experience in Scotland, reported by various bodies including the Royal Society of Edinburgh (2008), National Farmers Union (2008) and the Scottish Agricultural College (2008) of 'farming's retreat from the hills' gives rise to serious concern about the future of livestock farming in the English uplands should such trends move southwards.

Respondents were asked their opinion regarding the future of farming. Similar to other recent studies in England (for example, Burton et al. 2005), their responses revealed a range of emotions, from bitterness and cynicism through to guarded optimism. Farmers frequently saw their future role related much more closely to environmental management (I01): *'Really we're just paid to be park keepers aren't we. Keep the place looking nice ... we're just paid to have it look nice for the tourists, but the thing is they've made us take all the sheep off the fell.'*

There was a corresponding sense of being unwanted and undervalued, particularly in relation to WE (I03): *'I think the farming activity in the valley is now considered to be fairly peripheral to the general sort of aim of Wild Ennerdale ... we all get the impression that they would quite like us to go away.* But there was also evidence of resilience and a determination to continue (IO3): *It looks as though we've got one more generation that's going to keep it going and I feel quite strongly about it really.'*

There was also a view that if farmers were to be involved in public goods provision, they would need to be paid appropriately (I05): *'And if they want to look at this landscape they'll have to pay for it ... and pay well. I know my generation a lot of them they're sick to the back teeth of what's going on.'*

[11] In reality the 'upland community' comprises a number of diverse groups, with different sets of interests and varying capacities to voice those interests.

The interviews also revealed the complexity of farming households in the valley, indicating changing gender roles and the importance of off-farm employment (I04): *'My husband actually works away from the farm so I may be in a situation where I choose to do that more. I'm a trained X so I do have other things that I can do, so it's balancing up the time involved. You get less situations now where you've got your farmer's wife at home – in the past the farmer, the man, would be out working and the woman, the wife was in the house so there was always somebody there for eventualities such as bed and breakfast or people popping in or whatever, but more and more ... people have part-time jobs or the other partner has a full-time job even in many cases. Purely because of the financial uncertainty of farming, not many people go into farming now if they haven't got some sort of back-up.'*

12.3.2 Farming Perceptions of WE

The WESP (WEP 2006) states that *'Wild Ennerdale is not about abandoning land, excluding people or trying to create a past landscape. On the contrary, human activity is a crucial part of the process, along with the need to provide quantifiable economic, social and environmental benefits which are sustainable'*. The strategic plan outlines the long history of human influence in Ennerdale valley, stating that *'Ennerdale has provided for people's needs for many centuries. The range of monuments and features within the valley demonstrates how the landscape has been influenced and altered by man for over 3500 years ... during the Bronze Age low intensity pastoral farming was introduced to the fells.'* However, the WESP also states that *'historic grazing levels have reduced species composition on the fells and suppressed upland heath communities'* (WEP 2006). Thus whilst Ennerdale has a long history of farming,[12] overgrazing has been identified as an important management issue by the WEP.

Respondents were asked how they viewed WE and its vision for the future of the valley and were also invited to discuss the implications of WE for their farm management. The views expressed indicate both scope for collaboration and compromise with WE integration and concerns over practical difficulties related to merging forest with pasture. Some farmers who use forest tracks for access to

[12] Archaeological evidence suggests that the Ennerdale valley has been subject to low intensity farming since the Bronze Age (WEP 2006) though the present-day farming landscape is largely the product of evolution since the Norse colonization of Cumbria around AD 900. Oxford Archaeology Unit (2003) state that the post-medieval period saw the increasingly intensive pastoral exploitation of the valley sides (in common with other upland areas), adding that the increasing numbers of sheep on the fell inevitably had a considerable impact upon the vegetation, preventing the proliferation of heather moors. The practice of transhumance (summer grazing of stock on the common fell land) continued into the sixteenth and seventeenth centuries but was often first documented in the twelfth and thirteenth centuries (Oxford Archaeology Unit (2003) provide a detailed historical account of human settlement, agriculture & land use in the valley).

pasture expressed concern over proposals to *'allow sections of the forest track network to revert to vegetated tracks'* (WEP 2006). One farmer (I02) stated that *'I would prefer from my point of view that they didn't start blocking off the access roads.'* However, the main focus of farmers concern was linked to a proposal in the (WEP 2006) to *'remove redundant boundary fencing to move towards extensive grazing regimes within existing forest boundaries.'* There was widespread anxiety of serious implications for shepherding if the boundary fences between forest and open fell were allowed to deteriorate:

> *'The bit that potentially affects us in a major way would be the taking down of boundary fences ... They haven't been terribly sympathetic to our concerns about the breakdown of our heafing and shepherding systems, if we haven't got some physical barrier to keep them* [sheep] *out of the woods ... you can only shepherd them if you can find them.'* (I03)
>
> *'... if they take them fences down, all the stock will go out of that valley, it'll go tomorrow, because you can't shepherd it. A lot of stock would harm their selves walking through* [the woods] *... the knock-on effect of that is if you don't gather you start getting parasites and you can't get in to treat your animals. But you can't get that through to them* [WE]. *Well, once they take them fences down they'll push the farmers out ...'* (I05)

12.3.3 Working Together

There was, however, also a sense that WE and farming could co-exist. One farmer (I03) offers a pragmatic view as to striking a balance, which again focused on the importance of maintaining boundary fences:

> *'I think that the two things* [sheep farming and wilding] *can run side by side, but they're going to have to make certain concessions to farming activity ... from our point of view as long as the boundary fence remains fairly sound it shouldn't necessarily affect us to a great extent in the near future.'* (I03)

Another farmer (I02) suggests that if the boundary fences do come down, there was still the possibility of reconciling the interests of farming and wilding if WE pursued a more active policy towards removing the Sitka spruce, retaining (and actively planting) semi-natural oak woodland and opened up glades within the woodland:

> 'Inside the forest ... there's one or two real old oak woods up there, hundreds of years old, now they're nice ... that's natural, they want kept. And if they go back to that ... get that spruce cleaned out, but don't let a lot grow in the woods, make it so it's green underneath and then you can eat it with stock and it'd be like a parkland sort of ... now that'll blend in with the valley.'

I03 also offered grounds for optimism, stating *'it* [WE] *must be positive if at least one farming business has benefited considerably* [through working with WE on a cattle grazing initiative] *if it's increased the profitability of at least one farm in the valley that gives a better chance of that farm surviving in the future, so anything that does that has got to be positive really.'*

12.3.4 Defining Wilding

A debate with clear relevance for WE is whether wild areas should be left untrammelled or be manipulated toward a more 'natural' state (Cole 2001). There are difficulties associated with defining the concept of wilding (Alexander 1996), and more specifically re-wilding, which as Fenton (2004) indicates, risks falling into a 'value trap.' The WE use of 'wild' denotes 'a philosophical approach to managing the valley' (WEP 2006) encompassing two key areas: (1) the degree to which natural processes influence the environment (physical attributes) – *which might be broadly interpreted as leave it and see what happens*; and (2) the sense of wildness which people experience/perceive (emotive reactions) – *the social representation of wilderness.*

There is clearly some resonance with Nash (1982) and his 'if it looks wild it is wild' definition of wilderness: *'to accept as wilderness those places people call wilderness ... not so much on what wilderness is but what men think it is.'* Carver et al. (2002) describe Nash's views on wilderness as one extreme on a continuum from the paved to the primeval. The position along the continuum at which wilderness occurs has more to do with perceptions than it does with ecological conditions.

Thus the WE process will not create wilderness *per se*, but rather an addition to the wide range of altered or non-natural ecosystems found in England. Nevertheless, the resulting landscape will contain attributes of wildness including remoteness (from settlement and access) and naturalness (both perceived and ecological). To some this will be wild, to others it will be a glorified picnic site.

The wilding approach adopted by WE was perceived by several respondents as being unclear, with one farmer (I02) stating that *'the planners are just taking a step back and say we'll suck it and see which way it's going, cos they don't know exactly what's going to happen.'* Others were suspicious about the factors behind the establishment of WE:

> *I would say that quite a lot of the driving force behind Wild Ennerdale is the fact that none of the timber up there is commercially viable and to me it seems like an awfully good way of not doing anything else, you know, not spending more money on it really ... it's quite a nice way of getting rid of a bit of a liability to be honest, just badge it as something else and walk away and leave it ... they've just created this Wild Ennerdale, gone barging right into it.* (I03)

12.3.5 Tidy Up the Mess

Farming respondents were particularly concerned about the perceived 'mess' of the WE area, and how this might look in the future. The potential for conflict over landscape preferences, particularly between a wild/managed landscape, has been highlighted by van den Berg and Koole (2006), who note that there are broadly two sub-populations with relatively extreme landscape preferences: 'environmentalists' and 'farmers'. Environmentalists have been found to display relatively strong preferences for wilderness settings as compared to more managed natural settings. By contrast, farmers have been found to display relatively strong aesthetic preferences for managed settings. Interviews with farming respondents in Ennerdale supported this position, as the following extracts indicate:

> *If you go up that valley now and you look at the topside of the fences on both sides that we have and we have sheep on, that's the nice bit that everybody looks at, down in the scrow* [mess] *that the forestry have, that's the bit that needs tidies up to encourage people to come if that's what they want to do. Like our bit is all right, they want to get up off their backsides, tidy the mess up.* (I05)
>
> *'Tidy up the scrow that they've left behind now, which again is a complete contradiction, we were always told not to make any mess and the Forestry Commission left a nuclear landscape behind them when they'd finished.'* (I03)

12.3.6 Project Engagement

Respondents were asked about their current level of engagement with WE. Many of the respondents did not feel that there had been adequate communication. For example, I04 stated that *'they've kept us up to date with what they've decided but I don't feel they've necessarily open about what they were planning to do at the start ... I didn't feel necessarily included in decisions.'* In the focus group meeting, a tourism provider noted that *'I'm amazed as a newcomer to the area, I've only been here since March, I seem to know a lot more about it than people who've been here all their lives!'*

A farmer (I03) highlighted the need for deeper, farm-level engagement with the project, stating that *'if they've got any ideas of what they're wanting to do they need to put the actual proposals in their entirety to each farmer and how it might link in their business, rather than just decide what they're doing, get one volunteer to do it.'* However, another respondent (I02) felt that there had been relatively good communication, and argued that *'all the farmers in the valley were invited to an initial consultation where the idea was put forward. And from then it was taken forward, and as far as I'm aware, everyone was included or had the opportunity to be included, so, we can't all say that we didn't know that it was happening.'*

As stated earlier, the WEP (2006) outlines a vision which aims 'to allow the evolution of Ennerdale as a wild valley for the benefit of people, relying more on natural processes to shape its landscape and ecology'. The vision is qualified with an assertion that 'the valley will sustain the livelihoods of local people in keeping and enhancing the valley's special qualities and that a broader section of local people will have a greater sense of involvement in its future.' This vision is undermined by a sense of exclusion felt by some farmers, who saw themselves as important stakeholders in the valley. One farmer (I07) noted bluntly that *'it's between the Forestry Commission, National Park, United Utilities and the National Trust isn't it and they're not involving* [farmers]'.

There was consensus amongst respondents that WE should include what might be broadly termed the 'farming cultural landscape' in their vision for the future of the valley. For example, I03 stated that '*I think that it would be really good if this Wild Ennerdale partnership actually included within the partnership the farming activities as well as just the wilderness activity, because it's all part of the whole picture isn't it?* But he also thought that WE presented opportunities for community development: '*I think if I was going to see it developed for the benefit of the community it would be good if any sort of development involved the locals, get the people from the village involved ... because there's too many of these things are actually developed by people from outside the area.*'

Höchtl et al. (2005) assert that decision makers should be aware of the positive and negative aspects of (re)wilding and all stakeholders, especially those affecting local communities, should be included in any process that concerns the establishment of protected areas which are left to develop without human control. As a member of the Parish Council noted in the focus group meeting '*Any project in this valley has got to be hot-wired into what's going on in the valley, into the community, it has to celebrate it and sustain it, not cut across it.*'

12.3.7 The Cultural Landscape of Ennerdale

Whilst livestock–farming relations may be socially constructed and dynamic, thus engendering particular sets of farming practices at particular times and places, they nevertheless form lifescapes of social, cultural and economic interactions between humans, livestock and landscapes and is what Gray (1998) refers to as 'consubstantiality'. There was a deep sense of connection between the farmers of Ennerdale interviewed for this project and the physical environment of the valley – a sense of being part of the evolving cultural landscape. Höchtl et al. (2005) highlight how the main impacts on inhabitants of change from rural landscape to wilderness are psychological and economic in nature. They describe how the wilding process can lead to a perceived loss of historical experience, cultural knowledge and local identity. One farmer (I03) stated that '*you know I'll be disappointed, I'll be bloody annoyed actually if something caused the farming activities that's gone on for ... certainly in my case for 5 generations, pretty much the same thing you know, I probably now wear polartec and they used to wear woolen long-johns or summat* [something] *but you know the activity is just the same sort of stuff ... the heritage side of the thing is as important to me as the actual physical state of things.*'

This link to heritage and cultural landscape was also emphasized by I07, who noted that '*agricultures been here for hundreds and hundreds of years, it's what makes it. I mean we've been here in this valley about 128 years, how many years has agriculture been in the fells now? Before the National Park and the Wild Ennerdale initiative that's for sure.*' Another farmer (I01) noted simply that '*I feel like they* [WE] *own it but it's our heritage.*' As Slee et al. (2009) note, in the past the uplands have been sustained by people whose actions are not driven exclusively by narrow economic motives. Custodianship, tradition, socially and culturally valued actions have all influenced decision-making.

12.3.8 Farmers as Interpreters of Landscape

The role of farmers as interpreters of landscape has been highlighted by a number of projects in Cumbria (Burton et al. 2005). Most recently, the Flora of the Fells project (Flora of the Fells 2006) has involved farmer-led walks to 'explore the biodiversity' of the Lake District National Park (indeed, one of the interview respondents had participated in this scheme). The Fells & Dales local action group of the European Union rural development project LEADER is also keen to explore the potential for this role in the future. A farmer interpretation role could therefore potentially offer opportunities for collaboration between the farming community and WE, and farmers were asked whether they would be interested in participating in such projects. One farmer (I02) noted that he already spent time informally discussing his job with the public: *'They'll lean over wall and* [say] *'good morning' and 'what are you doing?' and they're interested in what you're doing. You don't mind spending 10 min with people to do that if they're interested in what you're doing.'* I04 agreed, stating that *'I think they're most of them are always interested when you're gathering sheep out of the fell, they'll want to talk to you.'* I06 remarked that *'they want to come and see how we live, they want to see our way of life they want to come and look and they ask you ... sometimes you get nothing done for telling them!'*

Convery and Dutson (2006) indicate how such conversations and insights were an enriching part of a visitors' experience of an area, but that the informal nature of such interactions was often important. The transition to a more formalized arrangement presents a number of obstacles, and whilst there was a clear willingness expressed during most interviews for farmers to engage with visitors in this way, there were also concerns expressed regarding time demands and perceived health and safety problems.

The attitude of the farming respondents towards a future interpretation role is neatly summed by one farmer (I02), who stated that whilst *'I'm farming, my goal is farming...* it would still be possible to get involved in tourist activities, "He suggested that *'you could certainly do a walk a month, something like that, or a couple of walks a month. And it's surprising how many people that came in March would probably come again in summer or the back end to see what's changing. To see the seasons change. Because it is a beautiful valley with all the different colours of the different trees and everything.*"

12.4 Discussion

The evidence from this small study suggests that the social consequences of a policy to create wild land require careful consideration. In a worst case scenario, MacDonald et al. (2000) argue that such policies risk creating a continuing cycle of increasing rural depopulation, deprivation, further land abandonment and loss of traditional land management skills. To counter such problems, Höchtl et al. (2005) and Kirby

(2003) recommend a wide ranging consultation process amongst stakeholders regarding the establishment of wild areas. As Matouch et al. (2006) indicate, *'a participatory approach is of particular significance in upland areas in Europe where there tends to be an intimate association between communities and the areas in which they live...* [and which support] *their livelihoods – either directly through agriculture and forestry or indirectly through upland tourism. Any project initiative which does not fully consider the aspirations, welfare and economic activity of upland communities has little chance of success.'*

There is evidence from this study that WE has not fully considered the sensibilities and complex livelihoods of the farming community, and as a consequence, a group of significant stakeholders feel alienated from the project. Of course, even with meaningful participation (for example, full joint decision-making) it still may not be possible to keep all stakeholders happy, particularly in relation to wilding projects, which tend to be difficult, contentious and complex.

Whilst terms such as 'public goods', 'ecosystem services' and 'multifunctionality' are gradually gaining credibility within the British hill farming community, many farmers still 'just want to farm' (Convery and Dutson 2006); a 'post-productivist' position for upland agriculture is far from universally accepted. Despite finding it ever more difficult to sustain their livelihoods, upland farmers and (other land managers) are increasingly being called to respond to new policy imperatives and manage the environment for the benefit of wider public goods (Long et al. 2009). For example, in recognizing the importance of uplands for biodiversity, the Government's Public Service Agreement strives to upgrade 90% of upland SSSIs into favorable or improving condition by 2010. This has driven significant recent investment in upland restoration, but in places, some would argue, has compromised effective farming practice.

Land managers whose natural inclination is often to want to produce food or timber have been asked to make a sometimes uneasy transition to managers of a suite of environmental services and public goods. There is no magic bullet to ease this transition; ultimately, this is likely to come down to local level dialogue, patience, cooperation and trust. There also needs to be an economic incentive ('what's it in for us?' as one respondent remarked) for the farmer. Nevertheless, the involvement of one farmer in WE, through grazing cattle in the forest, has demonstrated that this project can successfully involve the farming community and this gives some optimism for the future.

12.5 Conclusions

Whilst many aspects of the natural economy of Ennerdale have been subject to research and investigation, there is a need to deepen understanding of the nature of enterprises linked with this valley and in particular to consider how best to develop a model of management that is in harmony with the special qualities of the valley and the aspirations of those who live in and depend upon it.

The farming lifescape of Ennerdale represents a complex interrelationship of people, place and production system, and the research detailed in this chapter reveals a wide and interrelated set of themes and issues that include cultural landscape, social capital and farming/landscape lineages. Ennerdale valley has a long history of management; some members of the farming community have ties to the land spanning several generations. In contrast, the extensive conifer plantations are relatively recent, yet form the starting point for the WE initiative. WE has developed to such as stage that the partnership must fully consider its future role and impacts. In particular, the partnership should embrace wider partnership working with the various communities of Ennerdale and a much greater appreciation of the role farmers have played, and continue to play, in shaping the cultural landscape of the valley. This is true for WE as it is for any multi-stakeholder wilding project.

12.6 Management Implications

- *A socio-temporal perspective is important* – Ennerdale valley has a long history of agriculture and management. The extensive conifer plantations are relatively recent, yet form the starting point for the WE initiative
- *Cultural landscape* – the farming lifescape of Ennerdale represents a complex interrelationship of people, place and production system, creating a specific cultural landscape. How will the project affect this landscape? In developing project goals the cultural landscape needs to be fully considered
- *Effective participation* – wilding projects need to fully consider, and work with, the aspirations, welfare and economic activity of all local stakeholders if there are to have any chance of success

References

Alexander M (1996) A guide to the production of management plans for nature reserves and protected areas. Countryside Council for Wales, Bangor
Blackshall J, Manley J, Rebane M (2001) The upland management handbook. English Nature, Peterborough
Burton R, Mansfield L, Schwarz G, Brown K, Convery I (2005) Social capital in hill farming. Report for International Centre for the Sustainable Uplands, Cumbria
Burton R, Schwarz G, Brown K, Convery I, Mansfield L (2008) The future of hefted upland commons in areas of high public goods provision: learning from the Lake District experience. In: Bonn A, Hubacek K, Stewart J, Allott T (eds) Drivers of change in uplands. Routledge, London
Carver S, Samson P (2004) Eee, it's wild up North! Ecos 25:29–33
Carver S, Evans A, Fritz S (2002) Wilderness attribute mapping in the United Kingdom. Int J Wilderness 8:24–29
Cockbain W (2006) NFU claim victory on HFA. NFU Press Release (December 14)
Cole DN (2001) Management dilemmas that will shape wilderness in the 21st century. J For 99:4–8
Convery I, Dutson T (2006) Sense of place in Northern England. Biodivers Sci Manage 2:213–218

Convery I, Bailey C, Mort M, Baxter J (2005) Death in the wrong place? Emotional geographies of the UK 2001 foot and mouth disease epidemic. J Rural Stud 21:99–109
Convery I, Mort M, Bailey C, Baxter J (2008) Animal disease & human trauma. Palgrave Macmillan, Basingstoke
Conyers S, Garthwaite D, Bishop J, Boatman N (2008) Uplands entry level stewardship (UELS) final report. Central Science Laboratory, York
Cumbria County Council (2006) Cumbria's parishes – people. http://www.cumbria.gov.uk/factsandfigures/Statisticsal-summaries/parish-profiles.asp. Accessed 14 Dec 2006
Curry D (2002) Policy commission on the future of farming and food: farming and food a sustainable future. HMSO, London
Dutson T, Convery I (2007) Sense of place and community development in the Northern English uplands. Int J Interdiscip Soc Sci 1:37–48
Fenton J (1996) Wild land or wilderness – is there a difference? Ecos 17:12–18
Fenton J (2004) Wild thoughts followed up. Ecos 25:18–20
Fielding AH, Haworth PF (1999) Upland habitats. Routledge, London
Fisher M (2003) Self-willed land – the rewilding of open spaces in the UK. http://www.self-willed-land.org.uk/rep_res/SELF.pdf. Accessed 18 Nov 2006
Flora of the Fells (2006). http://www.floraofthefells.com. Accessed 5 Dec 2007
Forestry Commission (1998) A new focus for England's woodlands. England Forestry Strategy, Cambridge
Franks J, Lowe P, Phillipson J, Scott C (2003) The impact of foot and mouth disease on farm businesses in Cumbria. Land Use Policy 20:159–168
Gray J (1998) Family farms in the Scottish borders: a practical definition by hill sheep farmers. J Rural Stud 14:341–356
Green B (1995) Plenty and wilderness? Creating a new countryside. Ecos 16:3–9
Habron D (1998) Visual perception of wild land in Scotland. Landsc Urban Plan 42:45–56
Hall CM, Page SJ (2002) The geography of tourism and recreation environment, place and space. Routledge, London
Höchtl F, Lehringer S, Konold W (2005) 'Wilderness' – what it means when it becomes a reality – a case study from the southwestern Alps. Landsc Urban Plan 70:85–95
Jerram R (2004) The Bowland initiative: a lifescape project a review of environmental outputs. Lancashire Rural Futures, Lancaster
Kirby K (2003) What might a British forest landscape driven by large herbivores look like? English Nature research report 530. English Nature, Peterborough
Kollock P (1998) Social dilemmas: the anatomy of cooperation. Annu Rev Sociol 4:183–214
Long M, Convery I, Simmons E (2009) Stakeholder perspectives on the uplands entry level stewardship scheme and its significance for the future of upland farming. J R Agric Soc Engl 170:1–8
Lowe P, Edwards S, Ward N (2001) The foot and mouth crisis: issues for public policy and research. Centre for rural economy working chapter 64. University of Newcastle upon Tyne, Newcastle
MacDonald D, Crabtree JR, Wiesinger G, Dax T, Stamou M, Fleury P, Gutierrez Lazpita J, Gibon A (2000) Agricultural abandonment in mountain areas of Europe: environmental consequences and policy response. J Environ Manage 59:47–69
Mansfield L (2008) The Cumbria hill sheep initiative: a solution to the decline in upland hill farming community in England? In: Robinson G (ed) Sustainable rural systems. Ashgate, London
Marsden T (1999) Rural futures: the consumption countryside and its regulation. Sociol Rural 39:501–520
Matouch S, Forster J, Scott Jones P, Mitchell R, Rebane M, Rose P, Scheidegger C, Soier D, Stoll-Kleemann S (2006) Uplands thematic working group report. http://www.nbu.ac.uk/bioforum/Uplands.htm. Accessed 19 Dec 2006
Ministry of Agriculture, Fisheries and Food (1999) A new direction for agriculture England rural development plan. MAFF, London
Nash R (1982) Wilderness and the American mind, 3rd edn. Yale University Press, New Haven
National Farmers Union (2008) Manifesto for the hills. NFU Scotland. Ingliston, Edinburgh

National Trust (2008) Annual report 2007/08. http://www.nationaltrust.org.uk/main/w-anual-report08-printversion.pdf. Accessed 2 Sept 2009
Northwest Regional Forestry Framework Partnership (2005) The agenda for growth regional forestry framework for England's north west. http://www.iwood.org.uk/. Accessed 5 Jan 2007
Oxford Archaeology Unit (2003) Historic landscape survey. http://www.wildennerdale.co.uk/images/library/Historic%20Landscape%20Survey%202003.pdf. Accessed 2 Sept 2009
Pope C, Ziebland S, Mays N (2000) Analysing qualitative data. Br Med J 320:114–116
Potter C, Tilzey M (2005) Agricultural policy discourses in the European post-fordist transition: neoliberalism, neomercantilism and multifunctionality. Prog Hum Geogr 29:581–600
Royal Society (2002) Inquiry into infectious diseases in livestock. Royal Society, Plymouth
Royal Society of Edinburgh (2008) Committee of Inquiry into the future of Scotland's hills and islands. Royal Society of Edinburgh, Edinburgh
Scottish Agricultural College (2008) Farming's retreat from the hills. SAC, Edinburgh
Slee B, Reed M, Convery I, Dutson T (2009) Supporting the Council for Rural Communities in its inquiry into the future of England's upland communities. Tender document produced by The Macaulay Institute; Aberdeen Centre for Environmental Sustainability; University of Leeds Sustainability Research Institute & University of Cumbria
Smyth AM (2006) Activity tourism in an upland landscape: dudes and hippies in the hills. Biodivers Sci Manage 2:223–226
van de Berg AE, Koole SL (2006) New wilderness in the Netherlands: an investigation of visual preferences for nature development landscapes. Landsc Urban Plan 78:362–372
Waitt G, Lane R, Head L (2003) The boundaries of nature tourism. Ann Tour Res 30:523–545
WEP (2006) Wild Ennerdale Stewardship plan. http://www.wildennerdale.co.uk/managing-%20stewardship%20plan.htm. Accessed 23 June 2007

Part IV
Integrated Perspectives

Chapter 13
The Role of Forest Landscape Restoration in Supporting a Transition Towards More Sustainable Coastal Development

Peter R. Burbridge

13.1 Introduction

Coastal areas are under great pressure from development throughout the world. Various estimates indicate that some 60% of the world's population lives within 100 km of the sea and some 75% of urban areas with more than 1.5 million inhabitants are located in estuaries or open coastlines (Bodungen and Turner 2001). Poorly planned and managed development in coastal areas has led to wide-scale degradation of coastal systems, including forests. Estimates suggest that more than 70% of Europe's coastal areas have been degraded (CEC 1999). In tropical regions major pressures to develop coastal areas have, in many cases, proved non-sustainable.

It is anticipated that coastal regions will be the focus for future development and that the resulting pressures will increase competition for access to and use of coastal areas. Unless major advances are achieved in coastal area planning and management, and in watershed management, there will be increased stress on coastal systems.

Coastal forest systems in the tropics, and to a lesser extent in temperate regions, have been a major focus for conversion to other uses. Many of these developments have proven non-sustainable. Examples include conversion of tidal swamp forests for irrigated rice cultivation in Sumatra (Burbridge et al. 1981; Collier 1979), conversion of *Melaleuca* wetland forests for agriculture in Vietnam (Maltby et al. 1996; Safford et al. 1997, 2009), and conversion of mangrove forests to shrimp mariculture in areas of Asia and Latin America (Stevenson and Burbridge 1997). In Thailand some 70% of the mangrove found along the Gulf of Thailand has been degraded through the development of shrimp ponds, urban development, infrastructure development, and placer mining of tin deposits in mangrove systems (Kongsangchai 1984; Stevenson et al. 1999). The destruction of these coastal forests has led to the

P.R. Burbridge (✉)
School of Marine Science and Technology,
University of Newcastle-upon-Tyne, Newcastle, UK
e-mail: pburbridge@gmail.com

J. Stanturf et al. (eds.), *Forest Landscape Restoration: Integrating Natural and Social Sciences*, World Forests 15, DOI 10.1007/978-94-007-5326-6_13,

loss of a wide range of economic and environmental goods and services that could sustain many different forms of economic and social development (Burbridge 1994; Burbridge et al. 1981; Dixon and Burbridge 1984; Turner et al. 2000, 2003).

The destruction and mismanagement of coastal forests and other coastal ecosystems has resulted in many coasts being in a state of almost continuous dis-equilibrium (Burbridge and Pethick 2003). As a result of this dis-equilibrium combined with impacts resulting from poorly planned and mismanaged development in watersheds upstream, including the clearance of forests to accommodate alternative forms of development, the vulnerability of human settlements in coastal areas to natural hazards such as flooding has increased. Although increased international funding has been directed toward better understanding of the relationships between watershed management and coastal development, forest landscape restoration (FLR) does not appear to play a central role in such efforts.

There are initiatives to rehabilitate tropical coastal forest systems. However, the success of these initiatives will depend heavily upon a wide range of factors, including:

- Competition for access to and use of coastal systems
- Public support from stakeholders
- Adverse interactions among coastal development activities
- Sectoral based management of coastal development
- Poor watershed management
- Failure to link watershed management with coastal area development
- Global change, including climate change and the influence of man on hydrology, sediment budgets, energy and nutrients reaching coastal systems from upland areas

Ecosystem management concepts and principles, including FLR, could help to sustain the rehabilitation of coastal forest systems and reduce the vulnerability of coastal development from natural and human-induced hazards, such as flooding.

13.2 Concepts and Principles of Integrated Coastal Management

During the past 35 years increasing recognition of problems associated with poorly planned and managed development in coastal regions has led to the development of concepts and principles to support more integrated forms of coastal development, including the conservation of coastal ecosystem functions. A new discipline termed Integrated Coastal Management (ICM)[1] has been adopted by many countries as a means of promoting more sustainable use of coastal areas and natural resources

[1] The term Coastal Zone is also used to define the zone of transition from purely marine to purely terrestrial environments, however this has lost favour as a result of growing recognition of the trans-boundary management considerations, such as changes in hydrology due to development in river catchments, that have a major impact on the management of coastal areas and human activities. The term Integrated Coastal Management (ICM) is used instead of Integrated Coastal Zone Management in recognition of the linkages between catchments, coastal areasand ecosystems, and marine areas and ecosystems. The term ICM will be used in this paper.

(Chua 2004; Sorenson 2000, 2002). This has been complemented by the development of watershed management concepts and principles and more recently by the principles and good practices of Forest Landscape Restoration (Aldrich et al. 2004; ITTO 2002; Maginnis and Jackson 2005).

13.2.1 The Concepts of ICM

Integrated Coastal Management (ICM) describes a framework and process for formulating and implementing plans and management strategies to promote wise and sustainable use of coastal areas and resources. The basic objective of ICM is to maximise long-term economic and social benefits from the wise use of coastal resources. At the same time, it is very important that ICM be seen as a practical means of meeting short-term development objectives, such as helping to diversify economic activities in rural areas.

13.2.2 ICM as an Iterative Process

ICM is based on an iterative cycle of actions designed to help develop a robust and adaptive process for improving the planning and management of human activities in coastal areas. There is no Golden Rule or universal framework for promoting ICM and in application, ICM is an iterative process whereby it is possible to start with specific problems and issues and to develop increasingly comprehensive and sophisticated ICM initiatives.

The main steps in this process are:

- **Increase Awareness**-based on traditional knowledge and user friendly scientific information
- **Stimulate a Dialogue** among key stakeholders
- **Foster Cooperation** among stakeholders including sectoral governmental agencies
- **Encourage Coordination** among stakeholders in the formulation and implementation of policies, plans, investment and management strategies
- **Work towards Integration** of policies, plans, investment and management initiatives for coastal areas and human activities (Humphrey and Burbridge 1999)

The main elements of the management process include:

- Comprehensive assessment of environmental, social and economic issues that influence the sustainable development of coastal areas and associated natural resources
- Setting of objectives that meet local, regional and national social and economic needs and aspirations
- Formulation of plans and management strategies designed to anticipate and respond to issues as well as the development needs and aspirations of coastal societies
- Monitoring the success of ICM plans and management strategies and incorporating improvements to those plans and strategies as part of a cyclical process.

13.2.3 Adopting a Functional Approach to ICM

The most common causes of stress and degradation in coastal tropical forests are changes in hydrology- mainly in the form of reduced surface and ground water flows into the forest system, changes in materials fluxes (sediments and nutrients) and changes in the energy required to maintain the health and productivity of coastal forest systems.

The Land Ocean Interactions in the Coastal Zone (LOICZ) project published a synthesis of its first 10 years of research (Crossland et al. 2005) in which a number of conclusions concerning coastal systems and their relationship with river basin systems, marine systems and the global ecosystem have a bearing on FLR; namely:

- The coastal domain is the most dynamic part of the global ecosystem and the realm most subject to natural and man-induced global change
- At a global scale, coastal systems (including forests) play a significant role in regulating global change
- Although major river systems have a profound influence on coastal and near-shore marine systems at a regional level, the mounting pressures from human development and their effects on coastal systems are felt most acutely at small to medium catchment scales
- Changes to coastal systems cannot be confined within administrative boundaries. Instead, studies need to be oriented towards watershed and catchment based perspectives to understand coastal dynamics and to integrate the results into human management activities (Crossland et al. 2005)

These findings from the LOICZ project reinforce the emerging concepts of integrated coastal management where the "coastal zone" is treated as part of a dynamic continuum linking terrestrial and marine components of the Earth's ecosystem, rather than an isolated "zone" in which systems such as coastal forests can be managed without reference to natural and man-induced changes in hydrology, or fluxes of materials in upland and oceanic systems (Burbridge et al. 2005).

Adopting a functional landscape approach logically links river basin systems and coastal systems. Key factors to consider are the standards of planning and management of human activities in the broader landscape and their influence on the hydrology, sediment budgets, nutrient fluxes and energy fluxes that maintain the health and productivity of coastal forests downstream. The vulnerability of coastal forest systems to poor planning and management of development in upland catchments is illustrated in Table 13.1.

13.3 Complementarity Between ICM and FLR

It is therefore critically important to integrate coastal forest systems into the framework of FLR and related efforts such as River Basin Restoration into a broader framework that couples the management of human activities within river basin systems and the management of forest systems in coastal areas. A more detailed

Table 13.1 Vulnerability of coastal tropical forest systems to development in coastal areas and from upland development

	Coastal forest type		
Development activity	Mangrove	Melaleuca	Tidal swamp forest
Agriculture and farming	Likely	Likely	Likely
Aquaculture and mariculture	Likely	Likely	Possible
Dredging and filling	Likely		Likely
Harbours	Likely	Likely	
Roadways and causeways	Likely	Likely	Likely
Shipping	Likely	Possible	Possible
Electric power generation	Likely	Possible	Possible
Heavy industry	Likely	Likely	Likely
Upland mining	Possible	Possible	Possible
Coastal mining	Likely	Possible	Possible
Oil and gas development	Likely	Likely	Possible
Military facilities, training and testing	Likely	Likely	Possible
Sanitary sewage discharges	Possible	Possible	Possible
Solid waste disposal	Likely	Likely	Likely
Water development and control	Likely	Likely	Likely
Coastal Defence and Shoreline management	Likely	Likely	Possible
Tourism and Recreation	Possible	Possible	Possible

Potential impacts:
Significant adverse effect Likely (pink symbol = "Likely")
Adverse effects possible (yellow symbol = "Possible")

Sources: Abdeldayem et al. (2005), Bassoullet et al. (1998), Burbridge (1988), Burbridge (1996), Burbridge et al. (2005), Maltby et al. (1996)

examination of FLR is provided elsewhere in this volume (Lamb et al. 2012) and the main principles of FLR can be summarized as:

- Restoration of a balanced and agreed package of forest functions
- Active engagement, collaboration and negotiation among a mix of stakeholders
- Working across a landscape
- Learning and adapting

Because FLR is a flexible concept it has been applied at various scales from the individual site through to larger geographic areas and a variety of principles have developed (see Elliot 2002; Lamb et al. 2012; Maginnis and Jackson 2005; Mansourian et al. 2005; Oregon Department Forestry 2001). Pfund and Stadtmüller (2005) identified similarities in the goals and components between FLR and other innovative approaches to development, including the Ecosystem Approach, the Sustainable

Table 13.2 Examples of complementary principles of ICM and FLR

ICM principles	FLR principles
Water is the major integrating force in coastal resource systems	Restoration of the hydrologic linkages among natural systems
Coastal zone land and water uses must be jointly planned and managed	FLR emphasises integration of land uses
Sustainable development of coastal resources is a major purpose of coastal management and planning	Sustainable use of forest systems and forest functions is emphasised
Multiple use of renewable coastal resources is emphasized by coastal management and planning	Restoration of the multiple functions of forest ecosystems
The focus of coastal management and planning is on common property resources	FLR attempts to rehabilitate degraded common property resources
Integrated, multiple-sector involvement is essential to coastal management and planning	Is a multi-sectoral approach extending the decision-making process to all key stakeholders
Coastal management and planning boundaries are issue based and adaptive	FLR boundaries are adaptive
Coastal management and planning is structured for incremental implementation	FLR can be applied in an incremental manner, moving from small scale to larger scales within a functional landscape
Coastal management and planning emphasises the nature-synchronous approach to development	FLR seeks to work with natural system processes, including seasonal variations in hydrology
Ensure the support and involvement of all relevant bodies	FLR seeks to integrate stakeholders in developing a consensus on restoration priority actions
Use participatory planning to develop consensus	FLR involves stakeholders and encourages participation as well as it secures the long-term existence of the benefits for the society
Adaptive management is the basis of the ICM cycle	FLR embodies adaptive management principles and practices

Sources: Aldrich et al. (2004), Oregon Department Forestry (2001), IUCN (2005), Pfund and Stachmuller (2005)

Livelihoods Approach, Farming Systems ("Gestion de Terroir"), and Integrated Natural Resources Management. There are also potentially powerful links between FLR and Integrated Coastal Management (ICM), as indicated in Table 13.2.

13.4 Integrating FLR and ICM Processes

The focus of FLR is on restoring forest functionality in terms of enabling the forest system to generate goods, services and ecological processes at a broader landscape level rather than promoting increased tree cover within a particular location (FRIS 2006; ITTO 2002; IUCN 2005; Lamb et al. 2012; Maginnis and Jackson 2005).

The ICM process described above also seeks to promote sustainable use of the diverse functions and natural resources generated by coastal ecosystems at a broad landscape level. The ICM process is built around the concept that the coastal zones represents a transition between marine and terrestrial components of the earth and gives emphasis to developing specially adapted "integrated planning and management measures" that respect and work with the natural processes that support the inter-linkages among coastal ecosystems and the human activities they sustain. The ICM process can also be applied at different scales within the coastal landscape from a specific site through to the mosaic of natural systems and human activities. At the same time, the process of developing ICM is embodying a broader landscape perspective that links river catchments and near-shore marine systems.

As illustrated in Table 13.2, there are a number of principles that are essentially the same and/or complement one another that are central to both the FLR and ICM development planning frameworks. This provides major opportunities to strengthen the integration of the two approaches. To some extent this is already being attempted in international initiatives such as the Hill to Ocean (H2O) programme, BASIN, EUROBASINS, LOICZ and the EU Water Framework Directive.[2]

The question I would pose is "Are we giving enough emphasis to stronger integration of FLR and ICM?" The answer is NO. The coastal domain is the most dynamic part of the global ecosystem, and is the realm most vulnerable to natural and man-induced global change. Human development in coastal regions represents a powerful catalyst for direct changes in coastal systems, including coastal forests, and more broadly at a global system scale. At a global scale, coastal systems play a significant role in regulating global change. By developing stronger integration of FLR at a catchment scale with ICM, there is a major opportunity to substantially reduce the negative impacts of human activities and so have a positive influence on Global Change.

13.4.1 Examples of Coastal Forest Degradation

Two examples of non-sustainable development of coastal forest systems are presented in Boxes 13.1 and 13.2 below to illustrate how FLR would have served the conservation and sustainable management of these coastal forests and other inter-related coastal ecosystems if it had been employed. Each example demonstrates the need to understand the forest ecosystem, the functions they perform and the activities they can sustain. In both examples the integration of FLR and ICM could help to resolve issues affecting their sustainable use.

---

[2] BASIN (http://na-basin.org/)

EUROBASINS (http://eurobasin.dtuaqua.dk/eurobasin/index/index.html)

LOICZ (http://www.loicz.org/)

EU Water Framework Directive (http://ec.europa.eu/environment/index_en.htm)

Box 13.1 *Melaleuca quinquenervia* Wetland Forests in the Mekong Delta, Vietnam

Vietnam has two major river systems the Red River and the Mekong. There has been a long history of agricultural development of the floodplains and other alluvial lands bordering the Red River. Today, the Red River and associated river basin lands form the "Food Basket" of Vietnam.

The Mekong river system and its extensive delta have very different soils from the Red River system. For example there are extensive areas of potential acid sulphate soils (PASS) in the Mekong that are not common in the Red River delta. The difficulty of exploiting the PASS for agriculture has meant that extensive areas of *Melaleuca quinquenervia* (Cav.) S.F. Blake forest remained until the 1970s (Fig. 13.1). During that period war led to the draining of the wetland forest areas, use of defoliants and use of fire to destroy the forests and deny a safe haven for combatants. Following cessation of hostilities, the Vietnamese attempted to extend the drainage canals and to convert large tracts of *Melaleuca* wetland forest into irrigated agricultural lands (Fig. 13.2). Most of these agricultural conversions failed due to exposure of the PASS to the atmosphere and the resulting high acidity in the groundwater, irrigation water and soils once the water table was lowered. There was subsequent large-scale abandonment of the agricultural areas.

Fig. 13.1 Intact *Melaleuca quinquenervia* swamp forest in the Mekong Delta, Vietnam (Photo Peter Burbridge)

(continued)

Box 13.1 (continued)

Fig. 13.2 Drainage canals constructed in the *Melaleuca* forest for access and to drain the swamp to allow agricultural development (Photo Peter Burbridge)

In the early 1980s the World Conservation Union (IUCN) was asked to help the Vietnamese Government find ways of rehabilitating some 260,000 ha of degraded former *Melaleuca* forests in the Long Xuen Quadrangle and find ways of helping landless farmers to make a living. The Vietnamese officials had come to realise that the rehabilitation of the wetland forest would sustain a wide variety of economic goods and services as well as restoring valued environmental services. This presented a major challenge as any management solution would require restoration of the hydrology of the wetland forest sites to submerge the PASS and curtail the exposure of these soils to the atmosphere. The complex network of drainage and irrigation canals had been adopted as major means of access by boats which made it difficult to fill in or block canals and other man-made water channels (Fig. 13.3). At a broader scale, there are major plans to dam the Mekong and divert waters for agriculture in countries upstream.

The solution proposed by IUCN was to modify an agrosilviculture model adopted by the Vietnamese Forestry authority, which had 10 ha units with a core area of 7.5 ha for Melaleuca replanting and 2.5 ha for agriculture, including a house for the farmer and family. It was reasoned that if agriculture based on field crops failed because of the PASS, then it would make sense to reduce the areas for crops and to develop alternative forms of livelihood for the farmers participating in the agrosilviculture system.

(continued)

Box 13.1 (continued)

Fig. 13.3 Drainage canals serve as boat access to the area (Photo Peter Burbridge)

The modified model incorporated an area devoted to a nursery for stocks to be replanted in the expanded reforestation area, and the area set aside for agriculture was reduced to to 0.5 ha. A system of raised beds covering 0.25 ha was developed to leach out the acid from the soils and to form the basis for acid tolerant cash crops, such as coffee, citrus and pineapple, and for the cultivation of medicinal plants and food crops for the household (Burbridge 1995).

Arrangements were made with the regional agricultural cooperatives to help market the cash crops. New forms of income generation were developed based on on-site distillation of essential oils from the leaves and bark of the *Melaleuca*, and honey production. Seed trials were introduced to find early flowering varieties of *Melaleuca* to accelerate the production of honey.

There remained the problem of restoring the hydrology while maintaining access to and from the rehabilitation sites. A simple system of hand-operated locks was suggested as an alternative to filling in canals and using pumps to re-establish the water table. The ideas and preliminary development work for the rehabilitation of the *Melaleuca* wetland forests by the IUCN was then taken over by an Australian Aid (AusAid) project. A Darwin initiative was also undertaken to help refine the work by the IUCN (see Maltby et al. 1996; Safford et al. 1997, 2009).

Major progress has been achieved in the rehabilitation of a large area of the wetlands and in refining the rehabilitation model and techniques. However,

(continued)

Box 13.1 (continued)

this success at the local scale must be seen within the broader context of government plans for major dams on the Mekong to store and abstract water throughout the Mekong river system. Such changes in the regional hydrology of the river system will likely reduce base water flows locally that will adversely affect the rehabilitation of the wetlands.

The challenges faced in rehabilitating the *Melaleuca* wetland forests in the Mekong Delta are common to those faced today in lowland and forest systems. The principles and practices that have been developed for FLR would have been of great help to those of us struggling to integrate watershed management, restoration of local hydrologic conditions as a basis for rehabilitation of soils and re-creation of flooding and humidity conditions conducive to the successful replanting of *Melaleuca*, as well as developing agro-silviculture systems that landless farmers could adopt and manage as the basis for sustainable use of the wetland forests.

Sources: Beilfuss and Barzen 1994; Burbridge 1995; Burbridge and Scott 1990; Le 1989, 1993; Maltby et al. 1996; Safford et al. 1997, 2009; Takahashi et al. 2004.

Box 13.2 Freshwater, Tidally Influenced Peat Swamp Forests in Indonesia

Indonesia has extensive areas of tidally influenced, freshwater swamp forests (Fig. 13.4). These forests are capable of yielding some 25 m^3 ha^{-1} of high value timber and timber products from Dipterocarps and other species (Abell 1979).

During the 1960s and 1970s these forests were considered suitable for conversion to agricultural lands to support the Transmigration Programme where landless farmers and people displaced by the development of dams, industry and urban development in Java could be relocated (Figs. 13.5 and 13.6). An elegant scheme for using the tidal regime in estuaries to "pump" freshwater into irrigation channels during rising tides and then drain the wastewater as the tide fell was developed by the Dutch in the 1930s and adopted by the Transmigration Programme (Fig. 13.7).

Large-scale clearance of tidal swamp forests and excavation of primary, secondary and tertiary canals was undertaken to form transmigration sites. However, advice from soil scientists and water resources specialists was largely ignored in the drive to meet political targets for the resettlement of people. This resulted in the development of areas of deep peat and potential acid sulphate soils (PASS) and consequent problems of release of sulphurous

(continued)

Box 13.2 (continued)

Fig. 13.4 Freshwater, tidally influenced peat swamp forest in Indonesia (Photo Peter Burbridge)

Fig. 13.5 The swamp forest was drained and developed under the Transmigration Programme to provide land for landless farmers and others who needed to be relocated due to development pressures in Java (Photo Peter Burbridge)

(continued)

Box 13.2 (continued)

Fig. 13.6 Landless farmers were moved to the Transmigration Sites from elsewhere in Indonesia but farms on deep peats and potentially acid sulphate soils have proven unsustainable (Photo Peter Burbridge)

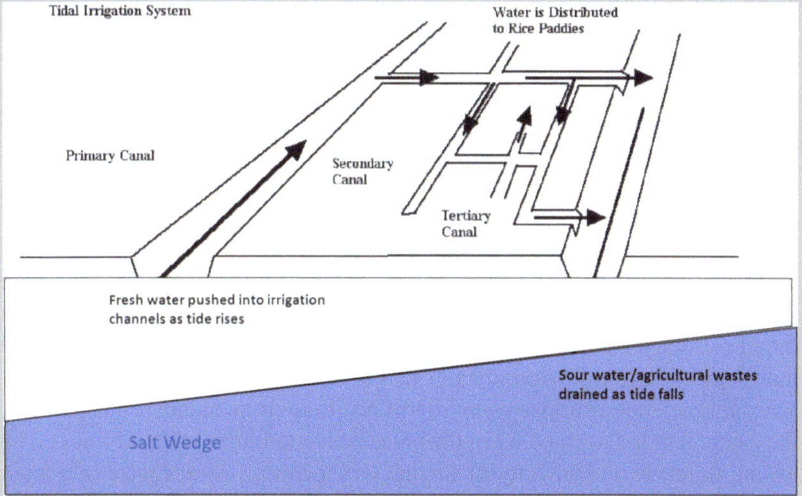

Fig. 13.7 The Transmigration Programme was based on an irrigation and drainage scheme using tidal movements to "pump" freshwater in and drain wastewater out. Development in the upland catchment basin has increased seasonality of water flows causing a landward shift in the saltwater wedge in the dry season and flooding during the wet season

(continued)

Box 13.2 (continued)

Fig. 13.8 Poor rice production on potentially acid sulphate soils has led to land abandonment (Photo Peter Burbridge)

acids, carbonic acids, subsidence of soils, and shrinkage of the interstitial spaces in the soil matrix. This led to poor crop production and large areas became unsuitable for the continuation of agriculture (Fig. 13.8) and consequently it resulted in the abandonment of agriculture as well as growth of illegal logging in surrounding production forests and areas set aside for nature conservation.

A further set of problems arose as the watershed above the estuaries was developed for urban, industrial and agricultural purposes. These developments altered the hydrology of the river systems to the point where the tidal irrigation system was disrupted. In the dry season there was reduced base water flow in the river which resulted in greater penetration of saline waters into the estuary. Further, it forced brackish water into the irrigation canals and led to salinization of groundwater and soils (Fig. 13.7). During the rainy season, accelerated surface water runoff resulting from changes in land use in the catchment led to flooding in the areas converted from wetland forest. The extent, duration and severity of impact on crops and homes were increased due to the erosion, subsidence and shrinkage of the soils in the transmigration sites. The difficulties of maintaining any form of agriculture led to large scale abandonment of areas converted for transmigration.

A second stage of transmigration development was embarked upon in the early 1980s where consultants were asked to examine sites where agriculture

(continued)

Box 13.2 (contiuned)

had proven unsustainable and to make recommendations for their rehabilitation for new forms of agriculture, rehabilitation as forest systems, or conversion to an alternative use. The author was involved in developing the system for determining the causes that led to non-sustainable use transmigration sites, and to determine what options might be available for their rehabilitation. The system was based on a reversal of the environmental impact assessment process where the impacts were identified along with the causal factors and then examination of the transmigration planning and management arrangements to determine what led to the creation of the adverse impacts.

The findings from this process were used to improve the planning and management of transmigration, and to identify options where carefully planned and implemented measures could lead to successful rehabilitation of the transmigration sites (TSSD reports).

Sources: Abell 1979; Brown and Burbridge 1999; Burbridge 1989, 1990; Burbridge and Mar agos 1985; Collier 1979; Crisman 1990; Halls 1997; Hardjowigeno 1985; Houterman et al. 2005; ITTO 2002; Koesoebiono et al. 1982; Silvius and Giesen 1992, 1996; Silvius and Suryadiputra 2005; Sorensen 1993; Sukardjo and Toro 1988; Wignyosukarto 2008; World Bank 2006.

13.4.2 Potential Benefits from Integrating ICM and FLR

There are potential environmental, social and economic benefits that could be derived from integrating Forest Landscape Restoration into the principles and practice of Integrated Coastal Management. These include:

- Maintaining the hydrology and material fluxes essential to the health and productivity of coastal ecosystems;
- Restoring the functions and natural resources generated by riparian ecosystems, coastal ecosystems and marine ecosystems;
- Reducing the adverse effects of natural and man-induced hazards;
- Ameliorating Global Change;
- Supporting the expansion and diversification of economic activities in the coastal realm and within watersheds;
- Protecting the health and welfare of communities;
- Helping nations meet international obligations under conventions, protocols, treaties and other legally or morally binding instruments.

In the case of the rehabilitation of the *Melaleuca* wetlands in Vietnam, FLR in the upper reaches of the Mekong river system would help to maintain base water flows in the lower Mekong. This would help to maintain groundwater levels in the areas of Melaleuca that are being rehabilitated, reduce the risks of fire and damage to replanted sites, reduce the exposure of the potential acid sulphate soils, and reduce pollution of the river waters from acid drainage. In turn this would help improve riverine fisheries and the transition towards more sustainable use of the Melaleuca wetland forests. However, given the population pressures in the countries upstream of Vietnam and the demand for hydro-electric power and for the abstraction of water for irrigation in the region, there will be limited scope for linking FLR and ICM until the economic and social benefits that could result are built into development planning of the Mekong system.

FLR could have a series of powerful benefits in Indonesia. For example, improved watershed management upstream of the transmigration areas would reduce flooding of agricultural fields, maintain base water flows in the rivers and reduce the ingress of saline water into estuaries which is causing salinization of tidal irrigation waters, and would improve the economic and social welfare of the transmigrants. FLR would also help maintain the economic benefits of previous investment in Transmigration projects where sites have been abandoned, but could be rehabilitated. This would also reduce the burden of repayments of loans from international banks.

As in the case of Vietnam, the adoption of FLR as a means of supporting a transition towards more sustainable use of coastal areas and ecosystems in Indonesia will require some very effective economic analyses of the costs and benefits that could be derived from FLR as an economic, environmental and social development tool.

It will also be important to create the enabling conditions to support FLR. Pokharel et al. (2005) provide a comprehensive discussion of the conditions that help to enable effective rehabilitation of forest systems and related landscapes and the economic, social and environmental benefits that can be gained.

The examples of the conversion and both the *Melaleuca* wetland forest in Vietnam and the tidal swamp forests in Indonesia demonstrate the need for a much broader perspective on the relationship among coastal forest development and rehabilitation and the management of watersheds than is commonly adopted in making decisions for their development. These examples also demonstrate the potential benefit of integrating FLR concepts into the evolving concepts of ICM.

13.4.3 Management Implications

There are potentially powerful synergies between the concepts and principles of FLR and ICM that can be applied and developed at different levels of landscape from an individual forest system level, such as a mangrove, to a much broader river basin level. There are a number of good practices embodied in FLR and ICM that we should give stronger emphasis to in reducing the negative effects of human development in river basins and coastal regions. For example, the context for management for major development and conservation initiatives has shifted from doing things for people to engaging stakeholders in the conceptualisation, planning and management of those initiatives. A brief summary of these good practices include:

- Start at the local level and focus on priority issues in developing linked ICM and FLR initiatives
- Work at both the national and local levels with strong linkages between levels;
- Where political systems allow, develop and open, participatory process, involving all stakeholders in planning and implementation;
- Build programmes around issues that have been identified through the participatory process
- Build constituencies and political support for resource management through public education programmes
- Develop "ownership" of the ICM and FLR plans and management strategies, involve the stakeholders/constituents in the examination of issues and problems and so developing a common understanding of how each is affected and how they could benefit from working together in deriving a common solution
- Build capacity at the national, regional, and local levels to practice integrated, community based management of FLR and ICM through training, learning-by-doing and cultivating host country colleagues who can forge long-term partnerships based on shared values
- Treat FLR and ICM management processes as complementary learning by doing processes (adaptive management) and aim to complete the loop between the formulation of plans and management strategies and their implementation, monitoring and, where necessary, adjustment as quickly as is feasible. This will help to demonstrate the effectiveness of innovative policies, plans and management strategies
- Encourage the adoption of policies which lead to economically and ecologically sustainable and equitable resources management
- Strengthen existing mechanisms for cross-sector cooperation, coordination and integration of policies, management plans, investment and resources utilisation associated with both ICM and FLR
- Integrated Coastal Management must be seen as an iterative process in which incremental improvements in cooperation among agencies is one of the most important achievements we should look for. Cooperation then forms the basis for coordination and integration
- Adopt an incremental, adaptive and long-term approach to linking FLR and ICM, recognising that ICM initiatives must undergo cycles of development, implementation and refinement, building on prior success and adapting and expanding to address new of more complex issues
- Develop national ICM and FLR policies, strategies, and management guidelines
- Encourage improved coordination and integration of donor assistance programs

Based on Aldrich et al. 2004; Bodungen and Turner 2001; Burbridge 1998; Carter and Gronow 2005; Chua et al.1996; Clark 1992; DellaSala et al. 2003; UNFF 2004; Sorensen 1997, 2000, 2002.

13.4.4 Conclusions

With increasing understanding of the powerful forces driving and being driven by Global Change, it will be increasingly important to adopt a strategic perspective on the power of integrating FLR and ICM as essential elements in the search for more sustainable uses of terrestrial and near shore marine ecosystems. Given the increasing importance of coastal areas and natural resources in sustaining human populations, and the increasing hazards posed by natural and man-induced global change the integration of FLR and ICM at all scales could yield powerful long-lasting environmental, social and economic synergies and benefits, including:

- Rehabilitation and then maintenance of the hydrologic and other environmental linkages between upland, coastal and marine systems
- Reduction in the risk to human activities from natural and man-induced hazards, such as flooding
- Amelioration of climate change
- Increased effectiveness of public and private investment
- Improved achievement of international norms of environmental management

These and other benefits would all be effective in supporting a transition towards more sustainable forms of coastal development. However, the synergies and potential benefits that can be derived from improving the integration of FLR and ICM need to be clearly spelled out through appropriate economic analyses before they will be more broadly understood.

References

Abdeldayem S, Hoevenaars J, Mollinga PP, Scheumann W, Slootweg R, Van Steenbergen F (2005) Agricultural drainage: towards an integrated approach. Irrig Drain Syst 19:71–87
Abell TMB (1979) The forest resource. Republic of Indonesia and United Kingdom Transmigration Project: site investigation along the Trans-Sumatra Highway, Overseas Development Administration, Land Resources Centre, London
Aldrich M, Belokurov A, Bowling J, Dudley N, Elliott C, Higgins-Zogib L, Hurd J, Lacerda L, Mansourian S, McShane T, Pollard D, Sayer J, Schuyt K (2004) Integrating forest protection, management and restoration at a landscape scale. WWF Forests for Life Programme, Gland
Bassoullet P, Djuwansah R, Gouleau D, Marius C (1998) Hydrosedimentological processes and soils of the Barito estuary, South Kalimantan, Indonesia. Oceanol Acta 9:217–226
Beilfuss RD, Barzen JA (1994) Hydrological wetland restoration in the Mekong Delta, Vietnam. In: Mitsch WJ (ed) Global wetlands: old world and new. Elsevier Science BV, Amsterdam
Bodungen Bv, Turner RK (2001) Science and integrated coastal management: an introduction. In: Bodungen Bv, Turner RK (eds) Science and integrated coastal management, Dahlem Workshop Report 85, Dahlem University Press, Berlin
Brown BE, Burbridge PR (1999) Coastal management in Southeast Asia. DFID, London
Burbridge PR (1988) Sustainable development: the interplay among environmental, economic and social factors. In: Lugo AE, Clark JR, Child RD (eds) Ecological development in the humid tropics. Winrock International, Morrilton

Burbridge PR (1989) Multiple-use assessment for coastal management. In: Magoon OT, Miner D, Tobin LTh, Clark D (eds) Coastal zone '89: proceedings of the sixth symposium on coastal and ocean management. American Society of Civil Engineers, New York

Burbridge PR (1990) Tidal wetland resources in the tropics. J Resour Manag Optim 7:115–138

Burbridge PR (1994) Integrated planning and management of freshwater habitats, including wetlands. In: Sasekumar A, Marshall N, Macintosh D (eds) Ecology and conservation of south-east Asian marine and freshwater environments including wetlands, vol 98, Developments in hydrobiology. Kluwer, Dordrecht

Burbridge PR (1995) Case study-Vietnam, Mekong Delta: difficulty of repairing damaged wetlands. In: Clark JR (ed) Coastal zone management handbook. CRC Press, New York

Burbridge PR (1996) Social, cultural and economic factors that influence the sustainable development of peatlands. In: Maltby E, Immrizi CP, Safford RJ (eds) Tropical lowland peatlands in SE Asia. IUCN, Gland

Burbridge PR (1998) Coastal zones from the "water" point of view. Paper presented at the 34th international planning congress, International Society of City and Regional Planners, Azores, Portugal

Burbridge PR, Maragos JE (1985) Coastal resources management and environmental assessment needs for aquatic resources development in Indonesia. International Institute for Environment and Development, Washington, DC

Burbridge PR, Pethick J (2002) Sustainability and management: coastal systems. In: Wefer G, Lamy F, Mountoura F (eds) Marine science frontiers for Europe. Springer, Heidelberg

Burbridge PR, Pethick J (2003) Sustainability and management: coastal systems. In: Wefer G, Lamy F, Mantoura F (eds) Marine science frontiers of Europe. Springer, Berlin

Burbridge PR, Scott D (1990) Report of the mission to assess opportunities for the development of an IUCN wetlands programme in Vietnam. IUCN, Gland

Burbridge PR, Dixon JA, Soewardi B (1981) Forestry and agriculture: options for resource allocation in choosing land for transmigration development. Appl Geogr 1:237–258

Burbridge, PR, Dankers N, Clark JR (1989) Multiple-use assessment for coastal management. In: Magoon OT, Miner D, Tobin LTh, Clark D (eds) Coastal zone '89: proceedings of the sixth symposium on coastal and ocean management. American Society of Civil Engineers, New York

Burbridge PR, Buddemeier RW, Letissier M, Costanza R (2005) Synthesis of main findings and conclusions. In: Crossland CJ, Kremer HH, Lindeboom HJ, Crossland JIM, LeTissier MD (eds) Coastal fluxes in the Anthropocene. Springer, Heidelberg

Carter J, Gronow J (2005) Recent experience in collaborative forest management. CIFOR Occasional Paper No. 43. Center for International Forest Research, Bogor, Indonesia

CEC (1999) Lessons from the European Commission's demonstration programme on integrated coastal zone management (ICZM). Commission of the European Communities, Luxembourg

Chua T-E (ed) (1996) Lessons learned from successes and failures of integrated coastal management initiatives. MPP-EAS Technical Report No. 4, GEF/UNDP/IMO Regional Programme for the Prevention and Management of Marine Pollution in East Asian Seas, Quezon City, Philippines

Chua T-E (2004) Integrated coastal management in transition. PEMSEA Update, partnerships for environmental management of the seas of East Asia, Quezon City, The Philippines

Clark JR (1992) Integrated management of coastal zones. FAO Fisheries Technical Paper No. 327, Food and Agriculture Organization, Rome

Collier W (1979) Social and economic aspects of tidal swampland development. In: Symposium on Tidal Swampland Development, Palembang, Indonesia

Crisman TL (1990) Constraints to successful rehabilitation of subtropical and tropical wetlands. In: Streever W (ed) (1999) An international perspective on wetland conservation. Kluwer, Dordrecht

Crossland CJ, Kremer HH, Lindeboom HJ, Crossland JIM, LeTissier MD (eds) (2005) Coastal fluxes in the Anthropocene. Springer, Heidelberg

DellaSala DA, Martin A, Spivak R, Schulke T, Bird B, Criley M, van Daalen C, Kreilick J, Brown R, Aplet G (2003) A citizens' call for ecological forest restoration: forest restoration principles and criteria. Ecol Restor 21:14–23

Dixon JA, Burbridge PR (1984) Economic considerations in mangrove management. In: Hamilton LS, Snedaker S (eds) Handbook on mangrove area management. IUCN/UNESCO/UNEP, East–west Center, Honolulu

EC (2001) EU Focus on coastal zones: turning the tide for Europe's coastal zones. European Commission, Brussels

Elliot T (2002) Forest landscape restoration: working examples from five ecoregions. WWF, Gland

FRIS (2006) Forest landscape management, concepts and definitions update. Forest Restoration Information Service, UNEP World Conservation Monitoring Centre, Cambridge, England

Halls A (ed) (1997) Wetlands, biodiversity and the Ramsar convention: the role of the convention on wetlands in the conservation and wise use of biodiversity. Ramsar Convention Bureau, Gland

Hardjowigeno S (1985) Conversion of tidal wetlands into rice land in Indonesia. UNDP/UNESCO research and training pilot programme on mangrove ecosystems in Asia and the Pacific. UNESCO, New Delhi, India

Houterman J, Djoeachir M, Susanto RH, van Steenbergen F (2005) Drainage and integrated water resources management: two basins. In: Toward an integrated perspective on agricultural drainage, final report and case studies. The International Bank for Reconstruction and Development, Washington, DC

Humphrey SL, Burbridge PR (1999) Thematic Study D: planning and management processes: sectoral and territorial cooperation. European Demonstration Programme on Integrated Coastal Zone Management, European Commission, Brussels

ITTO (2002) Guidelines for the restoration, management and rehabilitation of degraded and secondary tropical forests. International Tropical Timber Organization, Yokohama

IUCN (2005) Forest landscape restoration: broadening the vision of West African forests. International Conservation Union, Gland

Koesoebiono, Collier WL, Burbridge PR (1982) Mangroves, land and sea, coastal resource use and management in Asia and the Pacific. In: Soepadmo E, Rao AN, Macintosh DJ (eds) Proceedings of the Asian mangrove symposium on mangrove environment and management. University of Malaysia and UNESCO, Kuala Lumpur, Malaysia

Kongsangchai J (1984) Mining impacts upon mangrove forest in Thailand. In: Soepadmo E, Rao AN, Macintosh DJ (eds) Proceedings Asian mangrove symposium mangrove environment and management. University of Malaysia and UNESCO, Kuala Lumpur, Malaysia

Lamb D, Stanturf J, Madsen P (2012) What is forest landscape restoration? In: Stanturf JA, Lamb D, Madsen P (2012) Forest landscape restoration: Integrating natural and social sciences. Springer, Dordrecht

Le DD (1989) Wise use of wetland in Viet Nam. Paper presented at the international workshop on Wetlands in Tam Nong, Dong Thap, Mekong Delta, Viet Nam

Le DD (1993) Rehabilitation of the Melaleuca flood plain forests in the Mekong delta Vietnam towards the wise use of wetlands. Ramsar Convention Bureau, Gland

Maginnis S, Jackson W (2005) Restoring forest landscapes: forest landscape restoration aims to re-establish ecological integrity and human well-being in the degraded forest landscapes. IUCN, Gland, 6p

Maltby E, Burbridge PR, Fraser A (1996) Peatland and sulphate soils: a case study from Vietnam. In: Maltby E, Immrizi CP, Safford RJ (eds) Tropical lowland peatlands in SE Asia. IUCN, Gland

Mansourian S, Vallauri D, Dudley N (eds) (2005) Forest restoration in landscapes – beyond planting trees. Springer, Heidelberg

Oregon Department of Forestry (2001) Northwest Oregon state forests management plan. Salem, Oregon

Pfund J-L, Stadtmüller T (2005) Forest landscape restoration. In: Wenger R, Sommer R, von Dach SW (eds) Inforesources focus 2/05. Swiss Agency for Development and Cooperation (available online http://www.inforesources.ch/pdf/focus_2_05_e.pdf)

Pokharel BK, Stadtmüller T, Pfund J-L (2005) From degradation to restoration: an assessment of the enabling conditions for community forestry in Nepal. Intercooperation, Swiss Foundation for Development and International Cooperation, Switzerland

Safford RJ, Duong Van Ni, Maltby E, Vo-Tong Xuan (eds) (1997)Towards sustainable management of Tram Chim National Reserve, Vietnam. In: Proceedings of a workshop on balancing economic development with environmental conservation. Royal Holloway Institute for Environmental Research, University of London, UK

Safford RJ, Maltby E, Van Ni D, Branch NP (2009) *Melaleuca* wetlands and sustainable development in the Mekong Delta, Vietnam. In: Maltby E, Barker T (eds) The wetlands handbook. Wiley Interscience, New York

Silvius MJ, Giesen W (1992) Integration of conservation and land-use development of swamp forests of East Sumatra. In: Proceedings of the workshop on Sumatra, environment and development: its past, present and future. BIOTROP Special Publication No.46, SEAMEO BIOTROP, Bogor, Indonesia

Silvius MJ, Giesen W (1996) Towards integrated management of swamp forests: a case study from Sumatra. In: Maltby E, Immrizi CP, Safford RJ (eds) Tropical lowland peatlands in SE Asia. IUCN, Gland

Silvius MJ, Suryadiputra N (2005) Review of policies and practices in tropical peat swamp forest management in Indonesia. Wetlands International, Wageningen

Sorensen KW (1993) Indonesian peat swamp forests and their role as a carbon sink. Chemosphere 27:1065–1082

Sorensen J (1997) National and international efforts at integrated coastal management: definitions, achievement, and lessons. Coast Manag 25:3–41

Sorensen J (2000) Building a global database of coastal management efforts. University of Massachusetts Press, Boston

Sorensen J (2002) Baseline 2000 background report: the status of integrated coastal management as an international practice (second iteration). Urban Harbors Institute, University of Massachusetts, Boston

Stevenson NJ, Burbridge PR (1997) Abandoned shrimp ponds: option for mangrove rehabilitation. Special edition on Mangroves, International Newsletter of Coastal Management, Narragansett, Rhode Island

Stevenson NJ, Lewis RR, Burbridge PR (1999) Disused shrimp ponds and mangrove rehabilitation. In: Streever WJ (ed) An international perspective on wetland rehabilitation. Kluwer, Dordrecht

Sukardjo S, Toro AV (1988) Guidelines for the protection and management of mangrove and estuarine wetlands in Indonesia. UNDP/UNESCO Research and Training Pilot Programme on Mangrove Ecosystems in Asia and the Pacific. UNESCO, New Delhi, India

Takahashi H, Usup A, Hayasaka H (2004) The importance of groundwater level and soil moisture of subsurface layer on peat/forest fire in a tropical peat swamp forest. In: Paivanen J (ed) Proceedings of the 12th international peat congress: wise use of peatlands. International Peat Society, Jyväskylä

Turner RK, van den Bergh CJM, Soderqvist T, Barendregt A, van der Straaten J, Maltby E, van Ierland EC (2000) Ecological-economic analysis of wetlands: scientific integration for management and policy. Ecol Econ 35:7–23

Turner RK, Paavola J, Cooper P, Farber S, Jessamy V, Georgiou S (2003) Valuing nature: lessons learned and future research directions. Ecol Econ 46:493–510

UNFF (2004) Ecosystem approaches and sustainable forest management. A discussion paper for the UNFF Secretariat prepared by the Forest Conservation Programme of IUCN, the Program on Forests (PROFOR), and the World Bank. (Available online at http://www.un.org/esa/forests/pdf/session_documents/unff4/add-docs/iucn-profor-worldbank.pdf)

Wignyosukarto BS (2008) The hydraulic performance of tidal irrigation systems in the reclamation of acid sulphate soils. Department of Civil Engineering, Faculty of Engineering, Gadjah Mada University, Yogyakarta, Indonesia (Available online at http://budiws.files.wordpress.com/2008/01/bs-wignyosukarto-icec2006_b28_.pdf)

World Bank (2006) Reengaging in agricultural water management, challenges and options, Directions in development series. World Bank, Washington, DC

Chapter 14
Broad-Scale Restoration of Landscape Function with Timber, Carbon and Water Investment

R.J. Harper, K.R.J. Smettem, P.V. Townsend, J.R. Bartle, and J.F. McGrath

14.1 Introduction

A number of major environmental issues currently confront Australia. These include the relentless salinization of agricultural land and water resources, recurrent wind and water erosion and the prospect of climate change due to increases in the concentration of greenhouse gases in the atmosphere. For example, in 2000 in Australia it was estimated that 5.7 million hectares of land was affected by shallow water tables or salinity, with this likely to increase to 17 million hectares by 2050 (National Land

R.J. Harper (✉)
Alcoa Chair in Sustainable Water Management, Environmental Science, Murdoch University, Murdoch, WA 6150, Australia

Forest Products Commission, Locked Bag 888, Perth Business Centre, Perth, WA 6849, Australia
email: r.harper@murdoch.edu.au

K.R.J. Smettem
Centre for Eco-Hydrology, School of Environmental Systems Engineering, The University of Western Australia, Nedlands, WA 6907, Australia

P.V. Townsend
Forest Products Commission, Locked Bag 888, Perth Business Centre, Perth, WA 6849, Australia

J.R. Bartle
Cooperative Research Centre for Future Farm Industries, The University of Western Australia, Nedlands, WA 6907, Australia

J.F. McGrath
Cooperative Research Centre for Future Farm Industries, The University of Western Australia, Nedlands, WA 6907, Australia

J. Stanturf et al. (eds.), *Forest Landscape Restoration: Integrating Natural and Social Sciences*, World Forests 15, DOI 10.1007/978-94-007-5326-6_14,

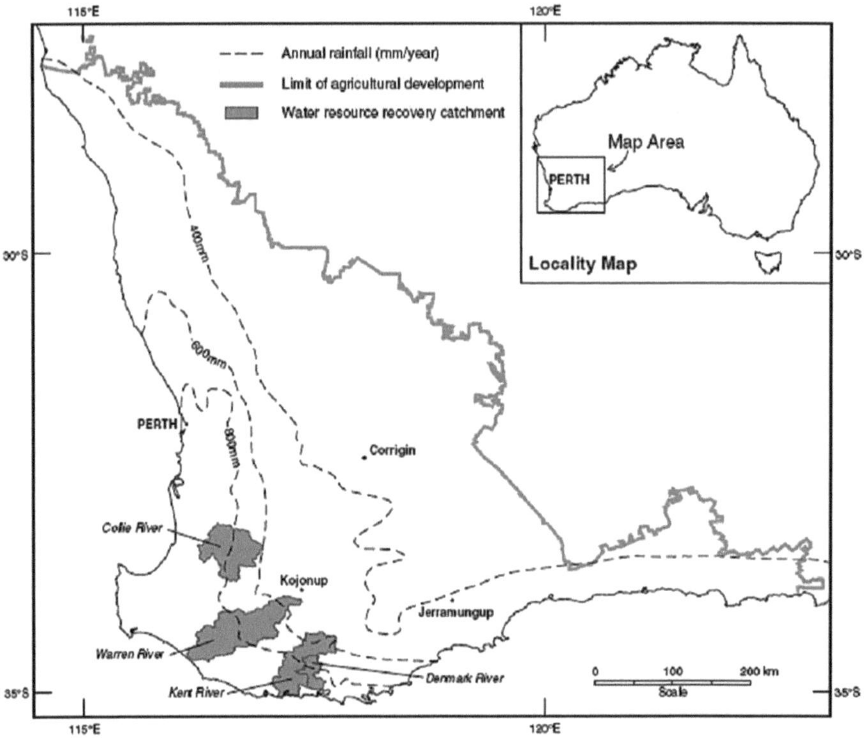

Fig. 14.1 Map of south-western WA, with locations described in the text

and Water Resources Audit 2001). Respective estimates for 2000 and 2050 for south-western Australia (Fig. 14.1) were 4.4 and 8.8 million ha. Various assessments have suggested that all inland watersheds will be salinized (National Land and Water Resources Audit 2001), up to 450 species are at risk of extinction (Keighery et al. 2004) and apart from significant loss of farmland productivity (Kingwell et al. 2003) there is significant threat to infrastructure (State Salinity Council 2000). Wind erosion is also a major concern, with regular recurrent events (Harper et al. 2010a; Select Committee into Land Conservation 1990). This region comprises one of 25 global biodiversity conservation hotspots (Myers et al. 2000).

Salinization in this region is caused by the remobilization of salts stored in deep regolith by groundwater that rises following the replacement of deep-rooted perennial plants with shallow rooted, annual agricultural species, and consequent groundwater recharge (Peck and Williamson 1987). There is evidence that salinity has been a cyclical phenomenon in this landscape, albeit at millennial timescales, with previous periods of salinization and recovery (Harper and Gilkes 2004). It has long been known that reforestation of watersheds will restore the hydrologic balance (Wood 1924) and thus reverse salinization; however the scale of investment required is immense, and likely to be limited from public funds.

In an early example of payment for environmental services, Shea and Bartle (1988) and Bartle and Shea (1989) advocated the approach of providing favourable economic and policy settings for afforestation. This was considered likely to result in forestry's widespread adoption by external investors, with consequent collateral environmental benefits. This has in fact occurred, with private timber companies establishing 280,000 ha of *Eucalyptus globulus* plantations in the region since 1990 (Parsons and Gavran 2007). Hydrological analysis suggests that where extensive afforestation has occurred in previously salinizing watersheds, such as those of the Denmark and Collie Rivers, that there has been either a stabilization (Mauger et al. 2001) or reduction (Bari et al. 2004) of salt loads.

Broadscale afforestation has, however, been restricted to the >600 mm rainfall zone of south-western Australia, as it is in this zone that forestry is profitable in its own right. Salinity is also a major issue in the 300–600 mm rainfall zone, a region that comprises around 15 million ha of cleared farmland (State Salinity Council 2000). The climate of this region is Mediterranean in nature, with rainfall occurring during the cooler winter months and an annual summer drought. Land use comprises rotations of cereal cropping and annual improved pastures, with farmers having replaced a range of natural vegetation communities including woodlands (Burvill 1979). Conventional forestry species and practices have not achieved widespread adoption in this region. Compared to the >600 mm rainfall zone, yields are lower, there are large transport distances to processing and export facilities and forestry is consequently not as profitable as the existing agricultural activities.

Two complementary strategies are being used to restore landscape function across this drier region, through increased afforestation. The first is to shift from the paradigm of forestry comprising tall trees grown in relatively long rotations and producing timber to the production of a range of biomass products (bioenergy, chemicals, sequestered carbon), and environmental services such as restoring watershed function to provide fresh water. The second strategy is to integrate these new systems into the dryland farming systems. These strategies are described in this chapter along with a case study that values the multiple products (wood, water, carbon) that would flow from the afforestation and repair of an agricultural watershed.

14.2 Strategy 1: New Species and Products from Trees

The commercially driven expansion of plantations in the >600 mm rainfall zone was catalysed by early Government investment in 7,000 ha of *Eucalyptus globulus* plantations, with this demonstrating the viability of the concept (Harper et al. 2009b; Shea 1999). Almost all of the subsequent investment in afforestation has been privately funded.

A similar approach has been taken in the areas of the region with 500–600 mm/year of rainfall. A pilot program, termed Strategic Tree Farming, commenced in 2005 with funding from the Australian and Western Australian Governments, and

Table 14.1 New tree species and products in the 300–600 mm rainfall zone

Species	Products	Target area (mm annual rainfall)
Pinus pinaster	Timber, carbon sequestration, bioenergy, water quality	>500
Eucalyptus saligna, E. cladocalyx	Timber, carbon sequestration, bioenergy, water quality	>500
Mallee eucalypts	Carbon sequestration, bioenergy, activated carbon, wood fiber, water quality	300–600
Various species	Carbon sequestration, biodiversity restoration	>300
Santalum spicatum	Sandalwood for fragrance products and oils	400–600

this has resulted in 18,000 ha of plantations being established on farms. These plantations are aimed at producing timber from species such as *Pinus pinaster*, *Eucalyptus saligna* and *E. cladocalyx*. Private investment has subsequently commenced in the region.

Developing forestry programs in regions with annual rainfall of between 300 and 500 mm/year is particularly challenging and a large investment has been made to evaluate potential new woody crop options and practices, both in Western Australia and similar regions elsewhere in Australia (Consortium 2001). Examples of products that have been evaluated include sawn-timber, firewood, biomass for electricity and eucalyptus oils (Zorzetto and Chudleigh 1999). Hobbs et al. (2007) describe the systematic assessment of the native flora of southern Australia to identify species with commercial potential.

Widespread afforestation will be required to restore landscape hydrology over broad areas, and the strategy has been to seek species for which large markets could potentially exist. Some species such as Western Australian sandalwood (*Santalum spicatum*) are often quite profitable in their own right, as they produce high value products including incense. Such species, however, may be limited as a treatment of broad-scale land-use problems as widespread planting may result in an oversupply of markets and a future reduction in prices.

The main approach to the problem of lower profitability for forestry in areas with limited rainfall has been to break the paradigm that forestry has to result in the harvest of single bole logs that are used for pulp or timber (Table 14.1). Investigations have consequently been made of:

1. Producing several new products from a single stand of trees (e.g. timber, carbon, biomass, eucalyptus oil),
2. Harvesting whole trees, independent of their form (e.g. biomass), and
3. Producing stands of trees with harvest cycles that are much shorter than normal (e.g. coppiced bioenergy crops), are phased with agriculture (biomass) or are not harvested at all (e.g. carbon, water, biodiversity plantings).

As a consequence of breaking this paradigm, silvicultural practices such as planting densities, thinning and pruning, can also be redefined. If, for example, the aim is to produce the maximum volume of biomass in the shortest possible time-frame (say 3–4 years), traditional approaches of managing a stand of trees within the constraints of long-term climate extremes may not apply. For other systems, however, they do, and the main aim of silvicultural management is to avoid deaths that occur from droughts, which regularly occur in this region. Not only is there an annual summer drought, but also periodic droughts associated with periods of lower rainfall (Harper et al. 2009a). Typical dryland forestry approaches have been to select species that are parsimonious with water, and avoiding the development of large leaf areas through reduced planting densities or fertilization.

Mallee-form eucalypts such as *Eucalyptus kochii* ssp. *plenissima, E. vegrandis* and *E. polybractea* have attracted most of the development effort so far, and 13,000 ha of tree belts have been established (URS Australia 2008). These are coppicing species with a large lignotuber, multiple stems and branches, with multiple products including carbon, biomass and eucalyptus oils (Bartle 2001; URS Australia 2008). The cropping systems in this region are often rotated with pasture phases and it is essential to protect young trees from grazing animals. As the cost of fencing is prohibitive, this is managed through the use of unpalatable species.

There have been preliminary attempts to develop environmental services markets based on payments for water restoration and biodiversity conservation both in this region and eastern Australia. The Strategic Tree Farming project contains an environmental service payment, recognising the contribution of these plantings towards restoring landscape hydrology.

Although there has been some consideration of payments for the restoration of biodiversity (Myers 1996; Stoneham et al. 2003), major issues include how to define the nature of the benefit and quantify their economic value. Purchase of biodiversity protection services is essentially voluntary and a problem of such markets is that they often do not generate significant amounts of funding (Jack et al. 2008) as many beneficiaries receive a ‘free-ride’.

14.2.1 Carbon Sequestration

Increased concentrations of greenhouse gases have been linked to global warming. The international response to this warming, the Kyoto Protocol to the United Nations Framework Convention on Climate Change, includes provisions that enable green-house sinks, or the sequestration of carbon in soils and vegetation to be used by Parties to fulfil their obligations (Watson et al. 2000). The Kyoto Protocol also allows for trading in emission reductions, and this opens the possibility that investment in carbon sinks may help underwrite broader natural resource management objectives (Harper et al. 2007; Koziell and Swingland 2002; Shea 1998).

Consequently, and in order to facilitate trading, the Western Australian Government passed the *Carbon Rights Act 2003* which establishes a statutory basis for the ownership and protection of carbon rights. Widespread carbon investment, however, will require a national approach. Australia ratified the Kyoto Protocol in late 2007 and has developed a blueprint for a national emissions trading scheme, the Carbon Pollution Reduction Scheme (CPRS) (Department of Climate Change 2008). This includes provision for producing carbon permits from farmland reforestation, broadly based on the treatment of forestry in Article 3.3 of the Kyoto Protocol. The final design of the scheme, and confirmation of reforestation as an allowable activity, will depend on passage of the legislation through the Australian Parliament.

Under the CPRS afforestation will only be considered if it occurs on land that was cleared prior to December 31, 1989. Forests are defined as being greater than 0.2 ha in size, have the potential to be more than 2 m high and comprise >20% land cover. Afforestation that would be eligible for inclusion can thus range from plantations where timber production is the primary aim, through to mixed plantings of native species for biodiversity protection (Table 14.1), as long as they meet the appropriate definitions of land-use change and forests.

The maximum amount of carbon that could be sequestered on cleared land in Western Australia by afforestation using Article 3.3 rules, is around 2.2 billion tonnes CO_2-e, with the amount sequestered decreasing in a general fashion with rainfall, and also depending on the amount of farmland actually planted (Harper et al. 2007). The profitability of carbon sequestration not only depends on the sequestration potential of a particular site and the carbon price, but also on the cost of the land. Thus, carbon farming is not necessarily most profitable in the highest rainfall areas (Harper et al. 2007).

Carbon investment in biodiversity plantings will provide a direct route to landscape and biodiversity restoration, although the rates of sequestration with native species may be less than those produced by plantation systems using exotic species. The underpinning data for such carbon sequestration in biodiversity plantings is generally poor, with most development work having occurred with exotic plantation species (Harper et al. 2005; Ritson and Sochacki 2003).

14.2.2 Biomass

Woody plant materials can be used either as a direct feedstock for electricity production or for liquid fuels (Bartle 2001; Schuck 2006). Where these replace fossil fuels they represent a means of reducing net greenhouse gas emissions. Australia has developed a national Renewable Energy Target (RET), with an initial target of 20% of generating capacity. This will provide an incentive for the use of biomass from forestry projects. Unlike the CPRS, legislation for the *Renewable Energy (Electricity) Amendment Bill 2010* has passed through the Australian Parliament.

A 5 MW trial plant has been built and evaluated near Narrogin, in south-western Western Australia, with this also producing other products such as eucalyptus oils and activated carbon (Schuck 2006). The technology for industrial-scale production of liquid fuels from woody materials is still under development (Yuan et al. 2008), as is the development of new products using biomass as a feedstock. The rate of adoption of biomass will depend on renewable energy policies, the cost competitiveness of biomass compared with fossil fuels and alternative renewable energy sources, and the demand for co-products that may provide greater revenue than energy products.

14.2.3 Water

Water in south-western Australia is derived from surface watersheds, groundwater and desalination, and is used for urban, agricultural and industrial purposes. The surface watersheds from which water is derived are almost completely forested. A major feature of the region is that deforestation of lower rainfall portions of watersheds results in the salinization of water, as it is in these areas that salt stores are the highest. As described earlier, afforestation restores water quality (Bari et al. 2004; Harper et al. 2001).

Water demand is increasing with increasing population and growth in the economy. Several watersheds that have been partially cleared for agriculture and several with slightly salinized water have been identified for recovery to potable levels through afforestation and other interventions (State Salinity Council 2000) (Fig. 14.1). A feature of these watersheds, such as those of the Collie, Warren-Tone, Denmark and Kent Rivers is that their headwaters have rainfalls of around 500–600 mm/year.

Afforestation in these watersheds could result in the production of fresh water, which could have a considerable economic value. The value of improved water quality is explored later in this chapter for the Warren-Tone River, where the estimated value of reforestation is measured in terms of the returns from water, wood and carbon sequestration.

Although afforestation may result in an overall reduction in water yield, the water will be less saline (Bari and Smettem 2006). This is in contrast to the situation in other regions of Australia where widespread afforestation of watersheds results in a reduction in run-off (Vertessy 2001) or reduction of recharge to freshwater aquifers (Benyon et al. 2006) and this competition for water between forestry and other land-uses is consequently seen as a dis-benefit.

14.3 Strategy 2: Integrating Forestry with Agriculture

The major approach in the 300–600 mm rainfall zone has been to integrate forestry with agriculture and thereby maintain agricultural production. This is in contrast to higher rainfall areas where in some cases there has been a replacement of farming with forestry. Hatton et al. (2003) question whether it is possible to obtain a landscape water balance similar to that which occurred prior to deforestation on the basis that the restored systems will not have leaf areas approaching those of the natural vegetation. George et al. (1999) similarly suggest that up to 80% of watersheds might have to be planted to restore landscape hydrology to the pre-deforestation state. A key question for southwest Western Australia is thus whether it is possible to restore or stabilize landscape hydrology, without resorting to very high proportions of afforestation in the landscape.

Four main approaches for integrating trees into farming systems appear promising. The aim of these systems is to obtain a similar aggregate leaf area from trees planted over a relatively smaller proportion of the farming landscape as:

1. Belts of trees interspersed amongst cropland,
2. Blocks of trees targeted to specific soils and landscape positions, such as planting trees on those sites that could be used to maximize the control of recharge or on identified zones of water accumulation,
3. Matching species to site conditions, and
4. Alternating short phases of trees with annual crops or pastures.

Paradoxically, for systems that are attempting to stabilize landscape hydrology, a major issue in drier areas is managing the supply of water to the trees. South-western Australia is a region where the Mediterranean climate is characterized by winter droughts with a return frequency of around 10 years (Harper et al. 2009a) and strong prospects of future climate change (CSIRO and Australian Bureau of Meteorology 2007). If inappropriate species or planting densities are used in relation to a site's water supply, the trees would be expected to die. Although this may not matter for the phase farming system, lessening the impact of drought is a major aim of systems that involve permanent afforestation with long rotations, such as with *P. radiata*. Strategies to achieve this include site selection procedures that avoid shallow soils and silvicultural management, particularly through the management of leaf area, and the selection and breeding of species with better tolerance to these conditions.

14.3.1 Belts of Trees

The first option is to use belts of trees, arranged across the landscape, with agriculture practiced in the alleys between. Belts open the option for spatial dispersal to achieve an efficient integration of land and water use. This is similar to agroforestry systems practiced elsewhere in the world.

Belt layout can be designed in a manner that is compatible with large scale cropping. The inter-belt or 'alley' width will usually be a multiple of the width of the largest machine used in cropping operations. Layouts can be rectangular, or follow the contours on sloping land. The flexibility of belt layout is able to accommodate emerging precision agriculture techniques.

Despite soil profiles that often comprise clays with bulk densities of up to 2.0 Mg/m^3 the roots of mallee eucalypts can penetrate to depths of 5–10 m within

7 years of planting and 12–40 m laterally (Robinson et al. 2006). Although this extensive exploration of soil profiles provides access to additional moisture and will boost tree productivity it may result in competition for moisture with crops, particularly in drier years (Oliver et al. 2005). Ripping of lateral roots should reduce this problem.

A key issue with belts is whether the extensive and deep root systems will intercept water moving through the regolith from the hills to the valley floors, and thus reduce groundwater pressures and saline discharge in valley floors. It would be presumed that belts planted on contour lines would be most effective in this regard, however there is no evidence that this is the case. The most difficult design objective is to arrange tree belts to optimise interception of both surface and sub-surface waters and other land management issues. For example, south-western Australia has large areas of sandy-surfaced soils that are susceptible to wind erosion (Select Committee into Land Conservation 1990) and tree belts will help reduce this problem by reducing surface wind speeds.

Species that can be used in this system include coppiced eucalyptus mallees (Bartle 2001), producing biomass and sequestering carbon particularly in the roots, or eucalyptus species such as *E. cladocalyx* or *E. saligna*. The latter species are suitable for both timber production and carbon sequestration and would be based on longer rotations than systems designed for maximising biomass production.

14.3.2 Targeting Areas of Greatest Recharge and Water Accumulation

This strategy involves identifying soils and landscape positions that contribute greater amounts of recharge to groundwater systems, or accumulate surface waters and allow greater productivity. Harper et al. (2005) for example combined soil and climate data with a simple water balance model (AgET) to identify areas in a watershed that contributed a disproportionate amount of recharge to groundwater, in relation to their areal extent. AgET (Raper et al. 2001) is a one-dimensional model and disregards lateral through-flow and capillary rise. This approach identifies those soils that are contributing most to groundwater recharge and these can be preferentially planted.

Relatively fresh water can accumulate in lower landscape positions, both as a result of surface flows in response to peak rainfall events (Gregory et al. 1992) or as seepage from perched groundwaters (George 1990). Higher growth rates often occur in lower slope positions relative to those upslope, such as reported in *Pinus radiata* plantations by McGrath et al. (1991).

Access to these water stores, in addition to that provided by rainfall, may extend the profitable range of different forest enterprises (Cooper et al. 2005). Similarly, it may be possible to enhance water supply to trees by the capture and diversion of water on slopes. However, care is required in this environment to ensure that these areas are not also affected by salinity. These are avenues for further investigation.

14.3.3 Matching Species to Sites

Restoration strategies can also be adjusted to match site conditions (Ryan et al. 2002). Some soils have intrinsic limitations and trees grow poorly or die. For example, some soils have water holding capacities inadequate to allow tree survival over the recurrent summer droughts. Harper et al. (2009a), for example, report markedly different survival of *E. globulus* on soils <2 m deep to bedrock compared to those >2 m deep (22% vs. 70%). Afforestation should be avoided on such soils.

Large areas of land have salinized soils and tree growth on these soils is strongly reduced compared to non-saline soils (Bennett and George 1995). Such soils can be relatively easily identified prior to establishment using techniques such as electromagnetic induction meters. Tree species adapted to salinized soil conditions can be preferentially planted (Marcar and Crawford 2004). Growth rates are likely to be lower on such sites, compared to non-saline areas, due to both water logging and salinity (Archibald et al. 2006). Such salinized land is often poorly productive for agriculture and may thus be readily available for afforestation. A benefit-cost analysis of such restoration, that includes bringing the carbon value of such plantings to account, has yet to be performed.

14.3.4 Phase Farming with Trees

Woody phase crops could provide an option for temporal dispersal to achieve efficient integration and water use. Phase farming with trees (PFT) is a concept that would involve inserting short rotations (3–5 years) of trees into existing agricultural systems on a 20–25 year cycle (Harper et al. 2000, 2010b). The premise is that the trees will rapidly de-water soil profiles to several metres depth and thus create a buffer of dry soil, with this being refilled during the subsequent agricultural phase. The buffer of dry soil would prevent recharge to groundwaters during the tree phase and indirectly during the subsequent cropping or pasture phase. This approach is expected to stabilise landscape hydrology. Deep soil profiles are widespread across the region (McArthur 1991).

This is analogous to phase farming systems with perennial legumes such as lucerne (*Medicago sativa*), differing in terms of likely rates of water use and depth of soil water depletion. The material from the forestry-cropping phase farming systems would be suitable as a bioenergy feedstock. Short rotations of tree plantations have been used for bioenergy production in North America and in green fallow systems in Africa, with the latter producing firewood whilst improving soil fertility through nitrogen fixation (Sanchez 2002).

This premise has been modelled (Harper et al. 2000) and subsequently evaluated at an experimental site near Corrigin, that has 300 mm annual rainfall and is typical of the broad-scale dryland farming systems of the region. Soil water content was depleted to wilting point to depths of 6.5 m beneath high density (4,000 stems/ha) plantings of *E. occidentalis* within 3 years of planting (Harper et al. 2010b). This was equivalent to soil water depletion of 440–780 mm, and assuming a recharge rate of 40 mm/year this could result in a rotational system with 3 years of trees followed by between 11 and 20 years of agriculture. With this approach, rates of total dry biomass production varied with species, planting densities and slope position. Biomass yields of 15–22 t/ha/3 year are possible (Sochacki et al. 2007), or around 8.6 million tonnes of biomass a year on a sustainable basis, when trees are planted on suitable land across the region.

To optimize biomass production and water use in this system a balance is required between planting density and water availability. Modelling suggests that the system is not suitable for soils that are relatively shallow (<5 m deep), contain saline or fresh regional water-tables, or have clayey surface textures (Harper et al. 2000; Hatton et al. 2002). In these situations the system will not work as the dry soil buffer is inadequate or soil water is replenished from the regional aquifer. For soils with clayey surfaces the rate of recharge is relatively small and the system is thus not required.

There are several unresolved issues that need to be addressed before phase farming systems using trees can be implemented. These include the development of cost-effective establishment and harvesting techniques and the realisation of markets for these products. While the lack of a payment for salinity benefits is common to all salinity treatments, the markets for biomass-based products, such as bioenergy, are currently uncertain. However in Australia, this should be resolved with the implementation of both the CPRS and national renewable energy target. Large future markets may also arise when enabling technologies, such as the conversion of woody biomass to liquid fuels or industrial feedstocks, are developed.

14.4 Case Study: The Role of Plantation Forestry in Watershed Recovery: The Warren-Tone Catchment, Western Australia

Various studies have assessed the profitability of establishing carbon sinks on farmland in this region (Flugge and Abadi 2006; Harper et al. 2007; Petersen et al. 2003). For example, Harper et al. (2007) show that the combined returns from timber and carbon can increase the overall profitability of a forestry enterprise. Moreover, this results in a lower cost per unit of carbon sequestered compared to situations where only carbon is produced.

Here, we extend this style of analysis by not only taking multiple products (wood, carbon, water) into account, but also the costs of externalities, such as the damages associated with agriculture, based on the study of Townsend et al. (2012). The concept of obtaining multiple economic products from reforestation has been explored for the Warren-Tone Catchment (Fig. 14.1) by comparing the returns from afforestation with agriculture.

The Warren-Tone Catchment covers an area of 408,000 ha and is made up of several base-flow and land surface discharge systems. The watershed's headwaters arise near the town of Kojonup, where average rainfall is approximately 500 mm/year, and flow in a south-westerly direction towards the coast, where the rainfall exceeds 1,200 mm/year. The end-of-watershed water yield and salinity are measured at the Barker Road Crossing gauging station, and are currently 260 GL/annum and 1,000 mg/L total dissolved solids (TDS), respectively. This watershed has been identified by the Western Australian Government as being recoverable (State Salinity Council 2000), with the aim being to reduce the end-of-watershed salinity to 500 mg/L TDS by 2030.

As elsewhere in the region, replacing deep-rooted native vegetation with annual, shallow-rooted species has caused groundwater recharge rates to increase, water tables to rise and the mobilization of salt stores within the soil. By the mid-1970s, the effects of salinity were evident in terms of increasing stream salinity and the salinization of soils throughout the farmed landscape. When clearing controls were introduced for the watershed in 1978, the total area of land converted from forest to agriculture was around 105,000 ha. Although the clearing controls prevented any further replacement of native forests with pastures and crops, the watershed's salinity levels continued to rise, as previously dry soil profiles filled with water and salt was mobilized.

Since 1990, agriculture in the middle portion of the watershed has been gradually replaced by plantation forestry, with 25,000 ha established by 2006. The impacts of plantations on the watershed water yield and the subsequent salinity benefits are described by the hydrological model, LUCICAT, that was initially developed for another watershed (Bari and Smettem 2006), but has since been calibrated for the Warren-Tone Watershed. The LUCICAT model predicts that plantations established between 1990 and 2006 will cause the salinity levels to fall to 700 mg/L with a water yield of 245 GL/year at Barker Road Crossing by 2035 (Smith et al. 2006). Based on the existing land uses, the watershed salinity will remain considerably higher than the original levels and well above the watershed recovery target.

The main options being considered to meet the watershed recovery target include (a) engineering solutions, (b) changes in landuse from annual to perennial species and (c) the establishment of more plantations. This case study focuses on the potential role of plantation forestry in the areas of the watershed with between 500 and 700 mm annual rainfall.

For the purposes of this exercise, two forestry regimes were evaluated in terms of the financial benefits of replacing agriculture with forestry. Included in the economic analysis are the potential returns from selling wood and carbon credits. These carbon credits are sold as plantations sequester carbon dioxide and bought back when plantations are thinned or harvested. The analysis also considers the value of improved water quality resulting from afforestation.

Economic input variables are summarized in Table 14.2, and these were developed into a linear regression model. The forestry regimes were *Pinus pinaster* and *Eucalyptus saligna* in the 500 and 700 mm rainfall zones (Table 14.3). For the purposes of this exercise these are assumed to be grown in block plantings with several thinnings from multiple rotations over 100 years, with growth rates of 10 and 16 m^3/ha/year of wood, respectively. It is estimated that the two sites would sequester

Table 14.2 Values for economic variables used in the forestry linear regression model

	Most likely value
Discount rate (real)	6%
Annual land rental (% of purchase price)	60%
Starting carbon price ($/t CO_2-e)	$16
Annual change in carbon price (%)	1.5%
Variation from standard silvicultural costs (%)	0%
Annual costs ($/ha)	$65
Change from reference log prices (%)	0%

Table 14.3 Agricultural returns and externalities and forestry returns from both timber and carbon ($/ha/year)

	Annual rainfall (mm/year)	
	500	700
Agricultural returns	150	190
Externality (salinity) costs of agriculture	–50	–30
Net value of agriculture	100	160
Timber return	–200	–113
Carbon return	354	357
Timber + carbon	154	244
Net benefit of forestry over agriculture	54	84

around 17 and 27 t CO_2-e/ha/year, respectively. It is further assumed that carbon would be pooled across many forestry projects.

The wood yield and carbon sequestration estimates were combined with the economic variables to project the annualised benefits for the two different forestry regimes. The opportunity costs of forestry were deducted from these values (Table 14.3). LUCICAT was used to predict the impacts of forestry on water yields and in-stream salinity at the end of the watershed.

In the Warren-Tone Catchment, the gross margins for agriculture are between $150 and $190/ha/year (Table 14.3). After taking into account the externality costs due to salinity, the net value of agriculture is reduced to $100 and $160/ha/year. The returns from afforestation are clearly identifiable for the watershed, with forestry being more profitable than agriculture, when both timber and carbon values are included. At the 500 and 700 mm/year sites, forestry had a net benefit over agriculture of $54 and $84/ha/year, respectively.

The water quality benefit to the watershed was estimated using the LUCICAT model. Each 1,000 ha of afforestation resulted in a decrease in stream salinity of approximately 7.2 mg/L at the 500 mm/year rainfall site and 4.0 mg/L at the

700 mm/year site. A further 27,800 ha of afforestation would be required in the drier parts of the watershed to return the stream salinity to a potable value of 500 mg/L. At this stream salinity concentration it would be possible to dam the river and market this water. With this level of afforestation, the watershed could produce 237 GL/year of water. If 100 GL of this water was sold each year, this value could be spread over the total 52,800 ha of plantations. With an assumed value of $150,000/GL, this would result in a net water value of $285/year for each hectare of afforestation. This is in addition to between $154 and $244/ha/year for timber and carbon (Table 14.3).

14.5 Discussion and Conclusions

Widespread forest landscape restoration has occurred in the higher (600 mm/year) rainfall areas of south-western Australia as a result of the development of various afforestation investment options. Approaches that consider the production of multiple products from afforestation, including carbon sequestration, appear promising for the 300–600 mm rainfall zone, given the current widespread concern with climate change and the emergence of an emissions trading scheme. Recent investigations of how Australia can reduce its net greenhouse emissions suggest a large role for afforestation (Garnaut 2008), and investment of this scale could also have an impact on other landscape-scale problems such as hydrological imbalance, erosion and loss of biodiversity (Harper et al. 2007). Afforestation represents a method of mitigation that can be implemented immediately, whereas other procedures such as geo-sequestration are awaiting technological innovation.

A restoration strategy aimed at restoring landscape function would thus combine several approaches by (a) identifying and planting those areas that contribute the most recharge and thus to the underlying cause of the salinity problems, (b) identifying those areas where water will accumulate and tree growth will be greatest and (c) identifying differences in soil conditions and tailoring species selection to take this into account. The strategy pursued will also depend on what products are being sought from the reforestation.

It is likely that increased afforestation will occur with financial mechanisms that recognise and value land, water and biodiversity conservation benefits. In the example from the Warren-Tone Catchment, the agricultural, forestry, water, salinity and carbon markets were all linked. Where there is a potential return from water, the value from afforestation ($285/ha/year) is clearly in excess of the returns that can be achieved from the existing agricultural land-use ($100–$160/ha/year) and even from a joint carbon and timber venture ($154–$244/ha/year). Land-use change is therefore socially desirable from both economic and environmental points of view.

Although the Warren-Tone Catchment case study shows that the value of water exceeds the value of both timber and carbon as products, not all watersheds have the prospect of being restored to produce fresh water (State Salinity Council 2000). This does not remove the imperative of tackling salinity across all watersheds in the region and an additional source of income for afforestation may come from payments for the public good salinity benefits that they provide.

Products from forestry appear to cluster around three broad groupings. In the first group is wood, for which yields can be readily predicted and there are existing markets. The second group comprises carbon, bioenergy and water. The yields of each can be quantified through existing models. The existence of markets for these products depends on regulation and the establishment of appropriate trading regimes. Importantly, there are emerging buyers for these products. The third group comprises the protection of land from salinity and biodiversity. Here the response to afforestation is harder to quantify and the development of a market may also require regulation. These environmental services have few direct buyers.

The role of Government can thus vary, based on the different types of products. For carbon and bioenergy, the markets are dependent on the Government setting emissions and renewable energy targets and establishing formal trading schemes. For other environmental services markets may not emerge and the Government may have to provide subsidies or payments. Such a market and regulatory framework may be important should the effects of climate change lead to existing plantations becoming commercially unviable in terms of the returns from wood production and carbon sequestration. In this case the benefits of retaining the existing plantations for water quality purposes could be quantified and appropriate payments made.

An emerging debate relates to the use of farmland for the mitigation of climate change, either through the production of biofuels or the sequestration of carbon, and the danger of displacing food production. Both the alley and phase farming systems offer the prospect of achieving climate change mitigation without either using food-grains or displacing farming production. Importantly, such afforestation could occur at a scale that will increase sustainability and lead to the restoration of landscape function.

Acknowledgements Project funding has been obtained from the Joint Venture Agroforestry Program (Projects CAL-3A, CAL-6A, CAL-8A, FPC-2A), the Cooperative Research Centre for Greenhouse Accounting, and Cooperative Research Centre for Future Farm Industries, and the Western Australian Government's Greenhouse Strategy (Action 3.1.11). We thank Kevin Goss, Don McGuire and Stan Sochacki for constructive comments on the manuscript.

References

Archibald RD, Harper RJ, Fox JED, Silberstein RP (2006) Tree performance and root-zone salt accumulation in three dryland Australian plantations. Agrofor Syst 66:191–204

Australia URS (2008) Oil mallee industry development plan for Western Australia. Forest Products Commission, Perth

Bari MA, Smettem KRJ (2006) A daily salt balance model for stream salinity generation processes following partial clearing from forest to pasture. Hydrol Earth Syst Sci 10:519–534

Bari MA, Mauger GW, Dixon RNM, Boniecka LH, Ward B, Sparks T, Waterhouse AM (2004) Salinity situation statement Denmark River. Water resource technical report no. WRT 30. Department of Environment, Perth, Western Australia

Bartle JR (2001) New perennial crops: Mallee eucalypts – a model, large scale perennial crop for the wheatbelt. In: Managing agricultural resources. Proceedings of outlook conference 2001, Canberra. ABARE, Canberra, pp 117–128

Bartle JR, Shea S (1989) Development of a pulpwood cropping industry in Western Australia. Land Water Res News 1:5–9

Bennett DL, George RJ (1995) Using the EM38 to measure the effect of soil salinity on *Eucalyptus globulus* in south-western Australia. Agric Water Manage 27:69–86

Benyon RG, Theiveyanathan S, Doody TM (2006) Impacts of tree plantations on groundwater in south-eastern Australia. Aust J Bot 54:181–192

Burvill GH (1979) The development of light lands. In: Burvill GH (ed) Agriculture in Western Australia. University of Western Australia Press, Nedlands, pp 157–172

Consortium (2001) The contribution of mid to low rainfall forestry and agroforestry to greenhouse and natural resource management outcomes: overview and analysis of opportunities. Australian Greenhouse Office/Murray-Darling Basin Commission, Canberra

Cooper D, Olsen G, Bartle JR (2005) Capture of agricultural surplus water determines the productivity and scale of new low-rainfall woody crop industries. Aust J Exp Agric 45:1369–1388

CSIRO, Australian Bureau of Meteorology (2007) Climate change in Australia-technical report 2007. CSIRO, Canberra

Department of Climate Change (2008) Carbon pollution reduction scheme green paper. Australian Government, Canberra

Flugge F, Abadi A (2006) Farming carbon: an economic analysis of agroforestry for carbon sequestration and dryland salinity reduction in Western Australia. Agrofor Syst 68:181–192

Garnaut R (2008) The Garnaut climate change review. Cambridge University Press, Melbourne

George RJ (1990) Reclaiming sandplain seeps by intercepting perched groundwater with eucalypts. Land Degrad Rehabil 2:13–25

George RJ, Nulsen RA, Ferdowsian R, Raper GP (1999) Interactions between trees and groundwaters in recharge and discharge areas – a survey of Western Australian sites. Agric Water Manage 39:91–113

Gregory PJ, Tennant D, Hamblin AP, Eastham J (1992) Components of the water balance on duplex soils in Western Australia. Aust J Exp Agric 32:845–855

Harper RJ, Gilkes RJ (2004) Aeolian influences on the soils and landforms of the southern Yilgarn Craton of semi-arid, south-western Australia. Geomorphology 59:215–235

Harper RJ, Hatton TJ, Crombie DS, Dawes WR, Abbott LK, Challen RP, House C (2000) Phase farming with trees: a scoping study of its potential for salinity control, soil quality enhancement and farm income improvement in dryland areas of southern Australia. Rural Industries Research and Development Corporation, RIRDC publication N° 00/48, RIRDC publication N° 00/48, Canberra

Harper RJ, Mauger G, Robinson N, McGrath JF, Smettem KRJ, Bartle JR, George RJ (2001) Manipulating catchment water balance using plantation and farm forestry: case studies from south-western Australia. In: Nambiar EKS, Brown AG (eds) Plantations, farm forestry and water. Joint Venture Agroforestry Program, Canberra, pp 44–50

Harper RJ, Smettem KRJ, Tomlinson RJ (2005) Using soil and climatic data to estimate the performance of trees, carbon sequestration and recharge potential at the catchment scale. Aust J Exp Agric 45:1389–1401

Harper RJ, Beck AC, Ritson P, Hill MJ, Mitchell CD, Barrett DJ, Smettem KRJ, Mann SS (2007) The potential of greenhouse sinks to underwrite improved land management. Ecol Eng 29:329–341

Harper RJ, Smettem KRJ, Carter JO, McGrath JF (2009a) Drought deaths in *Eucalyptus globulus* (Labill.) plantations in relation to soils, geomorphology and climate. Plant Soil 324:199–207

Harper RJ, Smettem KRJ, Reid RF, Callister A, McGrath JF, Brennan PB (2009b) Pulpwood crops. In: Reid RF, Nuberg I (eds) Agroforestry for natural resource management. CSIRO Publishing, Melbourne, pp 199–218

Harper RJ, Gilkes RJ, Hill MJ, Carter DJ (2010a) Wind erosion and soil carbon dynamics in south-western Australia. Aeolian Res 1:129–141

Harper RJ, Sochacki SJ, Smettem KRJ, Robinson N (2010b) Bioenergy feedstock potential from short-rotation woody crops in a dryland environment. Energy Fuel 24:225–231

Hatton TJ, Dawes W, Harper RJ (2002) Woodlots in rotation with agriculture. In: Stirzaker R, Vertessy R, Sarre A (eds) Trees, water and salt – an Australian guide to using trees for healthy catchments and productive farms. Rural Industries Research and Development Corporation, Canberra, pp 43–55

Hatton TJ, Ruprecht J, George RJ (2003) Preclearing hydrology of the Western Australian wheatbelt: target for the future? Plant Soil 257:341–356

Hobbs TJ, Bennell M, Huxtable D, Bartle JR, Neumann C, George N, O'Sullivan W, McKenna D (2007) Potential agroforestry species and regional industries for lower rainfall southern Australia. RIRDC, Canberra

Jack BK, Kousky C, Sims KRE (2008) Designing payments for ecosystem services: lessons from previous experience with incentive-based mechanisms. Proc Natl Acad Sci 105: 9465–9470

Keighery GJ, Halse SA, Harvey MS, McKenzie NL (eds) (2004) A biodiversity survey of the Western Australian agricultural zone, Records of the Western Australian Museum supplement 67. Western Australian Museum, Perth

Kingwell R, Hajkowicz S, Young J, Patton D, Trapnell L, Edward A, Krause M, Bathgate A (2003) Economic evaluation of salinity management options in cropping regions of Australia. Grains Research and Development Corporation and National Dryland Salinity Program, Canberra

Koziell I, Swingland IR (2002) Collateral biodiversity benefits associated with 'free-market' approaches to sustainable land use and forestry activities. Philos Trans R Soc Lond A Math Phys Eng Sci 360:1807–1816

Marcar N, Crawford D (2004) Trees for saline landscapes. Rural Industries Research and Development Corporation, Canberra

Mauger GW, Bari M, Boniecka L, Dixon RNM, Dogramaci SS, Platt J (2001) Salinity situation statement Collie River. Water resources commission, Water resource technical report no. WRT 29

McArthur WM (1991) Reference soils of south-western Australia. Australian Society of Soil Science Inc. (WA Branch), Perth

McGrath JF, Ward D, Jenkins PJ, Read B (1991) Influence of site factors on the productivity and drought susceptibility of *Pinus radiata* in the Blackwood valley region of Western Australia. In: Ryan PJ (ed) Productivity in perspective. Third Australian forest soils and nutrition conference, Melbourne. Forestry Commission of New South Wales, Sydney, pp 65–66

Myers N (1996) Environmental services of biodiversity. Proc Natl Acad Sci 93:2764–2769

Myers N, Mittermeier RA, Mittermeier CG, da Fonseca GAB, Kent J (2000) Biodiversity hotspots for conservation priorities. Nature 403:853–858

National Land and Water Resources Audit (2001) Australian dryland salinity assessment 2000. Extent, impacts, processes, monitoring and management options. National Land and Water Resources Audit, Canberra

Oliver YM, Lefroy EC, Stirzaker R, Davies CL (2005) Deep-drainage control and yield: the trade-off between trees and crops in agroforestry systems in the medium to low rainfall areas of Australia. Aust J Agric Res 56:1011–1026

Parsons M, Gavran M (2007) National plantation inventory Australia – 2007 update. Department of Agriculture, Fisheries and Forestry, Canberra

Peck AJ, Williamson DR (1987) Effects of forest clearing on groundwater. J Hydrol 94:47–65

Petersen EH, Schilizzi S, Bennett D (2003) An economic assessment of the role of commercial tree crops to achieve greenhouse gas neutrality in predominantly grazing systems of southwestern Australia. Aust J Agric Resour Econ 47:213–233

Raper GP, Guppy LM, Argent RM, George RJ (2001) Innovative use of simulation models in farm and catchment planning in Western Australia. In: Conacher AJ (ed) Land degradation: papers selected from contributions to the sixth meeting of the international geographical union's commission on land degradation and desertification, Perth, Western Australia, 20–28 Sept 1999. Kluwer Academic, Dordrecht, pp 319–331

Ritson P, Sochacki SJ (2003) Measurement and prediction of biomass and carbon content of *Pinus pinaster* trees in farm forestry plantations, south-western Australia. For Ecol Manage 175:103–117

Robinson N, Harper RJ, Smettem KRJ (2006) Soil water depletion by *Eucalyptus* spp. integrated into dryland agricultural systems. Plant Soil 286:141–151

Ryan PJ, Harper RJ, Laffan MD, Booth TH, McKenzie NJ (2002) Site assessment for farm forestry in Australia and its relationship to scale, productivity and sustainability. For Ecol Manag 171:133–152

Sanchez PA (2002) Soil fertility and hunger in Africa. Science 295:2019–2020

Schuck S (2006) Biomass as an energy source. Int J Environ Stud 63:823–835

Select Committee into Land Conservation (1990) Discussion paper N° 2. Agricultural region of Western Australia. Western Australia Legislative Assembly, Perth, Western Australia

Shea S (1998) Farming carbon. Landscope 14:17–22

Shea SR (1999) The development of tree crops for farms in Western Australia. In: Tree crops for farms. Landscope tree crops special edition. Conservation and Land Management, Perth, pp 2–5

Shea SR, Bartle JR (1988) Restoring nature's balance: the potential for major reforestation of south western Australia. Landscope 3:3–14

Smith M, Dixon R, Boniecka L, Berti M, Sparks T, Bari M, Platt J (2006) Salinity situation statement Warren River. Water resource technical series report no. WRT 32, Department of water

Sochacki S, Harper R, Smettem K (2007) Estimation of woody biomass production from a short rotation bio-energy system in semi-arid Australia. Biomass Bioenergy 31:608–616

State Salinity Council (2000) Natural resource management in Western Australia. The Salinity Strategy, Government of Western Australia

Stoneham G, Chaudhri V, Ha A, Strappazzon L (2003) Auctions for conservation contracts: an empirical examination of Victoria's BushTender trial. Aust J Agric Resour Econ 47:477–500

Townsend PV, Harper RJ, Brennan PD, Dean C, Wu S, Smettem KRJ, Cook SE (2012) Multiple environmental services as an opportunity for watershed restoration. Forest Policy Econ 17:45–58

Vertessy RA (2001) Impacts of plantation forestry on catchment runoff. In: Nambiar EKS, Brown AG (eds) Plantations, farm forestry and water. Joint Venture Agroforestry Program, Canberra, pp 9–19

Watson RT, Noble IR, Bolin B, Ravindranath NH, Verardo DJ, Dokken DJ (2000) Land use, land–use change, and forestry. Cambridge University Press, Cambridge

Wood WE (1924) Increase of salt in soil and streams following destruction of the native vegetation. J R Soc West Aust 10:35–47

Yuan JS, Tiller KH, Al-Ahmad H, Stewart NR, Stewart CN (2008) Plants to power: bioenergy to fuel the future. Trends Plant Sci 13:421–429

Zorzetto A, Chudleigh P (1999) Commercial prospects for low rainfall agroforestry. Joint venture agroforestry program, RIRDC Publication 99/15299/152

Chapter 15
Challenging a Paradigm: Toward Integrating Indigenous Species into Tropical Plantation Forestry

Anthony S. Davis, Douglass F. Jacobs, and R. Kasten Dumroese

15.1 Introduction

The world's forests are disappearing at an alarming rate. Between 2000 and 2010 the net annual loss of the world's forests was 6.4 million ha (0.13%). While the rate of forest cover loss has slowed compared to the period of 1990–2000 (net annual loss of 8.3 million ha $year^{-1}$) (FAO 2011), it is still severe enough to warrant a concerted effort to slow or reverse this trend. This is especially true in the tropical regions, where net forest losses increased from 6.3 million ha $year^{-1}$ in the 1990s to 8.0 million ha $year^{-1}$ between 2000 and 2005 (FAO 2011)

Tropical forests, which represent 44% of the world's forested area (FAO 2011) and contain much of the world's biological diversity (USAID 1992), are cleared mainly for agricultural purposes (García-Montial and Scatena 1994; Brothers 1997; Leopold et al. 2001; Pearce et al. 2003; Gibbs et al. 2010). Timber harvest (Islam and Weil 2000) and collection of fuel wood (Wilcox 1995; Islam and Weil 2000) also contribute to the removal of forest cover. Natural disturbances, including fires and hurricanes, add to the loss of forest cover, but forest recovery from natural causes tends to be more rapid than from anthropogenic disturbances (Myster and Pickett 1990; Finegan 1992). Deforestation is defined as the removal of forest cover

A.S. Davis (✉)
Center for Forest Nursery and Seedling Research, Department of Forest Resources, College of Natural Resources, University of Idaho, Moscow, ID 83844, USA
e-mail: asdavis@uidaho.edu

D.F. Jacobs
Hardwood Tree Improvement and Regeneration Center, Department of Forestry and Natural Resources, Purdue University, West Lafayette, IN 47907, USA

R.K. Dumroese
USDA Forest Service Rocky Mountain Research Station, Moscow, ID 83843, USA

J. Stanturf et al. (eds.), *Forest Landscape Restoration: Integrating Natural and Social Sciences*, World Forests 15, DOI 10.1007/978-94-007-5326-6_15,

and withdrawal of land from forest cover, whether deliberately or circumstantially (NRC 1995). Given estimates that over half of tropical forest area has been lost (Myers 2003), it is not surprising that tropical deforestation is considered the leading factor in worldwide biodiversity loss (Sánchez-Azofeifa et al. 2003).

In this chapter, the current state of tropical forestry is described, as are global experiences with the use of exotic species and monocultures in plantation establishment, and experiences to date of afforestation and reforestation in tropical countries. The purpose of this chapter is to draw attention to potential afforestation (establishing trees where forests have been absent for an extended period) and reforestation (regeneration of forests after harvesting) strategies that would improve the state of tropical forest management by conserving biological diversity in tropical regions and increasing potential economic benefits to local communities. The global and local benefits from adopting these strategies are economic and environmental; they can be accomplished through the cooperation of researchers and forestry-related practitioners working toward establishment of compositionally diverse, economically viable plantations of native species.

15.2 Land Conversion and Plantation Establishment

Land converted from forest to agriculture and pasture is often abandoned soon after conversion (Aide et al. 1995; Finegan 1996; Wright and Muller-Landau 2006). Reasons for abandonment include shifts in the economy (Deitz 1986), invasion of grasses (Aide et al. 1995), and degradation of soil (Aide et al. 1995; Simmons 1997). Soil degradation, which may cause an immediate and lasting reduction in productivity (Milham 1994), results from a decrease in microbial biomass associated with burning (García-Oliva et al. 1999), nutrient runoff (Malmer 1996), erosion (Kaihura et al. 1999), and/or hydrologically mediated nutrient loss (Malmer 1996). In some cases cultivation further degrades soil quality (Islam and Weil 2000) such that it becomes necessary for farmers to clear new land to maintain agricultural productivity.

Farmlands are often characterized by low species richness (Fujisaka et al. 1998; Zapfack et al. 2002), and are often dominated by exotic species (Rivera and Aide 1998; Chinea 2002). Slashing and burning can greatly alter the seed bank composition in tropical deciduous forests (Miller 1999). Consequently, the biological communities regenerating on these farmlands after abandonment do not resemble those on similar sites with low anthropogenic disturbance (Zou et al. 1995; Oosterhoorn and Kapelle 2000; Chinea 2002; Marcano-Vega et al. 2002). Planting trees and shrubs, both exotic and native, to create microhabitats that facilitate the establishment of native tree species may be effective in increasing native forest cover. Several authors (Aide et al. 1995; Parrotta 1999; Feyera et al. 2002) argue that plantations established with exotic species can act as nurse trees to facilitate natural forest regeneration. However, plantations established to accelerate natural succession must be done with careful consideration of species selection, as Healey and Gara (2003) found that plantations of non-native teak (*Tectona grandis* Linn.f.) limited the development of native species compared to natural succession on abandoned fields in Costa Rica. Additionally, Haggar et al. (1997) found that woody regeneration beneath plantations, in terms of species richness and abundance, was related to the species planted. A study conducted by Cusack and Montagnini (2004) affirmed that timber species established in plantations accelerated understory recruitment when compared to abandoned pasture in Costa Rica, and that the effectiveness of species at recruiting understory plants varied by site. Even so, such transformations may be slow; Honnay et al. (2002) determined that it takes at least a century for the understory of European temperate forest plantations to attain a similar understory composition as that found under a natural forest.

Approximately 5% of the world's forests are plantations, comprising a total area of 187 million ha (FAO 2001). While most plantations are located in industrialized, non-tropical nations, a growing number are found in developing tropical countries (Pandey and Ball 1998; Evans 1999). Plantations account for as much as 90% of the wood supply in countries including New Zealand and Chile (Park and Wilson 2007). In the future, an increased proportion of the world's wood supply will come from tree plantations (Pandey and Ball 1998; Hartley 2002). Based on high productivity and successful experiences with plantation management to date (Wright et al. 2000; Montagnini 2001), it is probable that tropical regions will host many of these plantations. While potential benefits of tropical plantation establishment, such as erosion control (Lugo 1997; McDonald et al. 2002), carbon sequestration (Wright et al. 2000) and efficient timber production (Montagnini 2001; Healey and Gara 2003) are well known, the continued high rate of loss of forest cover implies that additional economic incentives are required for many tropical landowners to maintain forests on their land. Further, if financial benefits are realized from the development of a carbon sequestration and credit system, then demand for, and motivation to establish, tropical forest plantations would increase (Wright et al. 2000). Evidence suggests that increased short-term financial incentives could be effective in increasing the number of plantations established, as well as ensuring their continued management, as Piotto et al. (2004) found with Costa Rican and Nicaraguan farmers. These landowners, with continued subsidization from their governments, were willing to reforest their properties.

Globally, plantation forestry is a controversial subject, with concerns about establishment of monocultures being a primary objection. Commonly stated risks include high susceptibility to damage by pests and pathogens (Jactel and Brockerhoff 2007), changes to hydrological processes, and the perceived inability of plantation forests, with their inherent compositional and structural characteristics, to facilitate ecological processes associated with naturally-regenerating forests. However, Gadgil and Bain (1999) argue there is no substantial evidence showing plantation forests are more susceptible to pathogen and pests than managed natural forests. In addition, research in the USA found that hydrological processes in young poplar plantations were similar to those of young natural forests (Perry et al. 2001) and, in China, that plantations on degraded sites decreased annual runoff and its coarse sediment content, which contributes to the restoration of natural hydrologic processes (Zhou et al. 2002).

Although horizontal and vertical structure (Cannell 1999) and species composition of flora (Thomas et al. 1999; Ross-Davis and Frego 2002) and fauna (Lambert and Hannon 2000; Lomolino and Perault 2000; Erdle and Pollard 2002) are significantly different in plantations and natural forests, establishment of plantations can ultimately lead to succession of natural understory communities (Haggar et al. 1997; Powers et al. 1997; Parrotta 1999; Otsamo 2000; Senbeta et al. 2002). Management for specific objectives in plantations can provide habitat for many plant and animal species (Cannell 1999; Cawsey and Freudenberger 2008), and incorporating indigenous tree species in plantation establishment helps maintain species and genetic diversity and allows for continued interactions between indigenous animal and plant species (Montagnini 2001). In fact, O'Neill et al. (2001) recommended plantation establishment to maintain genetic diversity in the Peruvian Amazon; plantations reduce high-grading in natural forests and well-planned plantations can decrease harvesting pressure placed on natural forests, which will continue to be exploited if needs for forest products are not met on lands managed for production of forest products (Buckman 1999).

15.3 Global Experiences with Exotic Species and Monocultures

Establishing plantations to help mitigate effects of deforestation is logical. Lamb (1998) and Paquette and Messier (2010) highlight the variable degree to which plantations can be successful at achieving restoration objectives; a sliding scale dependant on myriad factors ranging from ownership issues to environmental conditions. Afforestation and reforestation are important contributors to natural resource management through their provisions to local and global economies, and by alleviating pressures placed on natural forests (FAO 2011). To date, most afforestation and reforestation projects in tropical regions of the world have used exotic species with inherent fast growth rates (Leopold et al. 2001; Montagnini 2001) to establish monoculture plantations (Hartley 2002). Globally, *Pinus* and *Eucalyptus* are the most commonly-planted genera, comprising 20 and 10% of the total area in plantations, respectively (FAO 2001). In tropical regions, approximately 85% of forest plantations consist of the genera *Pinus*, *Eucalyptus*, and *Tectona* (Montagnini 2001).

Monocultures of exotic species in the tropics are often, at the time of establishment, viewed as a ready solution to the apparent need to cover the land. The selected exotic species frequently have exceptional growth rates and standardized management plans that can be easily implemented (Montagnini 2001). With the existing high rate of deforestation in tropical regions, many observers welcome plantation establishment as a means of returning stability to the land. The act of increasing forest cover worldwide to compensate for loss through agriculture, settlement, and other anthropogenic disturbances is an important and noble goal.

The benefits and costs associated with exotic species and monoculture plantation management have been experienced globally. Given an objective of maximizing timber and fiber production, forest plantations established as monocultures of exotic species appear to be most productive. It is important to recognize that most of the plantations now being established will differ from naturally-regenerated forests in both composition and structure. This means they will be sustained by different ecological processes and will generate different functional outcomes. Moreover, as conservation values change and new forest policies and management objectives are developed and refined, monocultural plantations of exotic species may no longer meet the needs of society. There are increasing numbers of examples where managers have sought to modify silvicultural practices to take account of these changing social goals. For example, forest management practices that associate harvest schedules with natural disturbances regimes are considered to be ecologically sustainable (Mönkkönen 1999), and mimicking natural systems in the humid tropical lowlands can lead to the design of sustainable land-use systems (Ewel 1999). In Israel, a country without extensive natural forests, forest management practices have shifted from the establishment of large, even-aged monoculture blocks to the use of small, uneven-aged multi-species blocks in an effort to increase ecological stability and biological diversity (Ginsberg 2002). Koch and Skovsgaard (1999) posited that the European approach to forestry has shifted from a single-use management objective (wood production) to management for multiple values, and that greater focus has been placed on protection of natural forests and minimizing stand conversion. Furthermore, plantations should be established in a manner that minimizes the time necessary to develop into a semi-natural forest.

These objectives aim to thwart the loss of biological diversity and authenticity that occurs with establishment of monocultures of exotic species, as has been noted in Britain (Peterken 2001). In Australia, the spread of Monterey pine (*P. radiata* D. Don) from plantations to natural forests has resulted in a decrease in native plant and animal diversity (Gill and Williams 1996). Increased understanding of the consequences of species introduction has resulted in many industrialized countries with well-developed forest management strategies now preferring forest compositions closer to those that preceded large-scale human intervention; forest restoration is often used to facilitate return of natural ecosystem components and/or processes (Harrington 1999). For example, forest restoration projects in the United States aim to reintroduce stand-driving events (i.e., fire) as a means of maintaining natural species compositions and facilitating natural processes (Blake and Schuette 2000; Bailey and Covington 2002), while in Central Europe projects are underway to re-convert non-native conifer forests to more characteristic deciduous forests (Zerbe 2002).

Lodgepole pine (*P. contorta* Dougl. ex. Loud.), extensively planted as an exotic species in Scandinavian countries, is being thoroughly studied given concerns over its potential to escape plantations and invade native forests (Sykes 2001), and subsequent strict management plans have been suggested to reduce the potential impact of this exotic species (Engelmark et al. 2001). In Sweden, for example, Engelmark et al. (2001) suggest management plans should include maintaining strict control over the locations and total area of lodgepole pine plantations established to reduce the incidence of this species spreading beyond plantations, and to define zones where lodgepole pine should not be planted as a means of promoting growth of native species. It has been argued that exotic species are free from natural pests and pathogens when first introduced, and that there is little evidence of native pests and pathogens adopting them as hosts (Gadgil and Bain 1999). However, one concern associated with the introduction of lodgepole pine in Sweden is the transfer of exotic pathogens along with the species, and the potential for these pathogens to spread from the exotic lodgepole pine to the native Scots pine (*P. sylvestris* L.) (Ennos 2001). The effect of such disease transmission can be devastating.

In the United States, the once-dominant American chestnut (*Castanea dentata* (Marsh.) Borkh.) was decimated by the introduction of an aggressive diffuse canker disease, *Cryphonectria parasitica* (Murrill) Barr (Anagnostakis 1987), which was likely introduced through the importation of chestnut seedlings from Europe (Marchant 2002). Since 1920, Dutch elm disease (*Ophiostoma ulmi* (Buisman) Nannf. and *O. novo-ulmi* (Brasier)), a wilt disease originally identified in Holland, has spread across Europe, North America, and Central Asia, and resulted in the death of most mature elms (*Ulmus* spp.) in the northern hemisphere (Brasier and Buck 2001). Introduction of Dutch elm disease into Britain and North America was likely by importation of infested timber (Brasier and Buck 2001). Lack of pests and pathogens adapted to exotic species as natural population controls is not always the case; in lowland humid tropics, pests and pathogens have limited the productivity of *Eucalyptus*, a genus comprising approximately ten million ha of tropical plantations (Turnbull 1999).

With these problems associated with monoculture plantations of exotic species, stricter controls are being placed on the new use of exotic species. For example, in the Forest Stewardship Council's (FSC 2012) Principles and Criteria, Principle 6.9 mandates that exotic species be 'carefully controlled and actively monitored to avoid adverse ecological impacts.' For this reason, it is paramount that recommendations for afforestation and reforestation strategies come from researchers and practitioners in countries with experience in plantation and exotic species management; their experiences can help land managers in developing countries avoid making similar mistakes. The best approach, however, may be to change the plantation paradigm and place more emphasis on plantations comprised of native tree species.

15.4 Establishing Tropical Plantations Using Native Species

Evidence shows that since 1995, the diversity of species planted globally has increased (FAO 2001); this is due in part to more research (e.g. Butterfield 1995, 1996; Haggar et al. 1998; Leopold et al. 2001; Montagnini 2001; McDonald et al. 2003; Pedraza and Williams-Linera 2003) investigating the suitability of indigenous tropical species for afforestation and reforestation. In many of these studies, native species have been identified as being at least as productive as exotic species. Montagnini (2001) reported a number of native Latin American species that achieved greater biomass than plantations of exotic species, and Leopold et al. (2001) found that growth rates achieved by mixed-species plantations of native hardwoods in Costa Rica compared favorably with those reported for exotic species. In Australia, mixed-species plantations including native species were found to be more productive than monocultures (Erskine et al. 2006). Haggar et al. (1998) determined that selected native species in the lowland humid tropics of Costa Rica could produce similar growth rates to exotic species grown in plantations, and that the native species showed higher survival and required less intensive site preparation for establishment. In the Jamaican Blue Mountains, McDonald et al. (2003) identified native species that, based on growth rate, would be preferable for reforestation. In Costa Rica, plantations established with native species are able to produce value-added products such as furniture and construction wood (Wightman et al. 2001). A comparison of monetary gains between native Indian rosewood (*Dalbergia sissoo* Roxb. ex DC.) and exotic eucalyptus (*E. tereticornis* Sm.) in India found that Indian rosewood had net annual gains of almost twice that of eucalyptus (Jalota and Sangha 2000). While in many cases exotics have shown to be better able to establish, under heavy competition, than indigenous counterparts (e.g. Otsamo et al. 1997), it is possible that further research into appropriate plant materials, or alternative nursery cultural practices (e.g. Dumroese et al. 2009), may lead to more acceptable indigenous species for plantation establishment in many regions for which exotic species are currently the primary component.

For the use of native species in afforestation and reforestation efforts to become accepted by landowners and government agencies, suitable species and their management requirements must be correctly identified and described. Both the management protocols and plant material must be made readily available for public consumption (McDonald et al. 2002). In Panama, the provision of materials and technology has led to increased tree planting by small producers (Simmons et al. 2002). Continued extensive research and dissemination of knowledge at the local level will help to accomplish these requirements. Species performance in plantations cannot always be predicted by that in natural forests, as evidenced by work in the Jamaican Blue Mountains by McDonald et al. (2003) where no relationship between the growth rates of native species in a plantation trial and their mean growth rates in natural forests was detected. As is the case in all plantations, site selection is important, as it is difficult to infer performance across a variety of sites (Turner et al. 1999). Further variation between development in natural stands and in plantations may be explained by interspecific interactions that are lacking in monocultures.

In tropical northern Thailand, where many native tree species have not been grown in nurseries, successful nursery production of native species is limited by lack of knowledge of propagation requirements (Elliott et al. 2002); minor changes in nursery cultural practices can have major implications for seedling quality and suitability for establishment (Dumroese et al. 2009). For production of seedlings, or other plant material where appropriate, of native species to become more common, it must be simple enough to be completed by community members with little or no training in this field. Research to identify suitable nursery practices to cultivate seedlings of native tropical species is needed before quality seedlings will become available (Blakesley et al. 2002). Increased availability of seedlings will advance afforestation and reforestation efforts. Traditional use and familiarity with the species, a benefit provided by working with native species as opposed to exotic species (Montagnini 2001), may help with the development of practical and effective propagation techniques.

15.5 Future Directions and Potential Benefits

Dependence on forest products for livelihood is common in tropical forest communities (Byron and Arnold 1999). Diversification of assets is a necessary practice in economic spheres to reduce risk. Application of this same concept to species diversity in tropical plantations would serve land managers well. In addition to decreasing the likelihood of catastrophic pest outbreaks, the effect of fluctuation in the market value of tree species would be minimized. As timber values can deviate temporally (Trømborg et al. 2000), it is important not to invest solely in one particular species. Plantations established with mixed species reduce economic risk by increasing the potential end-products (Butterfield 1995) and decreasing dependence on any one species. Plantations established in parts of Costa Rica with native species are used for value-added products of greater economic benefit than pulpwood production (Wightman et al. 2001).

Production of non-timber forest products (NTFPs) for use as food, medicinal treatments, and household items can be a major contributor to community income (Narendran et al. 2001). A stand managed for revenue beyond that provided by timber production will be of greater economic value (Tewari 2000). Added to revenue generated through timber production, NTFPs could increase interest in maintaining forest cover, as well as provide short-term revenue. Effective marketing to ensure fair prices and commercial accessibility could increase the economic benefits of NTFPs. Many factors, including the distance to market and product value, must be considered and can have a major influence on the value of the product to the local community (Shanley et al. 2002). Establishment of plantations with native species will enable production and collection of many of these NTFPs, many of which are culturally significant, which may be precluded through the use of exotic species.

As forests are cleared for agriculture or pasture, carbon is released into the atmosphere, an additional problem associated with deforestation (Deacon 1995). A strategy that may help to diminish the rise in atmospheric carbon is to increase global forest cover, which would use trees as 'biological scrubbers' (Richards and Andersson 2001). The issue of how a system of credits for carbon sequestration could affect the practice of forestry is important. Fearnside (1999, 2000) stated that stemming the rate of deforestation yields greater potential for arresting global warming, but acknowledged that increasing the area in plantations is more feasible. A meta-analysis conducted by Silver et al. (2000) identified that afforestation of abandoned fields in the tropics can serve effectively as a carbon sink for at least 40 years, not surprising given that forest ecosystems contain up to 100 times the carbon found in agricultural systems. Species selection for carbon sequestration is important as biomass accumulation and carbon uptake is known to vary by species (Schroth et al. 2002). Forests have the potential to store carbon beyond current levels (Harmon 2001), and capitalizing on this will help ameliorate climate change. Schroth et al. (2002) found that multi-strata plantations (i.e., agroforestry) had greater potential to store carbon than even-aged monocultures. Cannell (1999) reported that trees in plantations managed for maximum timber volume contain considerably less carbon than the same area of unmanaged forests, based on the strategy of harvesting plantations at the maximum mean annual increment.

Genetic modification of common tropical plantation species such as Monterey pine has led to an increase in carbon sequestration by as much as 22% (Jayawickrama 2001). Should future efforts to increase the sequestration ability be concentrated only on those species that are already commonly used for reforestation in tropical areas, a decrease in the diversity of species being planted is likely. Furthermore, should there be a financial benefit to having species that are more adept at sequestering carbon, this would likely lead to an increase in the percentage of forests containing those particular species. While this could be beneficial to conservation efforts if those species were native, it could hinder attempts to foster diversity in reforestation programs if they were exotic.

The potential economic benefits of the aforementioned practices could create great opportunities for tropical regions of the world through plantation establishment. Revenue from a carbon-credit system could provide tropical landowners with financial justification to maintain forest cover on their land. Ramirez et al. (2002) determined that forest conservation to sequester carbon in Costa Rica would be economically sensible. With adoption of the REDD+ program, or reducing emissions from deforestation and forest degradation, carbon sequestration has become a significant international policy objective (FAO 2011).

Because potential carbon sequestration is another example of why the rate of establishment of tropical forest plantations may increase, it is important that before added pressure to establish plantations is applied to tropical countries, a better understanding of species selection be developed. NTFPs in India's forests are largely unaccounted for in terms of the economic value they provide (Chopra 1993). In Indonesia, farm-households that participated in tree-growing projects benefited from diversified revenues and added cash incomes (Nibbering 1999). The incorporation of NTFPs with traditional farming and forestry practices appear to be leading to financial and environmental benefits that may provide a basis for sustainable land use (Leakey and Tchoundjeu 2001). A combination of management for NTFPs and carbon sequestration could provide long-term benefits to local communities as well as the environment. Furthermore, these practices would be more likely to gain certification, which could otherwise lead to a loss of market access if, for example, European producers chose to purchase certified temperate forest products over uncertified tropical forest products (Ruddell et al. 1998). In 2007, Norway banned tropical timber from public procurement programs (FAO 2011). Actions like these increase pressure on tropical forest plantation managers to consider the economic and environmental consequences of plantation practices.

15.6 Summary and Recommendations for Future Progress

To conserve biological diversity and increase potential economic benefits to tropical communities, a global focus on establishment of tropical forest plantations should consider the impacts of species selection and stand composition. Such a careful *a priori* consideration of species selection could increase the long-term ecological contribution of resulting plantations. Mixed stand management using native species is recommended to conserve biological diversity and increase local economic opportunities. The ability of native species to sustain natural ecological processes and provide proper habitat for native animal and plant species far exceeds that of exotic species. Furthermore, mixed stand management provides diversity to offset damage associated with pest outbreaks. Therefore, a plantation comprised of a mix of native species is far more likely to: (1) facilitate processes associated with natural ecosystems, (2) meet local cultural needs, (3) maintain traditional forest values, and (4) produce traditional forest products, while minimizing biological, economic, and environmental risks (i.e., pests and pathogens, and dependence on a single forest product). Increased species richness in plantations ensures the potential for future development and diversification of forest products that will benefit local communities.

An increase in global consumption of wood products, potential financial gains from carbon sequestration, and acceptance of the need for erosion control to maintain clean water and ground structure will lead to enlargement of tropical forest plantation area. Addressing concerns over future stand composition prior to plantation establishment facilitates the potential to maintain key components of natural ecosystems. Over time, ecological restoration by humans will become increasingly important toward the conservation of biodiversity (Dobson et al. 1997). Currently, practices are being developed to rehabilitate and restore habitat for a multitude of species. These practices, however, are expensive and experimental. By limiting the establishment of sites which, in the future, may need to be restored, future costs can be avoided through present actions.

We propose four key points of consideration for future tropical plantation establishment:

1. Plantations are a necessary part of resource management, both for wood supply and for ameliorating pressures placed on natural forests
2. Plantation composition and structure can greatly impact both monetary and ecological values
3. Diversification of species (i.e., multi-species management) may reduce risks associated with disease and market fluctuations, while increasing contributions to ecosystem diversity
4. Use of native species does not preclude obtaining the same remuneration as exotic species, increases provision of ecosystem services, and maintains cultural traditions associated with indigenous forest trees

As a method of conserving biological diversity, maintaining natural ecosystem processes, and increasing local and global benefits, we recommend the following guidelines for tropical forest plantation establishment:

1. Use native species, within their native range (with attention to matching species to suitable site conditions)
2. Use multiple species to allow for access to traditional (including timber and non-timber and non-timber products) and emerging (such as biofuels and carbon) markets
3. Select species that facilitate, or at minimum do not restrict, native understory development

Recent research in tropical regions toward identification of native species suitable for plantation establishment is encouraging and necessary. Dedication of researchers from around the world to assist with development of strategies to conserve tropical forests will decrease the chance of repeating mistakes made in boreal and temperate ecosystems. Partnerships like the Model Forest Network, which involves cooperation between government, indigenous people, practitioners, researchers, non-government organizations, and other stakeholders, appear to be more effective than traditional means in creating sustainable resource management practices (LaPierre 2002). Individuals and organizations with experience in developing new market products could prove to be an invaluable resource for the economic production of NTFPs. Implementing the aforementioned recommendations will likely lead to a more desirable future forest plantation composition, but it is paramount that landowners understand the reasons for the actions they are taking. Whitmore (1987) recommended that a global network of plantation researchers and practitioners would 'overcome barriers to plantation success.' This still holds true, and we would all benefit from striving to attain this goal. Through proper management, tropical afforestation and reforestation can benefit society locally and globally while maintaining key ecosystem components and providing important services currently threatened or degraded by anthropogenic use.

References

Aide TM, Zimmerman JK, Herrera L, Rosario M, Serrano M (1995) Forest recovery in abandoned tropical pastures in Puerto Rico. For Ecol Manage 77:77–86

Anagnostakis SL (1987) Chestnut blight: the classical problem of an introduced pathogen. Mycologia 79:23–27

Bailey JD, Covington WW (2002) Evaluating ponderosa pine regeneration following ecological restoration treatments in northern Arizona, USA. For Ecol Manage 155:271–278

Blake JG, Schuette B (2000) Restoration of an oak forest in east-central Missouri. Early effects of prescribed burning on woody regeneration. For Ecol Manage 139:109–126

Blakesley D, Hardwick K, Elliot S (2002) Research needs for restoring tropical forests in Southeast Asia for wildlife conservation: framework species selection and seed propagation. New For 24:165–174

Brasier CM, Buck KW (2001) Rapid evolutionary changes in a globally invading fungal pathogen (Dutch elm disease). Biol Invasions 3:223–233
Brothers TS (1997) Deforestation in the Dominican Republic: a village-level view. Environ Conserv 24:213–223
Buckman RE (1999) Planted forests, agriculture, and the environment – a perspective. New For 18:3–4
Butterfield RP (1995) Promoting biodiversity: advances in evaluating native species for reforestation. For Ecol Manage 75:111–121
Butterfield RP (1996) Early species selection for tropical reforestation: a consideration of stability. For Ecol Manage 81:161–168
Byron N, Arnold M (1999) What futures for the people of the tropical forests? World Dev 27:789–805
Cannell MGR (1999) Environmental impacts of forest monocultures: water use, acidification, wildlife conservation, and carbon storage. New For 17:239–262
Cawsey EM, Freudenberger D (2008) Assessing the biodiversity benefits of plantations: the Plantation Biodiversity Benefits Score. Ecol Manage Res 9:42–52
Chinea JD (2002) Tropical forest succession on abandoned farms in the Humacao Municipality of eastern Puerto Rico. For Ecol Manage 167:195–207
Chopra K (1993) The value of non-timber forest products: an estimation for tropical deciduous forests in India. Econ Bot 47:251–257
Cusack D, Montagnini F (2004) The role of native species plantations in recovery of understory woody diversity in degraded pasturelands of Costa Rica. For Ecol Manage 188:1–15
Deacon RT (1995) Assessing the relationship between government policy and deforestation. J Environ Econ 28:1–18
Deitz JL (1986) Economic history of Puerto Rico. Princeton University Press, Princeton
Dobson AP, Bradshaw AD, Baker AJM (1997) Hopes for the future: restoration ecology and conservation biology. Science 277:515–522
Dumroese RK, Jacobs DF, Davis AS (2009) Inoculating *Acacia koa* with Bradyrhizobium and applying fertilizer in the nursery: effects on nodule formation and seedling growth. HortScience 44:443–446
Elliott S, Kuarak C, Navakitbumrung P, Zangkum S, Anusarnsunthorn V, Blakesley D (2002) Propagating framework trees to restore seasonally dry tropical forest in northern Thailand. New For 23:63–70
Engelmark O, Sjöberg K, Anderson B, Rosvall O, Ågren GI, Baker WL, Barklund P, Björkman C, Despain DG, Elfving B, Ennos R, Karlman M, Knecht MF, Knight DH, Ledgard NJ, Lindlöw Å, Nilsson C, Peterken GF, Sörlin S, Sykes MT (2001) Ecological effects and management aspects of an exotic tree species: the case of lodgepole pine in Sweden. For Ecol Manage 141:3–13
Ennos RA (2001) The introduction of lodgepole pine as a major forest crop in Sweden: implications for host-pathogen evolution. For Ecol Manage 141:85–96
Erdle T, Pollard J (2002) Are plantations changing the tree species composition of New Brunswick's forest? For Chron 78:812–821
Erskine PD, Lamb D, Bristow M (2006) Tree species diversity and ecosystem function: can tropical multi-species plantations generate greater productivity? For Ecol Manage 233:205–210
Evans J (1999) Planted forests of the wet and dry tropics: their variety, nature, and significance. New For 17:25–36
Ewel JJ (1999) Natural systems as models for the design of sustainable systems of land use. Agrofor Syst 45:1–21
FAO (2001) State of the world's forests 2001. United Nations, Rome
FAO (2011) State of the world's forests 2011. United Nations, Rome
Fearnside PM (1999) Forests and global warming mitigation in Brazil: opportunities in the Brazilian forest sector for responses to global warming under the clean development mechanism. Biomass Bioenergy 16:171–189

Fearnside PM (2000) Uncertainty in land-use change and forestry sector mitigation options for global warming: plantation silviculture vs. avoided deforestation. Biomass Bioenergy 18:457–468
Feyera S, Beck E, Lüttge U (2002) Exotic trees as nurse-trees for the regeneration of natural tropical forests. Trees 16:245–249
Finegan B (1992) The management potential of neotropical secondary lowland rain forest. For Ecol Manage 47:295–321
Finegan B (1996) Pattern and process in neotropical secondary rain forests: the first 100 years of succession. Trends Ecol Evol 11:119–124
FSC (2012) FSC principles and criteria. Available online at: http://www.fsc.org/principles-and-criteria.34.htm. Accessed 18 June 2012
Fujisaka S, Escobar G, Veneklaas E (1998) Plant community diversity relative to human land uses in an Amazon forest colony. Biodivers Conserv 7:41–57
Gadgil PD, Bain J (1999) Vulnerability of planted forests to biotic and abiotic disturbances. New For 17:227–238
García-Montial DC, Scatena FN (1994) The effect if human activity on the structure and composition of a tropical forest in Puerto Rico. For Ecol Manage 63:57–78
García-Oliva F, Sanford RL Jr, Kelly E (1999) Effect of burning of tropical deciduous forest soil in Mexico on the microbial degradation of organic matter. Plant Soil 206:29–39
Gibbs HK, Ruesch AS, Achard F, Clayton MK, Holmgren P, Ramankutty N, Foley JA (2010) Tropical forests were the primary sources of new agricultural land in the 1980s and 1990s. Proc Natl Acad Sci 107(38):16732–16737
Gill AM, Williams JE (1996) Fire regimes and biodiversity: the effects of fragmentation of southeastern Australian eucalypt forests by urbanization, agriculture, and pine plantations. For Ecol Manage 85:261–278
Ginsberg P (2002) Planning and management of the afforestation process in Northern Israel. New For 24:27–38
Haggar J, Wightman K, Fisher R (1997) The potential of plantations to foster woody regeneration within a deforested landscape in lowland Costa Rica. For Ecol Manage 99:55–64
Haggar JP, Briscoe CB, Butterfield RP (1998) Native species: a resource for the diversification of forestry production in the lowland humid tropics. For Ecol Manage 106:195–203
Harmon ME (2001) Carbon sequestration in forests: addressing the scale question. J For 99:24–29
Harrington CA (1999) Forests planted for ecosystem restoration or conservation. New For 17:175–190
Hartley MJ (2002) Rationale and methods for conserving biodiversity in plantation forests. For Ecol Manage 155:81–95
Healey SP, Gara RI (2003) The effect of a teak (*Tectona grandis*) plantation on the establishment of native species in an abandoned pasture in Costa Rica. For Ecol Manage 17:497–507
Honnay O, Bossuyt B, Verheyen K, Butaye J, Jacquemyn H, Hermy M (2002) Ecological perspectives for the restoration of plant communities in European temperate forests. Biodivers Conserv 11:213–242
Islam KR, Weil RR (2000) Land use effects on soil quality in a tropical forest ecosystem of Bangladesh. Agric Ecosyst Environ 79:9–16
Jactel H, Brockerhoff EG (2007) Tree diversity reduces herbivory by forest insects. Ecol Lett 10:835–848
Jalota RK, Sangha KK (2000) Comparative ecological-economic analysis of growth performance of exotic *Eucalyptus tereticornis* and indigenous *Dalbergia sissoo* in monoculture plantations. Ecol Econ 33:487–495
Jayawickrama KJS (2001) Potential genetic gains for carbon sequestration: a preliminary study on radiate pine plantations in New Zealand. For Ecol Manage 152:313–322
Kaihura FBS, Kullaya IK, Kilasara M, Aune JB, Singh BR, Lal R (1999) Soil quality effects of accelerated erosion and management systems in three ecoregions of Tanzania. Soil Tillage Res 53:59–70

Koch NE, Skovsgaard JP (1999) Sustainable management of planted forests: some comparisons between Central Europe and the United States. New For 17:11–22
Lamb D (1998) Large-scale ecological restoration of degraded tropical forest lands: the potential role of timber plantations. Restor Ecol 6:271–279
Lambert JD, Hannon SJ (2000) Short-term effects of timber harvest on abundance territory characteristics, and pairing success of ovenbirds in riparian buffer strips. Auk 117:687–698
Lapierre L (2002) Canada's model forest program. For Chron 78:613–617
Leakey RRB, Tchoundjeu Z (2001) Diversification of tree crops: domestication of companion crops for poverty reduction and environmental services. Exp Agric 37:279–296
Leopold AC, Andrus R, Finkeldey A, Knowles D (2001) Attempting reforestation of wet tropical forests in Costa Rica. For Ecol Manage 142:243–249
Lomolino MV, Perault DR (2000) Assembly and disassembly of mammal communities in a fragmented temperate rainforest. Ecology 81:1517–1532
Lugo AE (1997) The apparent paradox of reestablishing species richness on degraded lands with tree monocultures. For Ecol Manage 99:9–19
Malmer A (1996) Hydrological effects and nutrient losses of forest plantation establishment on tropical rainforest land in Sabah. Malays J Hydrol 174:129–148
Marcano-Vega H, Aide TM, Báez D (2002) Forest regeneration in abandoned coffee plantations and pastures in the Cordillera Central of Puerto Rico. Plant Ecol 161:75–87
Marchant K (2002) Exotic forest pests – keeping them out. For Chron 78:272–274
McDonald MA, Healey JR, Stevens PA (2002) The effects if secondary forest clearance and subsequent land-use on erosion losses and soil properties in the Blue Mountains of Jamaica. Agric Ecosyst Environ 92:1–19
McDonald MA, Hofny-Collins A, Healey JR, Goodland TCR (2003) Evaluation of trees indigenous to the montane forest of the Blue Mountains, Jamaica for reforestation and agroforestry. For Ecol Manage 175:379–401
Milham N (1994) An analysis of farmers' incentives to conserve or degrade the land. J Environ Manage 40:51–64
Miller PM (1999) Effects of deforestation on seed banks in a tropical deciduous forest of western Mexico. J Trop Ecol 15:179–188
Mönkkönen M (1999) Managing Nordic boreal forest landscapes for biodiversity: ecological and economic perspectives. Biodivers Conserv 8:85–99
Montagnini F (2001) Strategies for the recovery of degraded ecosystems: experiences from Latin America. Interciencia 26:498–503
Myers N (2003) Conservation of biodiversity: how are we doing? Environmentalist 23:9–15
Myster RW, Pickett STA (1990) Initial conditions, history, and successional pathways in ten contrasting old fields. Am Midl Nat 124:231–238
Narendran K, Murthy IK, Suresh HS, Dattaraja HS, Ravindranath NH, Sukumar R (2001) Nontimber forest products extraction, utilization and valuation: a case study from the Nilgiri Biosphere Reserve, Southern India. Econ Bot 55:528–538
Nibbering JW (1999) Tree planting on deforested farmlands, Sewu Hills, Java, Indonesia. Agrofor Syst 46:65–82
NRC (1995) Silvicultural terms in Canada. Natural Resources Canada, Ottawa
O'Neill GA, Dawson I, Sotelo-Montes C, Guarino L, Guariguata M, Current D, Weber JC (2001) Strategies for the genetic conservation of trees in the Peruvian Amazon. Biodivers Conserv 10:837–850
Oosterhoorn M, Kapelle M (2000) Vegetation structure and composition along an interior-edge-exterior gradient in a Costa Rica montane cloud forest. For Ecol Manage 126:291–307
Otsamo R (2000) Secondary forest regeneration under fast-growing forest plantations on degraded *Imperata cylindrica* grasslands. New For 19:69–93
Otsamo A, Ådjers G, Hadi TS, Kuusipalo J, Vuokko R (1997) Evaluation of reforestation potential of 83 tree species planted on *Imperata cylindrica* dominated grassland. New For 14:127–143
Pandey D, Ball J (1998) The role of industrial plantations in future global fibre supplies. Unasylva 193:37–43

Paquette A, Messier C (2010) The role of plantations in managing the world's forests in the Anthropocene. Front Ecol Environ 8(1):27–34
Park A, Wilson ER (2007) Beautiful plantations: can intensive silviculture help Canada to fulfill ecological and timber production objectives? For Chron 83:825–839
Parrotta JA (1999) Productivity, nutrient cycling, and succession in single- and mixed-species plantations of *Casuarina equisetifolia*, *Eucalyptus robusta*, and *Leucaena leucocephala* in Puerto Rico. For Ecol Manage 124:45–77
Pearce D, Putz FE, Vanclay JK (2003) Sustainable forestry in the tropics: panacea or folly? For Ecol Manage 172:229–247
Pedraza RA, Williams-Linera G (2003) Evaluation of native tree species for the rehabilitation of deforested areas in a Mexican cloud forest. New For 26:83–99
Perry CH, Millar RC, Brooks KN (2001) Impacts of short-rotation hybrid poplar plantations on regional water yield. For Ecol Manage 143:143–151
Peterken GF (2001) Ecological effects of introduced tree species in Britain. For Ecol Manage 141:31–42
Piotto D, Montagnini F, Kanninen M, Ugalde L, Viquez E (2004) Forest plantations in Costa Rica and Nicaragua: performance of species and preferences of farmers. J Sustain For 18(4):59–77
Powers JS, Haggar JP, Fisher RF (1997) The effect of overstory composition on understory woody regeneration and species richness in 7-year-old plantations in Costa Rica. For Ecol Manage 99:43–54
Ramirez OA, Carpio CE, Ortiz R, Finnegan B (2002) Economic value of the carbon sink services of tropical secondary forests and its management implications. Environ Res Econ 21:23–46
Richards K, Andersson K (2001) The leaky sink: persistent obstacles to a forest carbon sequestration program based on individual projects. Clim Policy 1:41–54
Rivera LW, Aide TM (1998) Forest recovery in the karst region of Puerto Rico. For Ecol Manage 108:63–75
Ross-Davis AL, Frego KA (2002) Comparison of plantations and naturally regenerated clearcuts in the Acadian Forest: forest floor bryophyte community and habitat features. Can J Bot 80:21–33
Ruddell S, Stevens JA, Bourke IJ (1998) International market access issues for forest products. For Prod J 48:20–26
Sánchez-Azofeifa GA, Daily GC, Pfaff ASP, Busch C (2003) Integrity and isolation of Costa Rica's national parks and biological reserves: examining the dynamics of land-cover change. Biol Conserv 109:123–135
Schroth G, D'Angelo SA, Teixeira WG, Haag D, Lieberei R (2002) Conversion of secondary forest into agroforestry and monoculture plantations in Amazonia: consequences for biomass, litter, and soil carbon stocks after 7 years. For Ecol Manage 163:131–150
Senbeta F, Teketay D, Näslund B-Å (2002) Native woody species regeneration in exotic tree plantations at Munessa-Shashemene Forest, southern Ethiopia. New For 24:131–145
Shanley P, Luz L, Swingland IR (2002) The faint promise of a distant market: a survey of Belém's trade in non-timber forest products. Biodivers Conserv 11:615–636
Silver WL, Ostertag R, Lugo AE (2000) The potential for carbon sequestration through reforestation of abandoned tropical agricultural and pasture lands. Restor Ecol 8:394–407
Simmons CS (1997) Forest management practices in the Bayano region of Panama: cultural variations. World Dev 25:989–1000
Simmons CS, Walker RT, Wood CH (2002) Tree planting by small producers in the tropics: a comparative study of Brazil and Panama. Agrofor Syst 56:89–105
Sykes MT (2001) Modelling the potential distribution and community dynamics of lodgepole pine (*Pinus contorta* Dougl. ex. Loud.) in Scandinavia. For Ecol Manage 141:69–84
Tewari DD (2000) Valuation of non-timber forest products (NTFPs): models, problems, and issues. J Sustain For 11:47–68
Thomas SC, Halpern CB, Falk DA, Liguori DA, Austin KA (1999) Plant diversity in managed forests: understory responses to thinning and fertilization. Ecol Appl 9:864–879
Trømborg E, Buongiorno J, Solberg B (2000) The global timber market: implications of changes in economic growth, timber supply, and technological trends. For Policy Econ 1:53–69
Turnbull JW (1999) Eucalypt plantations. New For 17:37–52

Turner J, Gessel SP, Lambert MJ (1999) Sustainable management of native and exotic plantations in Australia. New For 18:17–32
USAID (1992) Tropical forests and biological diversity: USAID report to Congress 1990–1991. United States Agency for International Development, Washington, DC
Whitmore JL (1987) Plantation forestry in the tropics of Latin America: a research agenda. Unasylva 156:36–31.
Wightman KE, Shear T, Goldfarb B, Haggar J (2001) Nursery and field establishment techniques to improve seedling growth of three Costa Rican hardwoods. New For 22:75–96
Wilcox BA (1995) Tropical forest resources and biodiversity: the risks of forest loss and degradation. Unasylva 181:43–49.
Wright SJ, Muller-Landau HC (2006) The future of tropical forest species. Biotropica 38:287–301
Wright JA, DiNicola A, Gaitan E (2000) Latin American forest plantations: opportunities for carbon sequestration, economic development, and financial returns. J For 98:20–23
Zapfack L, Engwald S, Sonke B, Achoundong G, Madong BA (2002) The impact of land conversion on plant biodiversity in the forest zone of Cameroon. Biodivers Conserv 11:2047–2061
Zerbe S (2002) Restoration of natural broad-leaved woodland in Central Europe on sites with coniferous forest plantations. For Ecol Manage 167:27–42
Zhou GY, Morris JD, Yan JH, Yu ZY, Peng SL (2002) Hydrological impacts of reafforestation with eucalypts and indigenous species: a case study in southern China. For Ecol Manage 167:209–222
Zou X, Zucca CP, Waide RB, McDowell WH (1995) Long-term influence of deforestation on tree species composition and litter dynamics of a tropical rain forest in Puerto Rico. For Ecol Manage 78:147–157

Chapter 16
Forest Landscape Restoration: Restoring What and for Whom?

Agni Klintuni Boedhihartono and Jeffrey Sayer

16.1 Introduction

This chapter will argue that restoration science has proven far better at solving the problems of 'how' to restore various sorts of degraded land than at addressing the problem of "what" to restore. The establishment and maintenance of vegetation on degraded sites and the use of land preparation techniques, cover and nurse crops and the management of natural successional processes are all standard tools of foresters and watershed and range managers. There is a considerable literature on how to manage and restore landscapes and excellent reviews are provided by Mansourian et al. (2005), Perrow and Davy (2002) and many others. Current thinking of issues of landscape approaches are reviewed by Lafortezza et al. (2008). The literature on determining the 'how' of landscape scale interventions is less rich.

The interest in Forest Landscape Restoration (FLR) and the emergence of the Global Partnership on Forest Landscape Restoration (GPFLR) were to a large extent motivated by the fact that civil societies in various parts of the world have been critical of large scale restoration programmes undertaken by public forest agencies. Such public rejection of reforestation programmes has occurred in places as far apart as the Mediterranean basin, Scotland, the Uplands of Vietnam and the loess plateau in Northern China. The ethical issues surrounding reforestation at a landscape scale are discussed by Whisenant (1999).

The generic criticism has been that such schemes has used too few species, have planted them in even-age monocultures, have used excessively heavy mechanical land preparation techniques and have retained ownership of the trees in the hands of

A.K. Boedhihartono • J. Sayer (✉)
Centre for Tropical Environmental and Sustainability Science (TESS),
School of Earth and Environmental Sciences,
James Cook University, Cairns, Australia
e-mail: jeffrey.sayer@jcu.edu.au; agni.boedhihartono@jcu.edu.au

J. Stanturf et al. (eds.), *Forest Landscape Restoration: Integrating Natural and Social Sciences*, World Forests 15, DOI 10.1007/978-94-007-5326-6_16,

the government agencies who planted them. A lot of progress has been made in recent years in addressing these concerns of society. In all of the examples cited above there has been considerable progress in ensuring that restoration schemes do indeed restore conditions that yield the multiple benefit flows that local societies require.

This chapter will therefore review some recent initiatives to ensure that restoration programmes do indeed address the question of 'what' to restore and where. Our basic hypothesis is that multi-stakeholder mechanisms have to be in place to allow for consensus-based decision making on what the objectives of restoration should be. We believe that the landscape is an appropriate scale at which to conceptualize restoration programmes as it is at this scale that trade-offs are best addressed and it is at the landscape scale that it is important to get the distribution, dispersal and nature of tree cover correct in order to optimize the balance of benefit flows. Our concepts are those that have driven the work of the GPFLR in recent years and our examples come from the network of landscapes where the GPFLR members have been active.

16.2 Why Restore at Landscape Scales?

We are using the term 'landscape' to describe a mosaic of different land-cover types that have properties that differ from the simple sum of the properties of the individual cells of the mosaic. The cells are in general areas of land under private management or under the control of sectoral administrative agencies. These owners or sectoral agencies legitimately focus on narrow production functions. Landscape approaches attempt to mediate the activities of the individual owners and managers in ways that lead to an optimization of the broader flow of environmental, social and economic benefits that emerge at the landscape scale and underpin the quality of life of society. We therefore like to understand the word landscape not as a spatial planning unit but rather as the scale at which it is necessary to intervene in order to balance trade-offs and optimize all benefit flows. There are numerous formal definitions of the word landscape but the one that we have used and that best describes our concepts is adapted from Farina (2006).

> A landscape is a geographical space in which the process or object of interest is completely expressed or functions. It includes not only the biophysical components of an area but also social, political, psychological and institutional attributes (Sayer 2007)

We are thus defining the landscape in functional terms. It is the area within which one has to intervene in order to achieve some desired outcome or set of outcomes. In its simplest terms this might be the maintenance of forests on hillsides to improve water supplies and prevent erosion of agricultural lands lower down those slopes. However in our experience, and even in relatively uniform landscapes, there are usually many more values and benefits that have to be addressed and the difficulty comes from the fact that there are almost always trade-offs between them. An example would be the conservation of an endangered forest dependent bird or mammal in a landscape where poor people are practicing subsistence agriculture. The needs of the endangered species would be best met by conserving as much contiguous, undisturbed forest as possible whereas the needs of the poor people would be best met by maximizing the area available for their agriculture and providing them with access to the best quality land. To achieve a landscape-scale reconciliation of this trade-off would require that representatives of the conservation interest group negotiate with representatives of the poor farmers to find a compromise. An equitable compromise might require that the conservationists pay the farmers for the opportunity costs that they incur through forgoing cultivation of some areas of land.

The examples of upland watersheds and lowland agriculture and of endangered species and extensive agriculture are quite simple. However in our experience in real life there is always a high diversity of people with different interests in what happens in landscapes. This is for the self-evident reason that people's lives are impacted by far more than what happens on the land that they own or where they have their primary economic activity. They are also impacted by their access to areas for recreation, by visual amenity, by the infrastructure and level of economic activity in their community and a wealth of other landscape attributes. This means that an initiative to restore large areas of land may impact upon the landscape which is the living space of many people with contrasting and conflicting interests.

We have used a variety of techniques to bring people together to explore scenarios for mediating landscape changes. In our experience the initial stages of this process are critical to success and creating trust and collegiality must be a major objective of this process of stakeholder engagement. A number of approaches that have been used by the Worldwide Fund for Nature as part of its New Generation Plantations initiative are appropriate for dealing with landscape restoration projects (WWF 2009). One recent example does not strictly concern restoration but rather the establishment of extensive forest plantations in an area where forest has not been present in the recent past. The example is a plan to establish industrial plantations in grassland areas in Uruguay and of the social and environmental issues that this would raise. The techniques used to initiate a multi-stakeholder process could however be applied in any situation where large scale landscape change is being planned.

16.3 An Example of Afforestation in the Pampas of Uruguay

A multi-national pulp and paper company began buying land in central Uruguay in 2005 and began establishing plantations in 2006. Operations were on a small scale and up until 2008 70,000 ha had been purchased, 3,500 ha leased and 12,000 ha planted, with a relatively even split of eucalyptus and pine. The medium-term target was to plant 13,000 ha per year but up until the present the actual area planted has been somewhat less. The project is still in its initial stages and no investment decision regarding the construction of a mill has been made. However one long-term scenario is that 118,000 ha of plantations would be established of which 75% would be eucalyptus and 25 % pine and that this would feed a single line pulp mill that would be established in the centre of the Country. The company has proceeded slowly in order to build local capacity in the area of operations and to ensure local support for its plans.

An exhaustive environmental and social impact study of the potential plantation area has been conducted. It identifies few sensitive issues although there are local concerns about enhanced fire risk, possible impacts to water sources and the disruption of the open pampas landscapes by plantation blocks. Biodiversity concerns are modest and could easily be addressed in plantation design and by use of set-asides. There is strong local support for both the establishment of plantations and of a mill as people see these as providing jobs and local revenues that would support local infrastructure. Opportunities for young people in the area are at present limited. On the other hand, the controversy surrounding some other mills in the region has in general heightened political and civil society sensitivities to such investments in Latin America. There is an urban-based environmental lobby that is critical of the environmental and social impacts of large scale plantations of exotic species.

16.3.1 A Multi-Stakeholder Mediation Process

In September 2008 a team from WWF conducted a preliminary exercise at two locations in the potential plantation areas around the town of Durazno. The intention was to test approaches to stakeholder engagement in a plantation establishment situation. The company organized two workshops in two different parts of the plantation landscape, the first with 35 participants and the second with 25. The participants represented those people most likely to be impacted by the plantations and those best able to shape local opinion. They included farmers, teachers, local officials, media representatives, unemployed people, students, contractors, other local entrepreneurs and other rural workers.

It became apparent that obtaining the participation of such a broad group for more than a short time was going to be difficult hence the decision was made to limit the main workshop to a single day. Staff members of the company were treated as participants in the workshop which was run by a team of four external facilitators from WWF. Each of the workshops started with an evening dinner and social event during which the subject was briefly introduced and some ice-breaking exercises were used in order for participants to get to know each other. On the day of the workshop the following exercises were run:

- Heterogeneous groups of 5–7 people were asked to write on cards five opportunities that would be provided by plantation expansion and five threats that might result. The cards were posted on the walls, similar ideas clustered and the facilitators led a discussion of the issues raised. Participants scored the opportunities and threats by voting with sticky dots for the issues of greatest concern to them personally – we refer to this process as ‘dotmocracy’.

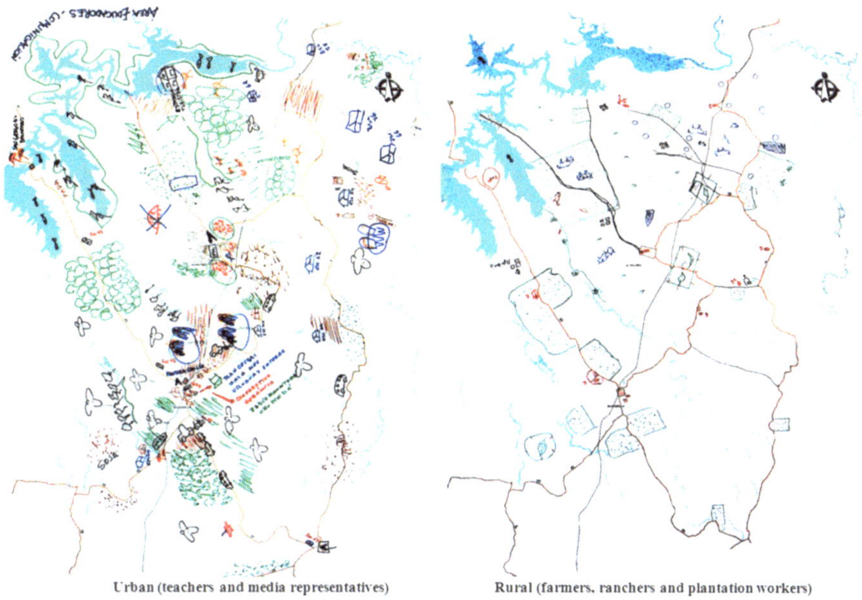

Fig. 16.1 Examples of visual representation of the present situation by a group of urban dwellers and a group of rural dwellers (**a**) Urban (teachers and media representatives) and (**b**) Rural (farmers, ranchers and plantation workers)

- Homogeneous groups of 5–7 people (farmers, teachers, workers etc) were asked to represent their present appreciation of the landscape on flip chart paper. A small number of geographic features were marked on the paper to provide scale and reference points. Groups were encouraged to indicate features of value or that were subject to threat. Each group then presented its assessment of the present situation to the entire group of workshop participants (Fig. 16.1).
- The same groups were then asked to draw their vision of an ideal landscape 10–20 years in the future and to present this to the plenary. In other exercises we have also invited participants in these exercises to present worst case scenarios but in this case we had insufficient time for this. The drawings produced by the groups were notable for their richness and for the fact that whilst the different categories of stakeholders produced different pictures the overall message from each group was similar. All saw scenarios with improved physical and social infrastructure and a mosaic of plantations and farmlands. The message was of a desire for balance and not for extreme scenarios although a group of younger workers and one of local officials clearly favored heavy investment and were strongly supportive of the construction of an industrial facility as it would provide jobs and tax revenues (Fig. 16.2).
- The drawings were photographed and manipulated using a variety of visual software and reproduced at the end of the day in ways that allowed different visions to be compared, contrasted and discussed. We did not encourage the meeting to

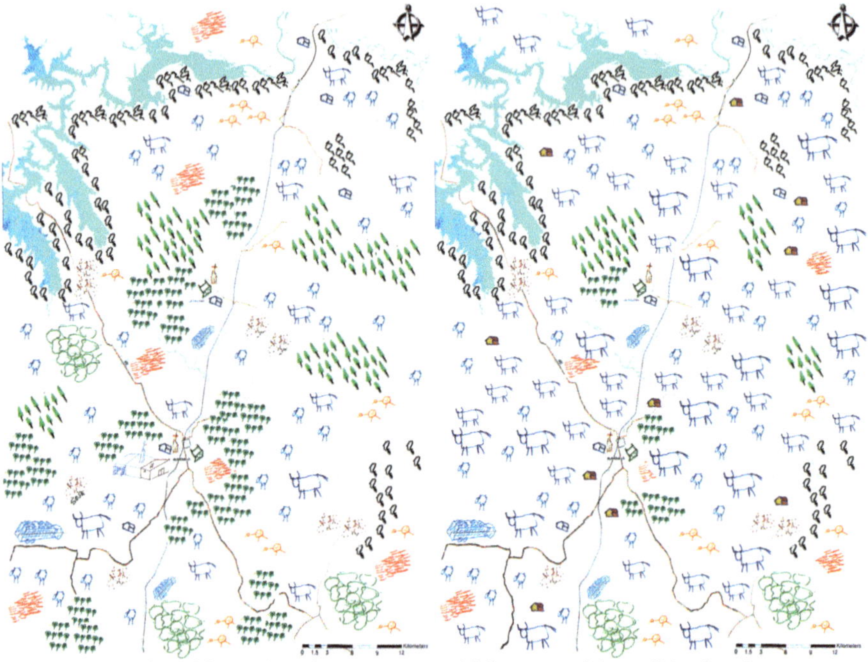

Expansion of forest plantations   Maintenance of the traditional cattle economy

Fig. 16.2 Examples of visual representation of scenarios for the future by heterogeneous groups. The first group favored expanded plantations and the second favored maintenance of the status quo (**a**) Expansion of forest plantations and (**b**) Maintenance of the traditional cattle economy

move towards a single desired scenario but rather used the exercise to provoke discussion of the variety of scenarios that were possible.

- Heterogeneous groups of 5–7 persons were invited to debate attributes of the landscape that might be used as indicators against which progress towards desired outcomes might be measured in a participatory way. Lack of time prevented this exercise being run to the point where a broad set of indicators could be established but in the short time available a number of ideas for indicators were identified and preliminary discussion held on ways in which they might be measured.

In the space of two 1-day workshops we were able to generate a rich discussion of future landscape scenarios reflecting the views of a wide variety of stakeholders in the landscape. Although it clearly was not possible to achieve consensus in such a short period it was possible to engage people in the debate and to lay the ground work for a continuing dialogue. The meetings defused some latent hostility to the industrial projects amongst local people and sensitized the company to the issues that it would have to address in order to have its plans accepted by the communities concerned.

16.3.2 Challenges and Difficulties

As in any exercise of this type it was difficult to select a group of participants that were genuinely representative of all potential stakeholder groups. By choosing certain individuals over others, one is inevitably empowering them as actors in the process and therefore disempowering others. Notwithstanding this, for a preliminary exercise the workshop groups were representative and were not constituted in a way that excluded any groups or issues.

If the company were to significantly expand its operations in the area then landscape scale issues would probably have to be addressed through more formal and democratically constituted stakeholder gatherings. However a continuation of small informal meetings such as those that we held would provide a good basis for gradually proceeding to a more formal arrangement that could provide a more continuous and systematic dialogue with communities. At an exploratory phase of the work these short informal workshops were probably appropriate for engaging a representative group in a debate and identifying the big issues.

It was generally accepted that the two workshops provided a good entry point for a broader stakeholder discussion of the social and environmental issues that would need to be addressed in a plantation programme of this nature. Participants clearly believed that this was an indication that the company was acting in "good faith". Such workshops should prevent concerns festering and should provide an escape valve for any frustrations that local actors might harbor. The events also sent a strong signal to the communities that the company was willing to listen to their views and to involve them in the process of developing the industry in the area.

The workshops enabled the balanced representation of various stakeholders from the local community. The company had an opportunity to learn from the workshops and identify actions that could help to strengthen local civil society support for future plantation and milling operations. A number of potential actions and interventions that the company might make to strengthen the integration of its activities with local communities were identified.

The use of independent facilitators was undoubtedly a large element of the success of these workshops but it was important that the facilitators were well briefed and had a good basic understanding of the issues that might arise in large scale plantation operations.

The use of small group exercises and facilitation techniques that encouraged social interaction resulted in the meetings being very friendly. Everyone had a chance to express themselves and all left feeling that their views had been heard. The use of drawings to visualize landscape scenarios is a very powerful tool that works well in a diversity of cultures and situations – but making full use of this technique requires skills in graphic design and computerized image manipulation. A fundamental element of success in these endeavors is not going too far or too fast. Taking time to listen, learn and share is vital. Moving cautiously from informal, ad hoc events to more formal ones is essential.

16.4 Restoring What and for Whom?

The example from Uruguay provides a simple account of how a process of stakeholder engagement might be initiated. We ourselves have used a variety of other techniques to mediate such stakeholder processes and to engage for the long-term with diverse stakeholder groups. We draw a distinction between our gradual long-term approaches to "engagement" and more conventional landscape planning where there tends to be heavy stakeholder involvement up-front in a process of defining a landscape plan – the plan then being implemented by a technical agency. We take the view that landscapes are complex systems that evolve continuously under the influence of many drivers of change. Plans will only enable us to influence a sub-set of all the potential factors that are driving change in the landscape. We therefore favor the concept that is explored by Easterly (2006) of not planning but rather 'seeking'. In the context of landscape restoration this means engaging for the long-term with the broadest possible range of people whose actions will impact on the landscape and attempting to influence their behavior in ways that will lead to better landscape scale outcomes. This logic has led us to avoid the word 'planning' that tends to be associated with formal ex-ante technically-driven processes and instead to practice 'muddling through' (Lindblom 1959, 1979). The way in which the muddling through approach applies to forest landscapes is explored in detail in Sayer et al. (2008).

The GPFLR has invested in the process of 'how' to influence forest landscapes and has documented a number of case studies of processes to influence rather than plan landscape change. Some of this work has formed part of an initiative of the International Union for Conservation of Nature (IUCN) known as the Landscapes and Livelihoods Strategy (LLS). Both the GPFLR and LLS are in the early stages of their landscape scale interventions but some lessons are already emerging. In the following section we review some of the techniques that have proven most useful and propose a generic conceptual framework for the application of these techniques in a coordinated way within restoration landscapes.

16.5 Techniques for Optimizing Landscape Scale Restoration Outcomes

16.5.1 Convene a Multi-stakeholder Platform

A transparent process through which different stakeholders can voice their opinions and concerns is fundamental and is widely recognized (Lynam et al. 2007). However the convening and facilitation of such platforms is challenging. People differ in their willingness to engage with these processes. The time costs of participation can be high. Extreme views tend to dominate the discussion and the differences in power of different stakeholders can make it difficult to ensure equity in the processes.

Our experience is that multistakeholder processes are best conceived of as 'coalitions of the willing' and that one usually has to be pragmatic and engage with a subset of motivated potential participants. As conservationists we tend to find it easiest to work with stakeholders who at least to some extent share our views of desirable future landscapes. We recognize therefore that it is essential to pro-actively seek the participation of at least some persons who can represent contrasting interests. We have found it easiest to work with groups of less than 40 people and preferably around 20. The groups that we have facilitated have often drawn heavily on participants from local NGOs and government technical agencies. We always seek continuity in participation but have found this difficult to achieve and have often worked with a core group enriched by the periodic involvement of an 'outer circle' of less engaged but nonetheless concerned persons. One such group that we have facilitated in a Congo Basin landscape has been meeting annually for 8 years and has its own list server and web site. Over 60 people have attended one or more of the meetings but each meeting has had less than 30 attendees. Considerable interaction amongst the stakeholders occurs between meetings. This group has built a strong shared vision of the desired future of the landscape and we believe that it has influenced the decisions and programmes of all participants.

There is almost unanimous agreement that local participation and multi-stakeholder processes are fundamental to any landscape scale intervention (Salt and Lindenmayer 2008). While not contesting this conclusion our experience has been that such processes are in reality difficult to conduct – they are always highly imperfect. At a landscape scale the diversity of stakeholders and interests is high and there are legitimate differences between people. Notwithstanding this such processes have to be engaged but they will inevitably consist more of 'muddling through' (Lindblom 1959) or 'seeking' (Easterly 2006) than conventional planning.

16.5.2 Visualization

Of all the techniques that we have used to attempt to build understanding and consensus around landscape scenarios by far the most powerful have been a variety of visual techniques. Participatory mapping and the use of GIS overlays as negotiation support tools are well tried and tested in rural development at local scales. Lewis and Sheppard (2005) and Sheppard (2006) have reviewed experience of using GIS techniques to visualize and negotiate landscape scenarios. Our own preferred approach to initiate the participatory exploration of landscape scenarios has been the use of simple drawing exercises such as those described above for the Uruguayan Pampas. We set small groups the task of drawing their understanding of the present state of the landscape and its values and then lead them to draw best and worst case future scenarios. The examples given in Fig. 16.3 are from an IUCN LLS landscape in the badly degraded areas in Eastern Sudan. We have on some occasions maximized

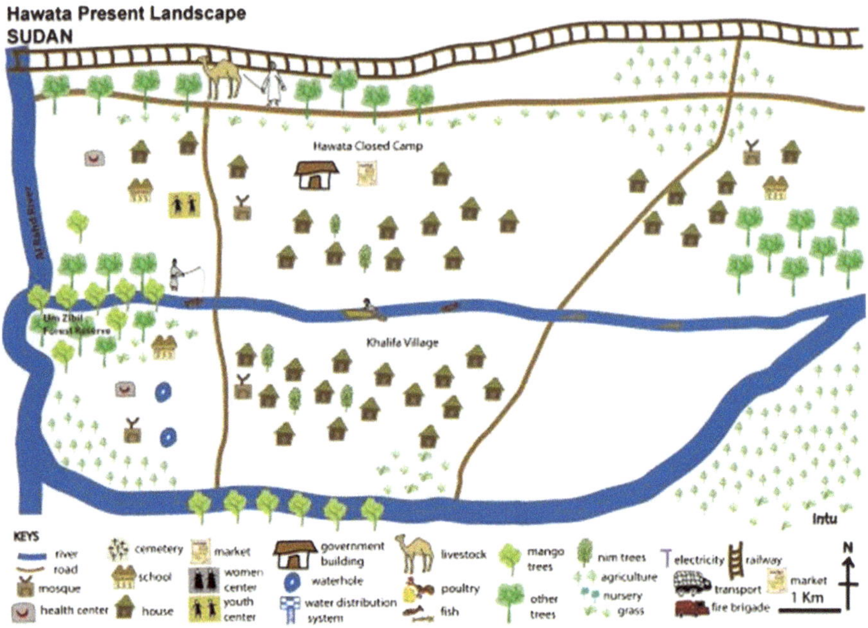

Fig. 16.3 A landscape at Hawata in Sudan visualized by a mixed stakeholder group and digitally manipulated

the internal heterogeneity of the groups who do the drawing in an attempt to move rapidly towards consensus. On other occasions we have divided stakeholders into interest groups in order to emphasize the divergence of interests amongst the stakeholders. On this occasion we separated men and women in order to explore the very different perceptions of landscape of the two gender groups.

As outsiders we have always found that these drawing exercises teach us a lot about the landscape. We are always surprised by the wealth and diversity of issues and features that emerge. The drawing exercises lead to rich and intense discussions about potential future situations. We sometimes simply use the drawings as a foundation for discussion and negotiation but on other occasions we have been able to move towards a degree of consensus amongst participants and to achieve drawings that genuinely provide a guide for landscape interventions. When this occurs we print the images and laminate them and use them as a reference point for tracking change in the landscape. These images define priorities and activities for the short term. Since the drawings are made by a diversity of people with varying levels of artistic skill and in short spaces of time we often scan the images and then manipulate them using a variety of visual software. An example of a "positive" scenario is given in Fig. 16.4. We have used visualization with multi-stakeholder groups and have found it useful to use the images from an initial consultation to initiate future rounds of negotiation. In general we attempt to convene these groups annually and

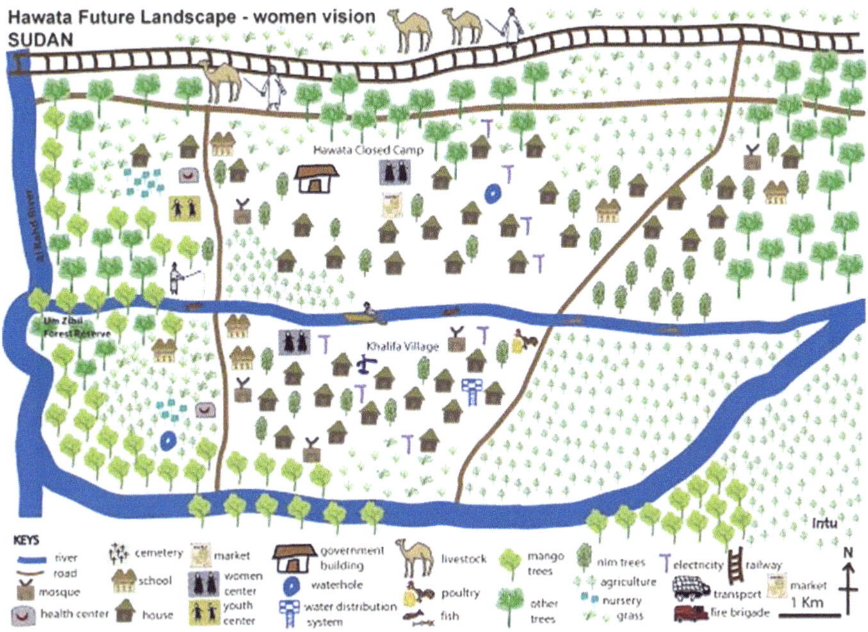

Fig. 16.4 A desirable future scenario for the Hawata area as represented by a women's group and subsequently manipulated digitally

to initiate each meeting with a review of the images from the preceding year. The techniques we use are similar to the 'Rich Pictures' used by Bell and Morse (2010).

16.5.3 Participatory Simulation Modeling

A third technique that has proven powerful in exploring landscape scenarios is the building of simple computerized simulation models of the processes of landscape change. A skilled facilitator discusses potential landscape change processes and works with local stakeholders to help them build models that simulate the changes that might occur. The models require quantitative data to enable them to run and so stakeholders have to have access to data on the main determinants of change and on the resulting flows of costs and benefits. An example might be changes in the price of timber or cost of transport driving changes in intensity of logging. Establishment of a protected area might drive down the income of households of people dependent on harvesting forest products etc. Such models have the value of making explicit the costs and benefits to different stakeholders of different landscape strategies. Such models are widely used in landscape scale conservation and development programmes but have not to our knowledge yet been used in restoration programmes.

Detailed discussions of case studies and basic principles of participatory modeling are given in Sayer and Campbell (2004) and Sandker et al. (2007).

16.5.4 Development of Outcome Indicators

Stakeholders must have a means of assessing the extent to which the landscape is evolving in the direction of the desired scenarios (Salt and Lindenmayer 2008). We have found it effective to draw such indicators out of the participatory activities outlined above. Visualization in particular provides a good basis for identifying indicators to which local participants can relate. Our usual approach is to ask individual participants to write on flash-cards the attributes of the landscape that are susceptible to change and which for them would constitute good indicators of the direction of change. The cards are then pinned onto charts and similar ideas are grouped. The facilitator then leads a discussion of the validity and measurability of each indicator. When the number of plausible indicators has been reduced to a manageable number and they have been expressed in measurable ways the participants are asked to vote for those that for them are most important and meaningful. Voting is with sticky dots – dotmocracy. The facilitator then leads a process of further defining the post popular indicators and elaborating a scoring matrix. An example of this process in a Central African landscape is given in Sayer et al. (2007). The process is also described in more detail in Aldrich and Sayer (2007). A variety of approaches to the development of such indicator sets are possible and there is a wealth of ways in which they can be aggregated and presented in the form of indices. The issues to consider in making choices are discussed in Bell and Morse (1999).

16.6 Conclusions

Many problems can be avoided and the effectiveness of landscape restoration programmes can be greatly enhanced if they are designed in ways that address the full range of interests and generate the support of the civil society stakeholders whose lives the programmes will impact. This appears to be such an obvious conclusion (Salt and Lindenmayer 2008) that it is surprising how often large scale restoration schemes miss out this fundamental step. Too many restoration programmes are designed by technical agencies or special interest groups and miss out on opportunities to provide their full potential range of benefit flows to society. We have found that facilitated participatory multi-stakeholder processes, with all their imperfections, are fundamental. These processes can be greatly enhanced by the use of a number of facilitation techniques of which visualization of scenarios and participatory simulation modeling have proven particularly effective. Both of these techniques require special skills on the part of facilitators. Visualization is greatly aided by having facilitators with good drawing skills and the ability to manipulate images using computer software. Participatory modeling can use simple off-the-shelf

Fig. 16.5 A desirable future scenario for the Hawata area as visualized by a men's group and subsequently manipulated digitally

software such as STELLA but still requires that the facilitator is familiar with the use of the software but also has sufficient understanding of the dynamics of the landscape to be able to define the parameters of the model. Lastly sets of indicators that enable stakeholders to track progress towards their desired scenarios and debate this progress are a valuable tool.

All of the above techniques must be seen as part of a process of shared learning by facilitators, experts and local stakeholders. The principles of adaptive management must be applied and it is essential to anticipate that stakeholders' views of desirable outcomes will change over time. There is rarely a final end point for restoration where no further management intervention is required. An effective multi-stakeholder process will reveal the constant need to adapt and redirect restoration efforts as the needs of society evolve (Fig. 16.5).

Figure 16.6 attempts to represent in the form of a flow chart the different steps that we have used in our landscape interventions. As outsiders we listen and learn in the early stages and practice the principles of appreciative enquiry (Cooperrider et al. 1995). We consider it important to ensure that we are fully aligned with national level development priorities. We then deploy the range of techniques for exploring scenarios and understanding the overall context of the intervention and we do this with a multi-stakeholder group. We establish general goals – a desirable future scenario and then work with stakeholders to establish indicators of the

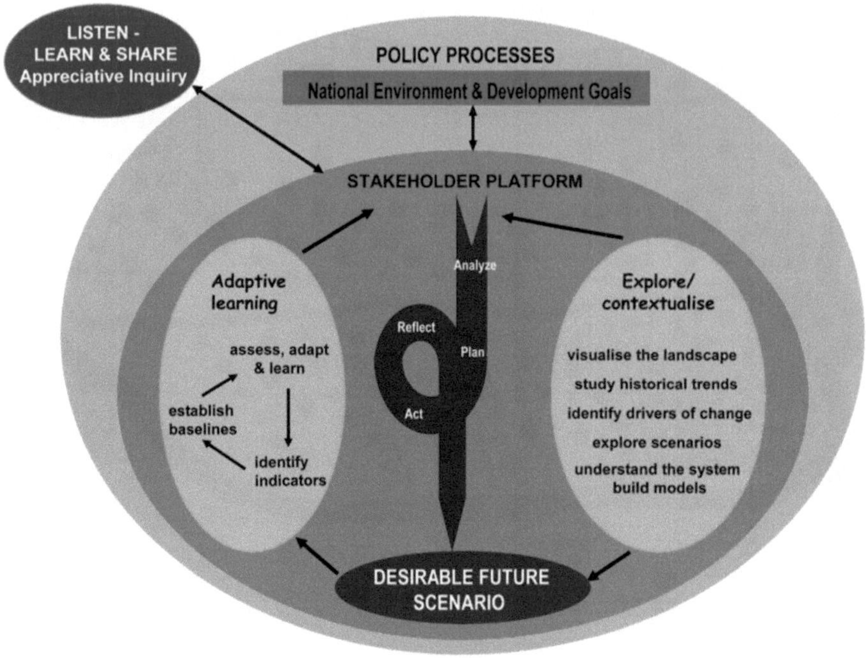

Fig. 16.6 Simplified general conceptual framework for landscape-scale restoration interventions. The stakeholder group has to be sensitive to national priorities and local concerns. On the *right* are approaches to building understanding of landscape dynamics and exploring scenarios. On the *left* are processes for active learning. An action learning cycle leads towards the desired future scenario

progress that is being made towards achieving the scenario. The stakeholder platform analyses and reflects upon the learning from the indicators and other participatory exercises and adapts its plans accordingly. The central arrow that turns back upon itself indicates that we are applying the principles of action research. Overall the concept is of a continuing process of reflection, action and observation. It recognizes that there is no end point to the process and that restoration is just one phase of a never ending process of landscape management.

References

Aldrich M, Sayer J (2007) LOAM in practice. WWF, Forests for Life Programme, Gland

Bell S, Morse S (1999) Sustainability indicators: measuring the immeasurable. Earthscan, London

Bell S, Morse S (2010) Rich pictures: a means to explore the "sustainable mind"? Sustain Dev. doi:10.1002/sd-497

Cooperrider DL, Barrett F, Srivastva S (1995) Social construction and appreciative inquiry: a journey in organizational theory. In: Hosking D, Dachler P, Gergen K (eds) Management and organization: relational alternatives to individualism. Avebury Press, Aldershot, pp 157–200

Easterly W (2006) The White Man's burden: why the West's efforts to aid the rest have done so much ill and so little good. Penguin, New York/London

Farina A (2006) Principles and methods of landscape ecology. Springer, Dordrecht

Lafortezza R, Chen J, Sanesi G, Crow TR (eds) (2008) Patterns and processes in forest landscapes: multiple use and sustainable management. Springer, Dordrecht

Lewis JL, Sheppard SRJ (2005) Culture and communication: can landscape visualization improve forest management consultation with indigenous communities? Landsc Urban Plan 77:291–313

Lindblom CE (1959) The science of muddling through. Public Adm Rev 19(2):79–88

Lindblom CE (1979) Still muddling, not yet through. Public Adm Rev 39(6):517–526

Lynam T, De Jong W, Sheil D, Kusumanto T, Evans K (2007) A review of tools for incorporating community knowledge, preferences, and values into decision making in natural resources management. Ecol Soc 12(1):5. Online URL: http://www.ecologyandsociety.org/vol12/iss1/art5/

Mansourian S, Vallauri D, Dudley N (eds) (2005) Forest restoration in landscapes: beyond planting trees. Springer, New York

Perrow MR, Davy AJ (2002) Handbook of ecological restoration, vol 1: Principles of restoration. Cambridge University Press, Cambridge

Salt D, Lindenmayer D (2008) The Bowral checklist: a framework for ecological management of landscapes. Land and Water Australia, Canberra

Sandker M, Suwarno A, Campbell BM (2007) Will forests remain in the face of oil palm expansion? Simulating change in Malinau, Indonesia. Ecol Soc 12(2):37. Online URL: http://www.ecologyandsociety.org/vol12/iss2/art37/

Sayer JA (2007) The peoples' forest: balancing local and global values. Doctoral thesis, Universidad Autonoma de Madrid

Sayer JA, Campbell B (2004) The science of sustainable development: local livelihoods and the global environment. Cambridge University Press, Cambridge

Sayer J, Campbell B, Petheram L, Aldrich M, Ruiz Pérez M, Endamana D, Nzooh Dongmo Z, Défo L, Mariki S, Doggart N, Burgess N (2007) Assessing environment and development outcomes in conservation landscapes. Biodivers Conserv 16:2677–2694

Sayer JA, Bull G, Elliott C (2008) Mediating forest transitions; grand design or muddling through. Soc Conserv 6:320–327

Sheppard SRJ (2006) Bridging the sustainability gap with landscape visualisation in community visioning hubs. Integr Assess J 6:79–108

Whisenant SG (1999) Repairing damaged wildlands; a process-oriented; landscape scale approach. Cambridge University Press, Cambridge

WWF (2009) Stakeholder engagement in plantations. Technical paper. WWF, Gland, Switzerland

Index

J. Stanturf et al. (eds.), *Forest Landscape Restoration: Integrating Natural and Social Sciences*, World Forests 15, DOI 10.1007/978-94-007-5326-6,

MIX
Papier aus verantwortungsvollen Quellen
Paper from responsible sources
FSC® C105338

If you have any concerns about our products,
you can contact us on
ProductSafety@springernature.com

In case Publisher is established outside the EU,
the EU authorized representative is:
Springer Nature Customer Service Center GmbH
Europaplatz 3, 69115 Heidelberg, Germany

Printed by Libri Plureos GmbH
in Hamburg, Germany